Interactions and Adaptation Strategies of Marine Organisms

Developments in Hydrobiology 121

Series editor
H. J. Dumont

Interactions and Adaptation Strategies of Marine Organisms

Proceedings of the 31st European Marine Biology Symposium, held in St. Petersburg, Russia, 9–13 September 1996

Edited by

A.D. Naumov, H. Hummel, A.A. Sukhotin & J.S. Ryland

Reprinted from Hydrobiologia, vol. 355 (1997)

Springer-Science+Business Media, B.V.

Library of Congress Cataloging-in-Publication Data

A C.I.P. Catalogue record for this book is available from the Library of Congress

ISBN 978-90-481-4988-9 ISBN 978-94-017-1907-0 (eBook)
DOI 10.1007/978-94-017-1907-0

Printed on acid-free paper

Hydrobiologia **355**: v–vi, 1997.
A.D. Naumov, H. Hummel, A.A. Sukhotin & J.S. Ryland (eds),
Interactions and Adaptation Strategies of Marine Organisms.

Contents

Adaptation Strategies in Marine Organisms

Hydrobiologia **355**: vii, 1997.
A. D. Naumov, H. Hummel, A. A. Sukhotin & J. S. Ryland (eds.),
Interactions and Adaptation Strategies of Marine Organisms.

Preface

The intention to organise a European Marine Biology Symposium (EMBS) in Russia arose in 1991 during UNESCO meetings on cooperation in coastal marine sciences with researchers from the Soviet Union. Glasnost and perestroika were coming to the foreground, and strong changes in the Soviet scientific community were appearing.

Although the first requests to organize a symposium in Russia were received by the EMBS committee with some hesitation, it was strongly felt by Konstantin Burdin (Moscow State University), Dirk Troost (UNESCO, Paris) and myself (Netherlands Institute of Ecology) that, in the midst of these societal changes and new developments, a major international symposium might have a positive impetus on the east European researchers involved in marine sciences. Finally, the Zoological Institute of St. Petersburg agreed to organize the 31st EMBS in 1996.

Adaptation strategies and interactions of marine organisms are scientific topics which correspond with both the subject matter of the major research tasks of Russian institutes, active in northern (sub-arctic and arctic) territories and with the first international (east and west European) cooperative projects. Irrespective of the limited number of topics, the central role of adaptations and interactions in the living environment was fully recognized and thereby attracted 190 participants to the symposium, from all over the world.

The book is an outline of the refereed reviews and papers, and presents a state-of-the-art view on some selected subjects of adaptational processes in marine organisms and interactions between them. Adaptations in ecophysiological parameters (at individual level) were studied mainly in response to changes in salinity, temperature, pollutants and food conditions. The interactions dealt primarily with intraspecific competition, predation and epibiosys at population level. From the contributions an interesting additional value arises from the differences in scientific approach between west and east European researchers, e.g. field surveys versus experiments or the practically restricted deterministic studies versus theoretical holistic treatises. Thereby the book may help in merging the best from the two sides with their different scientific and socio-cultural traditions.

Beside the scientific contributions, the splendour of St. Petersburg and the continuously enthusiastic input of so many colleagues from the Zoological Institute had their reflection on the atmosphere and intensity of the symposium, and made the 31st EMBS both a socio-cultural and a scientific success.

H. HUMMEL

Hydrobiologia **355:** 1–9, 1997.
A. D. Naumov, H. Hummel, A. A. Sukhotin & J. S. Ryland (eds), Interactions and Adaptation Strategies of Marine Organisms.

The importance of juveniles in structuring a littoral macrobenthic community

I. V. Burkovsky, A. A. Udalov & A. P. Stoljarov
Department of Hydrobiology, Faculty of Biology, Moscow State University, Moscow 119899, Russia

Key words: juvenile and adult macrobenthos, community structure, spatial distribution, life cycles

Abstract

The techniques of collecting, extracting and counting of meiobenthos were applied to the study of juvenile macrobenthos in soft sediments. The vertical and horizontal distribution of juvenile and adult macrobenthos, and the influence of various factors (i.e. tidal level, granulometric composition of sediments, plant biomass, moisture, organic content of sediments, etc.) on their distribution have been investigated. Combinations of these factors differ in their influence on species, depending on the animals' ontogenetic type (with or without pelagic larvae) and individual size. The spatial distributions of juveniles and adults were found to be different for macrobenthos with a pelagic stage in the life cycle. In contrast to this, juveniles and adults of species with direct development have more similar distribution patterns.

Introduction

Early juvenile stages of macrofauna are traditionally ignored in most benthic studies, and the distribution and abundance of these animals are treated in terms of interactions between adult organisms. The reason for this neglect is that the sieve mesh size used for the collection of macrobenthos is too large to collect these meiofaunally sized larvae and early juveniles. And traditional meiofaunal studies have also usually ignored this group by defining their subject taxonomically, not purely by size (Watzin, 1983).

Information about biology and life cycles of different macrobenthic species appears in the literature (Semenova, 1972; Kaufman, 1974; Mileikovsky, 1976; Sveshnikov, 1978), but data on multiple species and on the problem of community establishment and maintenance rarely occur (Santos & Simon, 1980; Watzin, 1986). Systematic observations on the development of community structure from juvenile to adult stages are lacking in spite of their importance.

In our research we aim to define the role of juvenile macrobenthos in community structure, to determine the factors affecting this group and to assess the changes in the benthic community during the growth of the organisms.

Materials and methods

The study was conducted on an intertidal muddy-sand beach (the Chernaya Gulf in the Kandalaksha Bay of the White Sea) 7 km away from the North Polar Circle. The littoral sediment is fine, slightly muddy sand (fraction <0.1 mm about 20%). Salinity varies from 16 to 24‰. Water temperature in July reaches +23 °C, sand temperature +34 °C. From late November till May the gulf is covered by ice. The mean tidal amplitude in this area is about 1.6 m. The macrofauna is dominated by *Mya arenaria*, *Hydrobia ulvae*, *Arenicola marina* and *Macoma balthica*. The site was described in detail by Burkovsky (1992).

An adequate description of juvenile macrobenthos is difficult because of the rapid temporal variation in appearence, growth and re-distribution of juveniles, and by their patchy distribution. Therefore sampling was designed to estimate both the temporal and spatial components of a species' distribution. The spatial distribution of juveniles and adults, and the effects of

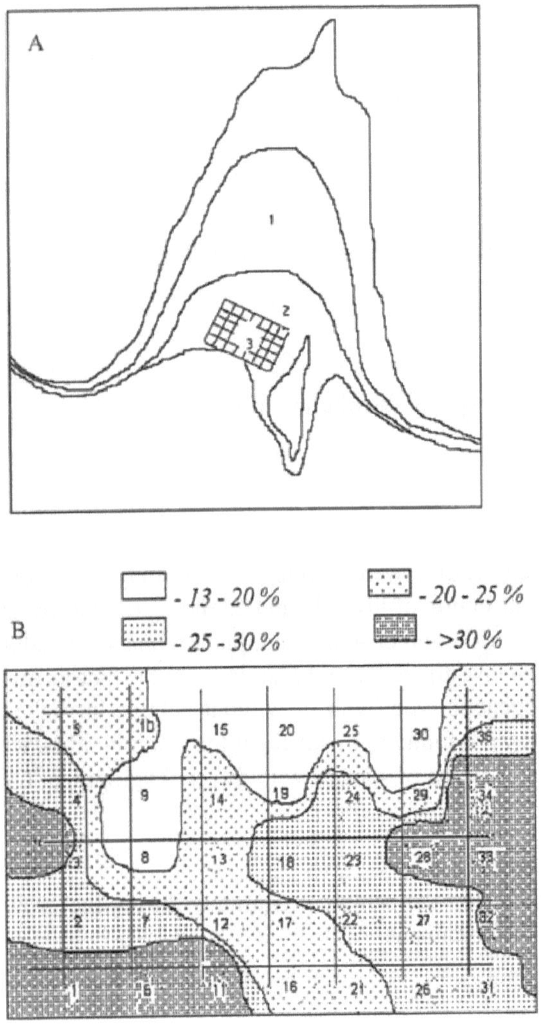

Figure 1. (A) Map of the sampling site (Chernaya Gulf; 1 = salt marsh, 2 = muddy-sand littoral, 3 = sampling area; dotted lines indicate the tidal level.). (B) Sediment composition of the investigated area (indicated is the percentage of sediment <0.1 mm)

abiotic factors, were determined during a single sampling survey from 5 to 25 July 1994 in order to minimize the influence of temporal changes. This survey was conducted in an area of about 10 000 m^2 (Figure 1a, b), embracing all littoral levels, with 35 stations at mutual distances of about 20 m. Plant biomass, the upper boundary of the H_2S and clay layers, tidal elevation, granulometric composition of sediments and the percentage of area covered by macrophytes were measured for each station.

To assess the temporal changes in community structure, macrobenthos was also investigated at weekly intervals from May to September 1995 on a transect consisting of 5 stations located from the upper littoral zone to the sublittoral.

Methods of collection and extraction of organisms from sediments, usually used for meiofauna, were applied to macrobenthic juveniles. Sediment samples of 25 cm^2 were taken to a depth of 5 cm. Four replicates were collected at every site. The sediment columns were subdivided in layers: 0–0.5, 0.5–2 and 2–5 cm. Adults of *A. marina* and *M. arenaria* were counted by standard macrobenthic methods.

None of the previously used methods of extraction yields quantitative data in all types of sediments and on all meio- and macrofaunal taxa. Therefore, in this study, the upper layer of the sediment was accurately divided and a shaking-decantation procedure was used to concentrate animals. The samples were then sieved through a 100 μm mesh sieve to separate the fine sediment particles. Because, in a pilot study, juveniles were found to be absent in the deeper (0.5–2 and 2–5 cm) layers, only mesh sizes of 250 and 500 μm were used to separate adult macrobenthos in the deeper sediments. All residues were sorted and larval, juvenile and adult macrobenthic animals were identified and measured.

For the analysis of spatial structure we used methods of cluster analysis and analysis of distribution of similarity indices. Pianka and Czekanovsky indices and Pearson's correlation coefficient were used as the measure of similarity. For identifying the factors affecting juvenile distribution Spearman's rank correlation coefficients (r) were used. Statistical procedures were provided using the ecological package ECOS and the statistical package SYSTAT.

Results

Taxonomic composition and life cycles of littoral species

During our survey a total of 26 macrofaunal species were found (Table 1). Juveniles of these species were found in nearly all cases. Adults of some species (*Musculus discors*, *Terebellides stroemi*, *Hormothoe imbricata*) were absent and the percentage of juveniles was too small (in <1% of the samples) despite their larvae being present in near-shore waters (Burkovsky, unpubl. data).

Two major groups of species were distinguished: species with pelagic larvae (less than 45% of all species) and species without planktonic larvae (Table 1). The larvae of the last group, more than 55%

Table 1. Taxonomic composition, types of development and abundance of the macrobenthos of the Black Gulf sand littoral community (the average density of organisms on the basis of spatial survey is presented) *: P= species with pelagic larvae in life cycle, B= species without pelagic larvae, but development within egg masses, maternal tubes or maternal organisms, P/B= life cycle includes very short pelagic stage (about several hours) and this stage usually does not affect the long-distance dispersion of organisms, F= species capable of fragmentation, ?= type of development is unknown or several different onthogenic types are described; **: Insects with benthic larvae.

Species	Average density, numbers/ m²		Ontogenic type (P, B , F) *
	Adults	Juveniles	
Peloscolex benedeni (Udekem)	3732	102	B, F
Hydrobia ulvae (Pennart)	2058	616	P/B
Paranais litoralis (Muller)	1529	29	B
Tubifex sp.	452	25	B, F
Pygospio elegans Claparede	161	25	P, B?
Macoma balthica (Linne)	154	79	P
Eteone longa (Fabricius)	126	45	P
Fabricia sabella (Ehrenberg)	68	218	B
Scoloplos armiger (Muller)	65	34	P, B?
Spio filicornis (Muller)	46	88	P
Jaera albifrons Leach	42	106	B
Halicriptus spinulosus Siebold	41	410	B
Arenicola marina Lamarck	33	35	P/B
Spio theeli Soderstrom	28	26	P
Priapulus caudatus Lamark	16	84	B
Gammarus duebeni Lilljebord	10	8	B
Littorina saxatilis (Olivi)	9	25	B
Mytilus edulis Linne	10	350	P
Littorina littorea (Linne)	3	?	P
Capitella capitata (Fabricius)	3	?	P, B?
Crangon crangon Linne	2	7	B
Musculus discors (Linne)	0	1	P
Balanus balanoides (Linne)	0	1	P
Chironomus salinarius Meigen	-**	59	B
Cricotopus vitripennis Meigen	-**	528	B
Muscidae gen.sp.	-**	1	B

of the species, develop within egg masses, maternal tubes or internally (Table 1). Some species that usually have pelagic larvae change their mode of development under littoral conditions; the pelagic stage is shortened or even eliminated from the life cycle. A striking example is *Pygospio elegans*, for which in the White Sea pelagic larvae were described (Sveshnikov, 1978) but in our community development takes place in maternal tubes containing up to 16–18 larvae. Their pelagic phase is less than several hours or absent. For *Scoloplos armiger* two types of development have been described: (1) development in egg masses, pelagic larvae being absent (Thorson, 1946; Rasmussen, 1973), (2) egg masses are absent, duration of the pelagic stage about three weeks (Sveshnikov, 1978; Plate & Husemann, 1992). We found neither pelagic larvae nor egg masses and therefore could not distinguish the type of development for *Scoloplos armiger* in our community (Table 1).

For some species with pelagic larvae (*Hydrobia ulvae*, *Macoma balthica*) the densities of juveniles were lower than for adults. On the other hand, for species having all stages of their life cycle in the benthos (*Fabricia sabella*, *Jaera albifrons*) the abundances were higher for juveniles than for adults (Table 1).

Juveniles of species with a relatively long pelagic stage in their life cylce, such as *Eteone longa* and *Spio filicornis*, had a constant density during the whole period of observation, because of prolonged period of settling. Species with a short pelagic stage of some hours (*Hydrobia ulvae*) or completing all stages of their life cycle in the benthos (*Jaera albifrons*, *Gammarus duebeni*, *Pygospio elegans*) have a maximal density of juveniles in July. These species reproduce at the same time, which is related to increasing temperatures (Kaufman, 1974). Exception to this are the priapulids, which are organisms with delayed larval development in the sediments, requiring several years. Priapulids have a reproductive output in winter (Shirley, 1990) and a constant density of juveniles.

Factors, affecting macrobenthic distribution

To estimate the effects of different factors on the macrobenthic distribution, Spearman rank correlation analysis was used. Correlation coefficients between animal abundance and the most important environmental factors are presented in Table 2.

Most species showed a positive correlation with depth and the submersion period. This correlation was more often significant for juveniles than for adults.

The abundance of *Zostera marina* was very closely associated with tidal level: the influence of the seagrass and of the level could not be separated. *Z. marina* holds water and fine organic matter, provides additional substrate for larval settling and protects larvae from drying out during low water. The juvenile macrofauna had a strong positive correlation with the biomass of *Z. marina*, but adults showed an insignificant correlation.

Most species showed an insignificant correlation with granulometric composition of sediments. The

Table 2. Spearman's rank correlation coefficients (r) for the effect of different factors on abundance of juvenile (juv.) and adult (ad.) macrofauna (Significance of r: $* = p<0.05$).

Species	Stage of organism's development	Spearman's ranc correlation coefficients (r) between species abundance and: Tidal level	Zostera's biomass	% of the sediment <0.1 mm	Depth of the H^2S layer	Amount of organic matter
Jaera albifrons	Ad.	0.002	−0.153	0.01	0.117	−0.015
	Juv.	0.205	−0.025	0	0.03	−0.099
Gammarus duebeni	Ad.	0.166	0.114	0.15	0.01	0.236
	Juv.	0.598*	0.589*	0.212	−0.121	0.172
Halicriptus spinulosus	Ad.	0.335	0.137	−0.139	0	0.014
	Juv.	0.592*	0.366*	−0.044	0.014	0.165
Macoma balthica	Ad.	0.134	−0.017	−0.183	−0.164	−0.018
	Juv.	−0.098	−0.097	0.427*	0.02	−0.158
Fabricia sabella	Ad.	0.442*	0.336	−0.015	0.132	−0.026
	Juv.	0.631*	0.627*	0.199	0.035	0.116
Scoloplos armiger	Ad.	0.298	0.249	0.083	0.196	0.189
	Juv.	0.091	−0.11	0.005	0.230	0.225
Eteone longa	Ad.	0.18	−0.101	0.002	0.001	0.047
	Juv.	0.614*	0.514*	0.038	−0.243	−0.132
Spio filicornis	Ad.	0.165	0.102	0.146	0.154	0.334
	Juv.	0.518*	0.422*	0.078	0.087	0.068
Hydrobia ulvae	Ad.	0.738*	0.575*	0.151	0.228	−0.02
	Juv.	0.515*	0.414*	0.197	0.163	−0.01
	Vel.	0.02	0.07	0.008	−0.047	0.13

only exception were the juveniles of *Macoma balthica*, that showed a significant correlation with the sediment fraction <0.1 mm.

A significant correlation between the upper boundary of the H_2S layer and juvenile abundance was not found. This can be explained by the occurrence of juveniles in the uppermost layer, where the effect of H_2S is very low. Spearman's coefficients between the amount of organic matter and species abundance were also low and insignificant.

Correlation coefficients between juvenile abundance and environmental factors were usually higher and more often significant than the coefficients for adults (Table 2). From this it might be assumed that the effects of environmental factors are stronger on juveniles, whereas during growth the pressure by abiotic factors becomes weaker and substituted by other factors, such as interactions between species.

Comparative analysis of spatial distribution of juvenile and adult macrobenthos

All macrofaunal juveniles occupy the uppermost 0.5 cm of sediments. Some burrowing species (*Macoma balthica, Mya arenaria, Arenicola marina, Scoloplos armiger, Halicriptus spinulosus, Priapulus caudatus*) move to the deeper layers during their growth. Other species remain in the upper 2 cm. Among adult macrobenthos *Hydrobia ulvae* dominated in the upper layer; more than 90% occur in the uppermost 1 cm and 73% appear on the sediment surface. Oligochaeta and small Polychaeta prefer the layer 0.5–2.0 cm, *S. armiger, A. marina* and *M. arenaria* reach down to a depth of 30 cm.

Earlier it was shown that the horizontal distribution of juvenile and adult macrobenthos is patchy (Burkovsky et al., 1996). Similarity indices were calculated to analyze the co-occurrence of adults and juveniles of species. Because the factors related with tidal level were the most important in affecting the community structure, the data from each littoral zone obtained on a transect from May to September 1995 were com-

Table 3. Average Pearson's correlation coefficients for macrobenthos.

Average similarity between species (juveniles only)	0.612 ± 0.016
Average similarity between species (adults only)	0.576 ± 0.019
Average similarity between adults and juveniles	0.556 ± 0.08

bined. Pearson's correlation coefficients were used as the measure of similarity sensitive to negative coincidences (i.e. including absences = double zero's). Therefore, a high similarity between species supposes the occupation by them of one littoral zone.

Similarity levels for juveniles significantly exceeded the mean similarity for adults (Table 3). The mean similarity between juveniles and adults was lowest. It shows that appearance and growth of juveniles occurred in a narrow band of the littoral zone, whereas during their development the organisms re-distribute over all littoral levels.

Analyzing the presence of juveniles at the transect, we conclude that during the first summer of their life, juveniles do not change their spatial distribution (Burkovsky, unpublished results). The establishment of the spatial pattern among adults may occur in the winter (Burkovsky, pers. observ.) and might be related to processes such as sediment disturbance by ice or melting. Therefore, we assume that the migration process demands more than one summer season, although in our research we have not observed migrations and have considered only the final result of it.

The similarity indices between juvenile and adult phases can be divided into two species groups (Table 4, first column): those considerably changing their dispersion during the life cycle (low similarity level: *Eteone longa*, *Hydrobia ulvae*, *Macoma balthica*) and those not changing their distribution (high similarity level: *Fabricia sabella*, *Mytilus edulis*, *Halicriptus spinulosus*, *Priapulus caudatus*). In the first major group, two sub-groups with different migration behaviour were distinguished on basis of the similarity indices between juveniles and adults on the one hand and all adult macrobenthos in general on the other hand (Table 4, second column). Some species with significantly increasing similarity indices from juveniles to adults (*H. ulvae*, *M. balthica*) migrate to the low and intermediate littoral zones during their growth. These zones are characterized by a maximal density and diversity of adult macrobenthos. In contrast to this, for the species of the other sub-group migrating during development (*E. longa*), the similarity indices significantly decreased, indicating that during development they migrate from the lower to higher littoral levels. The second major group consisted of species that did not change their spatial distribution and those without planktonic larvae (*F. sabella*, *P. elegans*, *H. spinulosus*, *P. caudatus*) and the fouling species *M. edulis*.

A cluster analysis by the average linkage method on the basis of the abundance of adults allowed us to assess the presence of groups of closely related species with a similarity level >0.5. Three groups were distinguished: the higher (group 1 in Figure 2), intermediate (group 2) and lower (group 3) littoral species. A similar classification by means of a cluster-analysis on the basis of juveniles was unsuccessful due to weak relations between juveniles of different species (Figure 3). Juvenile populations are characterized by low densities and therefore, if such groups exist, much more data are needed. It is possible, that for juveniles the groups of closely related species are absent because they occupy a narrow littoral zone.

Discussion

The results allow us to discuss some aspects of macrobenthic community establishment and the role of the juvenile pool in this process.

Firstly, we must note that in our littoral community the species composition of juveniles appears to be similar to the composition of adults. So, when adult stages of the species were not found on the littoral, juveniles were not found either, in spite of the presence of their larvae in near-shore waters. Our data do not allow us to decide how the taxonomic composition is maintained: by active habitat selection or by hydrodynamical processes in the near-shore waters. We only assume that the species composition of the adult community is not a result of random spatfall of all planktonic larvae present in the water column and subsequent elimination of some of them. On the contrary, it might be determined by mechanisms at the pelagic stage or at the moment of settling, as is well known for epibenthic species (Satchell & Farrell, 1993; Meenakumari & Balakrishnan, 1994).

Another interesting observation is that some species with pelagic larvae have lower densities of juveniles than of adults (Table 1). We suggest three possible explanations of this fact. Firstly, only part of the

Table 4. Similarity indices (Pearson correlation coefficients) between macrobenthic species.

Species	Similarity between juveniles of species (column 1) and adults of species	Similarity between juveniles of species and all adult macrobenthos	Similarity between adults of species and all adult macrobenthos
Fabricia sabella	0.868	0.867	0.944
Eteone longa	0.498	0.704	0.370
Pygospio elegans	0.765	0.785	0.886
Scoloplos armiger	0.717	0.482	0.516
Hydrobia ulvae	0.538	0.522	0.982
Macoma baltica	0.482	0.395	0.749
Mytilus edulis	0.814	0.709	0.623
Halicriptus spinulosus	0.842	0.947	0.858
Priapulus caudatus	0.804	0.834	0.800

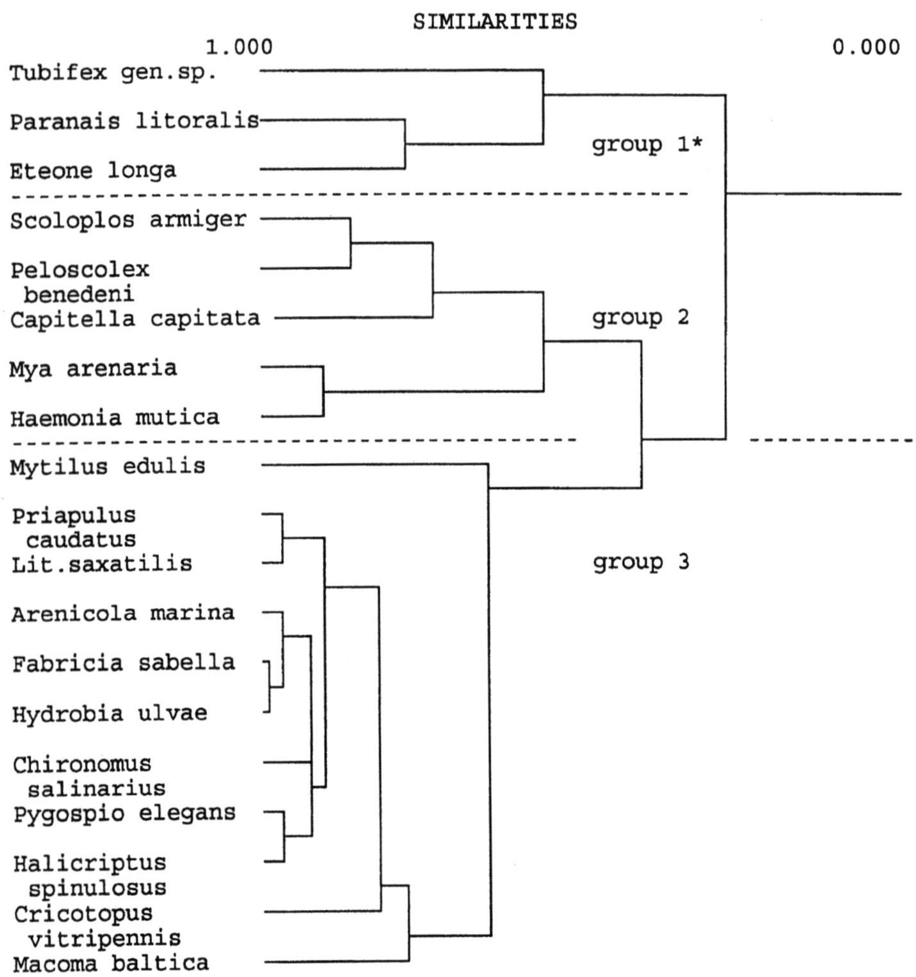

Figure 2. Dendrogram of similarity of adults (Czekanovsky similarity index; combined samples from one littoral level; * see text for explanation of groups)

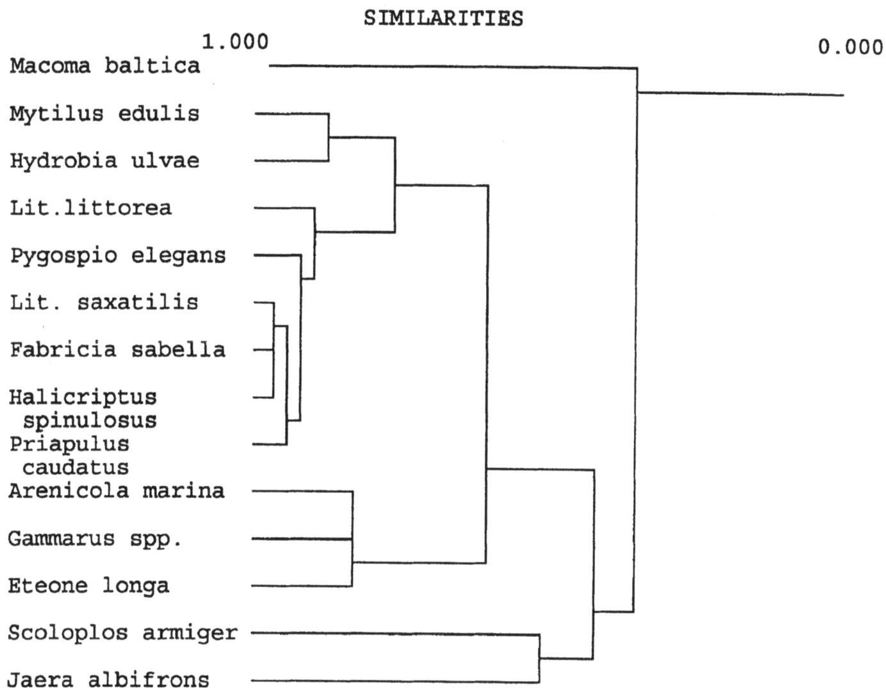

Figure 3. Dendrogram of similarity of juveniles (Czekanovsky similarity index; combined samples from one littoral level)

adult population is replaced by new recruits every year, because of a prolongation of a species' life by several years, as observed for the cold waters of the White sea in comparison with more southern waters (Semenova, 1972). A second cause could be the patchy distribution of juveniles that may lead to errors in counting. Thirdly, some larvae might settle on more attractive substrata in the upper sublittoral and appear later in the littoral. Dispersion of juveniles away from the adult populations until they have reached a certain size and their following (secondary) migration is well known for *A. marina* (Kaljakina, 1982; Reise, 1985) and *M. balthica* (Armonies & Hellwig-Armonies, 1992).

On the other hand, the abundance of juveniles of species passing all stages of their life cycle in the benthos is higher than adult abundance (Table 1).

Therefore, in the first group of species (with pelagic larvae) the adult density is determined especially by the dynamics of larvae and juveniles in the water column and adjacent areas, and by the moment of settling. In the second group the adult density is determined by survival of juvenile stages under littoral conditions. In the latter the processes most affecting population survival will be competition with other organisms, resistance of animals to environmental conditions, and elimination of young juveniles.

The marked differences in spatial distribution of juvenile and adult macrobenthos indicate strong changes in community structure during organisms' growth. These differences are determined by passive dispersion and active migration of organisms, elimination of juveniles by predators, mortality, and reaction patterns of organisms to environmental conditions. We can assume the importance of water flow in the formation of the larval pool as well as in the re-distribution of settled larvae and young juveniles in the littoral zone, as noted by Butman (1987, 1988). However, the lack of the possibility of estimating hydrological factors quantitatively prevents us from considering these factors. Among the environmental factors studied by us, such as tidal level, granulometric composition of sediments, plant biomass, moisture and organic content of sediments, those closely related with tidal level had a strong influence on macrobenthic species, especially on their juveniles. The effects of environmental factors on juveniles are stronger than on adults, and we can suggest that during growth the abiotic pressure becomes weaker, whereas for the distribution of adults, the impact of the other factors, such as species interactions, is more important.

In our littoral community we marked two tendencies in the character of spatial distribution of developing organisms. Some species which pass all stages of their life cycle in the benthos (*J. albifrons*, *G. duebeni*, *H. spinulosus*, *F. sabella*) did not change their dispersion pattern during development. On the other hand, the juveniles of species with pelagic larvae, *M. balthica*, *H. ulvae* and *E. longa* settled and developed mainly away from the adult populations. Juveniles of *E. longa* settled and developed in the low littoral zone, whereas their adults were found in the upper littoral levels. Such can be explained by a strong influence of the emersion period on newly settled larvae. Juveniles, as well as adults, of *H. ulvae* disperse by floating at the water surface, and aggregate in salt marshes, small depressions and in littoral puddles, as a consequence of which the numbers may vary by an order of magnitude within a few days (Armonies & Hartke, 1995).

Conclusions

In general, theories of benthic community organization explain observed community patterns in terms of distribution and abundance of adult organisms. Our research digressed from this tradition and focused the interest on the juvenile stage of macrobenthic community, because the main processes determining the adult composition occur in the earliest stages of an organism's development (Thorson, 1958; Watzin, 1986). The data allow estimation of the importance of juveniles for the formation and maintenance of community structure.

We have shown that the taxonomic composition of the benthic community is not a result of random spatfall of all larvae from the water column and their subsequent elimination. On the contrary, the species composition is determined at the moment of settling and then does not change, while the spatial distribution is determined, and may significantly change, by the elimination and migration of organisms during their growth and development from juvenile to adult.

Juvenile macrobenthos is more sensitive to environmental factors such as tidal level, granulometric composition of sediments and plant biomass than adult macrobenthos. We suggest that the distribution of juveniles is in general determined by environmental factors and that during growth the influence of abiotic factors becomes weaker whereby other factors, such as interactions between species, become more dominant.

Hardly any correspondence was found in the spatial distribution of juveniles and adults of macrobenthic species with a pelagic stage in their life cycle. Adults and juveniles of species with direct development show more similar distribution patterns. Differences in the spatial distribution of juvenile and adult macrobenthos allow to suggest the presence of structural changes in community organization during an organism's growth which may demand several seasons.

Acknowledgments

This research was supported by the Russian Fund of Fundamental Researches (grant No. 96-04-48367). We would like to express our gratitude Mr M. J. Kolobov, Mr M. V. Chertoprood, Ms Z. A. Zviaguintseva for help in collecting field material, Dr I. A. Jirkov, Dr A. B. Tzetlin, Dr N. V. Kucheruk for help with identification animals, Dr A. I. Azovsky, Dr V. O. Mokievsky for advice, Dr A. K. Kashunin for help in preparing this manuscript, two anonymous referees and Dr H. Hummel for editing.

References

Armonies, W. & D. Hartke, 1995. Floating of mud snails *Hydrobia ulvae* in tidal waters of the Wadden Sea, and its implications in distribution patterns. Helgolander wiss. Meeresunters. 49: 529–538.

Armonies, W. & M. Hellwig-Armonies, 1992. Settlement and migration of *Macoma balthica* spat. Neth. J. Sea Res. 29: 371–378.

Burkovsky, I. V., 1992. Structural-functional organization and stability of marine bottom communities. MSU Press, Moscow, 208 pp. [in Russian]

Burkovsky, I. V., A. A. Udalov & A. P. Stoljarov, 1996. Comparative study of the structure of the juvenile and adult macrobenthos of the White sea littoral. Zool. Zh. 10: 1452–1462 [in Russian].

Butman, C. A., 1987. Larval settlement of soft-sediment invertebrates: the spatial scales of pattern explained by active habitat selection and the emerging role of hydrodynamical processes. Oceanogr. Mar. Biol. annu. Rev. 25: 113–165.

Butman, C. A., J. P. Grassle & C. M. Webb, 1988. Substrate choices made by marine larvae settling in still water and in a flume flow. Nature 333: 771–773.

Kaljakina, N. M., 1982. Ecology of the *Arenicola marina* (L.) (Polychaeta) in the White sea. Ph.D. Diss., MSU, Moskow, 180 pp. [in Russian]

Kaufman, Z. S., 1974. Ecological rules of the reproductive output of the mass species of the White sea invertebrates. Zool. Zh. 55: 5–16. [in Russian]

Meenakumari, B. & N. Balakrishnan, 1994. Settlement and community interrelations of fouling organisms in Cochin Harbour, India. Fish. Technol. 31: 12–47.

Mileikovsky, S. A., 1976. The types of larval development of the bottom invertebrates. In: Ecology and biogeography of the plankton. Nauka, Moscow, 214–248. [in Russian]

Plate, S. & E. Husemann, 1992. An alternative mode of larval development in *Scoloplos armiger*. Helgol. Meeresunters. 45: 487–492.

Rasmussen, E., 1973. Systematics and ecology of the Isefiord marine fauna (Danmark). Ophelia 11: 92–93.

Reise, K., 1985. Tidal flat ecology. Springer, Berlin, 191 pp.

Santos, S. L. & J. L. Simon, 1980. Marine soft-bottom community establishment following annual defaunation: larval or adult recruitment? Mar. Ecol. Progr. Ser. 2: 235–241.

Satchell, E. R. & T. M. Farrell, 1993. Effects of settlement density on spatial arrangement in four intertidal barnacles. Mar. Biol. 116: 241–245.

Semenova, N. L., 1972. The distribution of bivalve *Macoma balthica* (L.) in some gulfs of the Kandalaksha Bay of the White Sea. Proc. White Sea Biol. Station 4: 87–102. [in Russian]

Shirley, T., 1990. Ecology of *Priapulus caudatus* Lamarck, 1816 (Priapulida) in the Alaskan subarctic ecosistem. Bull. Mar. Sci. 47: 149–158.

Sveshnikov, V. A., 1978. The morphology of Polychaeta larvae. Nauka, Moskow, 151 pp. [in Russian]

Thorson, G., 1946. Reproduction and larval development of Danish marine bottom invertebrates. Medd. Kommn. Fiskog. Havunders. Ser. Plancton 4: 1–523.

Thorson, G., 1958. Parallel level-bottom communities, their temperature adaptation, and their 'balance' between predators and food animals. In Buzzata-Traverso, A. A. (ed.), Perspectives in Marine Biology. University of California Press, Berkeley, California: 67–86.

Watzin, M. C., 1983. The effects of meiofauna on settling macrofauna: meiofauna may structure macrofaunal communities. Oecologia 59: 163–166.

Watzin, M. C., 1986. Larval settlement into marine soft-sediment systems: interactions with the meiofauna. J. exp. mar. Biol. Ecol. 98: 65–113.

Hydrobiologia **355:** 11–19, 1997.
A. D. Naumov, H. Hummel, A. A. Sukhotin & J. S. Ryland (eds), Interactions and Adaptation Strategies of Marine Organisms.

Epibenthic predation in marine soft-bottoms: being small and how to get away with it

Martin Thiel
Darling Marine Center, University of Maine, Walpole ME 04573, U.S.A.
Present address: Smithsonian Marine Station, 5612 Old Dixie Highway, Fort Pierce, FL 34946, U.S.A.

Accepted 30 May 1997

Key words: extended parental care, soft-bottoms, predation, protection, recruitment, amphipoda

Abstract

In intertidal soft-bottoms, epibenthic predation is one of the most important post-recruitment processes. Small juveniles are particularly susceptible to predation, and they often settle in the high intertidal where predation pressure is relatively low. Growth conditions in the high intertidal are, however, only suboptimal compared to the low intertidal. Juvenile amphipods *Leptocheirus pinguis* usually remain in their mother's burrow for extended time periods growing to average sizes of 4–6 mm during this extended parental care. In laboratory experiments, more juvenile amphipods survived in controls than in predator additions. In the predator treatments, most adult amphipods survived while many juveniles disappeared. Medium-sized juveniles (6–10 mm size) that had already established their own burrows emigrated in large numbers from the predator treatments whereas most of the adult females remained as residents in these trays. Juvenile *L. pinguis* survived periods where they are most susceptible to epibenthic predation in the protected burrows of their mother. Extended parental care enables juvenile amphipods to recruit immediately into the adult habitat with a good survival chance. It is hypothesized that some small soft-bottom infauna find protection in the burrows of other infauna, medium-sized infauna is most likely to engage in escape reactions, whereas large infauna build their own, deep, burrows, safe from epibenthic predators.

Introduction

Predation is an important post-recruitment process in soft-bottom environments (Ólafsson et al., 1994). Many soft-bottom inhabitants are potential prey of various other organisms during their life times. Their susceptibility to predation can vary during their life time. Some species become more attractive with increasing age and size by reaching profitable bite sizes for large predators. Others outgrow their most voracious predators becoming too big to be handled or building deeper burrows where they cannot be reached. Juvenile macrofauna are known to be primarily susceptible to epibenthic predation (Reise, 1985). Various strategies during the early life stages of juvenile macrofauna are attributed to the avoidance of predation. Two major categories of predator avoidance strategies can be identified in benthic marine organisms (Table 1). The first one might be termed separation strategy, where the potential prey either seeks temporal or spatial separation from its major predators. Prey organisms utilizing the second, the protection strategy, often live in the same general habitat as the predators but avoid predation by either hiding in crevices, burrows or by being untasty to predators (e.g. Reise, 1985; Hay & Fenical, 1996).

Both strategies have advantages and disadvantages. The separation strategy requires prey organisms to adjust their presence or activity to the predator presence. Prey organisms may be restricted to areas which are suboptimal with respect to resource supply (oxygen, food), but relatively safe from predation (see e.g. Beukema & de Vlas, 1989). The protection strategy might allow prey organisms to remain in optimal areas as long as they can avoid being eaten. Organisms able to find protection from predation within an optimal resource environment can achieve high growth rates. Those organisms that segregate from predators

Table 1. Predator avoidance strategies of benthic marine invertebrates.

Strategy	Prey behaviour	References (selected examples)
Seperation	Segregation of activity from predator activity	de Vlas, 1985; Kamermans & Huitema, 1994
	Recruitment to areas with low predator abundances	Günther, 1990; Beukema & de Vlas, 1989
	Preventive emigration in response to predator presence	Armonies, 1994
	Escape in response to immediate predator encouters	Ambrose, 1984; Olafsson & Persson, 1986; Thiel & Reise, 1993
Protection	Production or uptake of untasty substances	Kem, 1985; Hay & Fenical, 1986
	Selection of cryptic habitats	Jormalainen & Tuomi, 1989
	Shelter on, in or with other organisms	Ockelmann & Muus, 1978

and settle in areas with diminished resource supply have relatively low growth rates. Bivalves *Cerastoderma edulis* that were protected from predators by cages in low intertidal soft-bottoms with high predation pressure but optimal food supply grew much faster than their conspecifics in high intertidal soft-bottoms with few predators but only suboptimal food supply (Reise, 1985). Juvenile bivalves are extremely susceptible to epibenthic predators, however they outgrow most of their predators by the time they reach 30 mm in length (Reise, 1985). Many organisms in intertidal or shallow subtidal soft-bottoms seek temporal or spatial separation from the major epibenthic predators during the juvenile stages while small and susceptible. After reaching a certain size, they immigrate into the adult habitat, where growth conditions are usually more favorable than in the juvenile habitat. During these migrations via the water column (e.g. Armonies, 1994), organisms are exposed to other predators and may be misplaced by currents.

The ability to recruit immediately into the adult habitat, thus avoiding long migrations to predator-safe juvenile habitats would present an enormous advantage to recruiting soft-bottom fauna. Yet, few juvenile or small soft-bottom species seem to utilize the protection strategy successfully. A few small bivalve species brood early juvenile stages, and these recruit immediately into the adult habitat after being released from their mother's body (Gallardo, 1993). Some meiofauna organisms (e.g. Reise, 1985; Dittmann, 1996) as well as a variety of other organisms (e.g. Ockelmann & Muus, 1978; Ó Foighil & Gibson, 1984) are found in high numbers in the burrows of larger macrofauna where they might find protection from epibenthic predators. These guests cannot build deep burrows themselves, but rather rely on their hosts. Similarly, early larval and juvenile stages of some polychaetes remain in the relatively well protected burrows of their parents (e.g. *Nereis diversicolor*, Bartels-Hardege & Zeek, 1990). The juveniles of some amphipods remain in the burrows of their parents for extended time periods (Shillaker & Moore, 1985; Thiel et al., 1997) until they have attained almost half the adult size. We hypothesized that this form of extended parental care provides effective protection for the juvenile amphipods from epibenthic predation (Thiel et al., 1997). The amphipods *Leptocheirus pinguis* inhabit soft-bottoms at MLW (mean low water) where food supply is favorable (Thiel, in press), but epibenthic predators are also very abundant (unpubl. data).

The aim of the present study was to examine whether juvenile amphipods are more susceptible to predation than large adults. Only then would parental care be a useful strategy in avoiding predation. Amphipods were collected in the field to elucidate what sizes of amphipods did successfully establish individual burrows, and how deep these burrows were. The response of juvenile and adult amphipods to epibenthic predators was examined in predation experiments. The results of this study help to answer whether extended parental care can be a successful protection strategy for juvenile macrofauna in marine soft-bottoms.

Materials and methods

Field collection of amphipods Leptocheirus pinguis

In the summer of 1993 juvenile amphipods were collected from their habitat in Lowes Cove, Maine. Juveniles were collected from their mothers' burrows every two weeks from 8th June until 1st September 1993. Additionally, at each sampling date small amphipods were collected from their own individual burrows in order to determine the smallest size of amphipods which recruited into the adult habitat. In August 1996, a possible relationship between amphipod size and the burrow depths was examined. Individual amphipod burrows were sampled in Lowes Cove, the sediment was broken up along the burrow and the depth of the burrows measured with a ruler. All amphipods were preserved and their size measured with a computer-based video-analysis system along their dorsal surface from the rostrum to the base of the telson.

Predation experiments

Ovigerous female amphipods *Leptocheirus pinguis* were collected from mudflats in Lowes Cove, Maine. Eighteen females were introduced to each of 6 trays that contained natural sediments collected from the amphipod habitat. Another 18 females were preserved immediately after collection, and their eggs counted in order to determine the reproductive potential. The trays had a size of 900 cm^2 and the sediment in the trays had a depth of about 10 cm. Fresh seawater was flowing through the trays, and the females were allowed to establish burrows and rear juveniles within the burrows for approximately 5 weeks. During this time, the outflow was covered with a 500 μm screen, confining all amphipods to the trays. When the first juveniles started emerging from their mother's burrows, the screen was removed (= start of experiment), and the outflowing seawater passed through a trap with a 500 μm screen, collecting all organisms emigrating from the trays. For the next seven days, all emigrants were counted each morning, but then reintroduced to their respective trays. At day seven, predators (10 sandshrimp *Crangon septemspinosa* per tray) were introduced to three trays, while the three remaining trays were left predator-free as controls. Following the predator addition, all emigrants (= escapees) were collected and preserved every day. The experiment was terminated on day 14 and the contents of each tray were sieved over a 500 μm sieve. All juvenile and adult amphipods were collected, preserved, and measured.

Results

Sizes of juvenile amphipods Leptocheirus pinguis *in the field*

During the whole sampling period (June–September 1993) the average size of juveniles in their mothers' burrows varied between 4 and 6 mm (Figure 1a). Significant differences between consecutive sampling dates were only observed early in the summer (ANOVA, $p<0.05$; followed by Scheffé-test, $p<0.05$). The

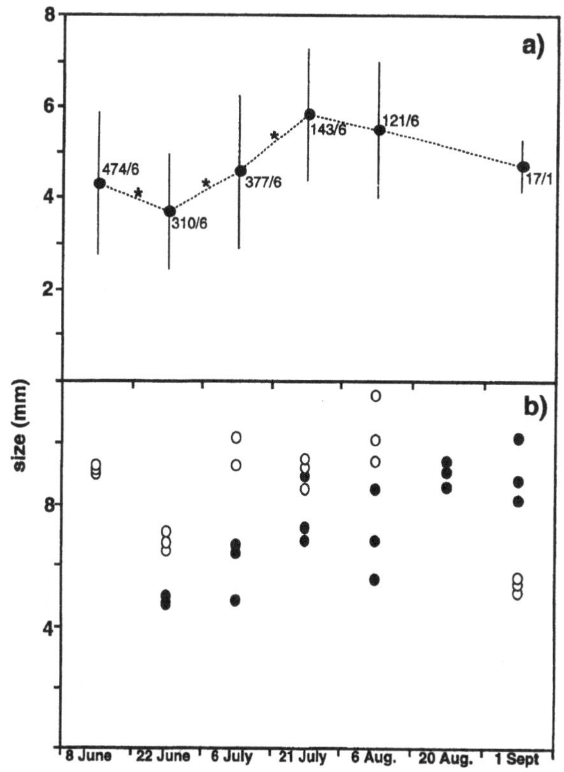

Figure 1. (a) Average size (mm; ± std err) of juveniles collected from their mothers' burrows in Lowes Cove during the summer of 1993 (n = number of juveniles/N = number of mothers from which juveniles were collected); significant differences (ANOVA, $p<0.05$; post-hoc Scheffé test, $p<0.05$) between consecutive sampling dates marked by $*$. (b) The three smallest juveniles *Leptocheirus pinguis* found in own burrows (filled dots) and the three largest juveniles in mothers' burrows (open dots) in Lowes Cove in the summer of 1993; no females with juveniles in their burrows were found on 20 August 1993.

size of the smallest juveniles in their own individual burrows slightly increased during the summer (Figure 1b) and most were at least 5 mm in length. There exists a significant correlation between the size of amphipods and the depths of their burrows ($y=3.662x+9.546$, $R^2=0.424$; t-test, $p<0.01$) (Figure 2). In August 1996, all amphipods collected were larger than 6 mm; only two individuals inhabited burrows which extended less than 20 mm below the sediment surface. All amphipods larger than 14 mm were found in burrows at least 40 mm deep.

Predation experiment

The 18 female *Leptocheirus pinguis* preserved at the beginning of the experiment contained on average 64 eggs in their brood pouch (mean ± s.e.: 63.67 ± 3.71)

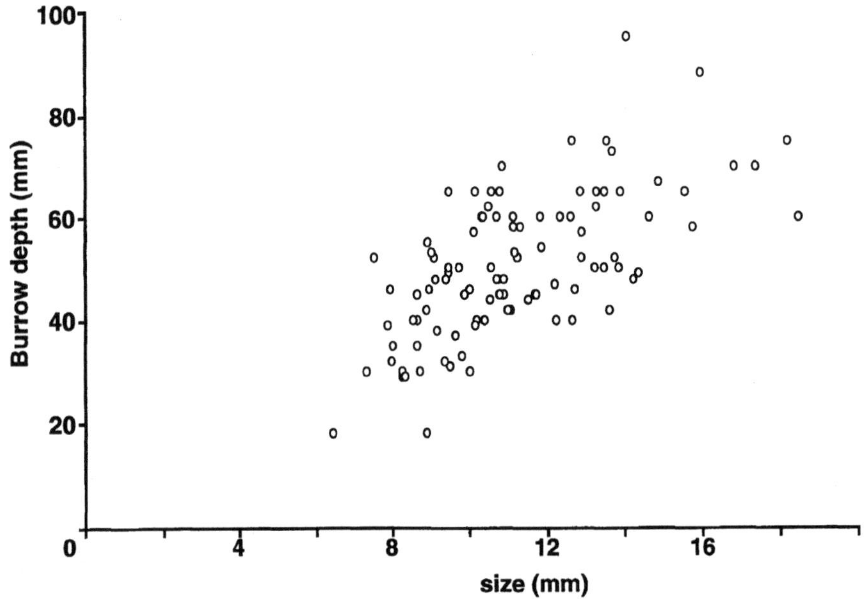

Figure 2. Relationship of the size (mm) of individual amphipods *Leptocheirus pinguis* and the depth of their burrows (mm) in Lowes Cove in August 1996.

resulting in a total reproductive potential of about 1150 juveniles for each tray (18 females × 64 eggs). The percentage of surviving juveniles in the controls (81.8%) was significantly higher than that of the predator treatments (61.6% surviving juveniles) (Mann-Whitney U-test, $p<0.05$). In the control trays without predators, an average of 941 juveniles survived at the end of the experiment (=81.8% of the total reproductive potential per tray) (Figure 3a). Those include 408 emigrants (=35.4% of total reproductive potential) and 533 residents per tray (=46.4%). In the predator treatment 584 emigrants (50.8%) and 124 resident juveniles (= 10.8%) per tray escaped from predation, adding up to 708 survivors (=61.6% of total reproductive potential). The percentages of females surviving in predator treatments and controls were not significantly different (Mann-Whitney U-test; $p>0.1$). Of the 18 large females in each tray an average of 100% survived in the control and an average of 78% survived in the predator treatment. No females emigrated from the controls, while about 20% escaped from the predator treatment (Figure 3b). Thus, in the predator treatment almost 60% of the large females but only 10% of the potential 1150 juveniles remained in the trays at the end of the experiment. A higher percentage of juveniles emigrated from the predator treatments (50%) than from the controls (30%).

Most juveniles between 6 mm and 10 mm size remained as residents in the predator-free control (Figure 4a). About 50% of each size class of the small juveniles (2–6 mm body size) emigrated from the predator-free control trays, but the other 50% immediately recruited to, and were residents in, the trays. The addition of predators resulted in a strong escape response in all juvenile sizes, particularly the large ones (6–10 mm body size) (Figure 4b). In the predator treatments, a relatively high percentage of small juveniles (2-6 mm body size) remained as residents in the trays (20 to >40% of each size class) (Figure 4b).

Discussion

Juvenile amphipods *Leptocheirus pinguis* are more susceptible to epibenthic predation than large adults. They spend at least part of the time during which they are particularly susceptible to predation in the protection of their mother's burrow. Juvenile amphipods in their own burrows react to epibenthic predators with a strong emigration response, whereas large adults remain in their deep burrows, indicating that the behavioral response of amphipods *L. pinguis* to these predators changes with size.

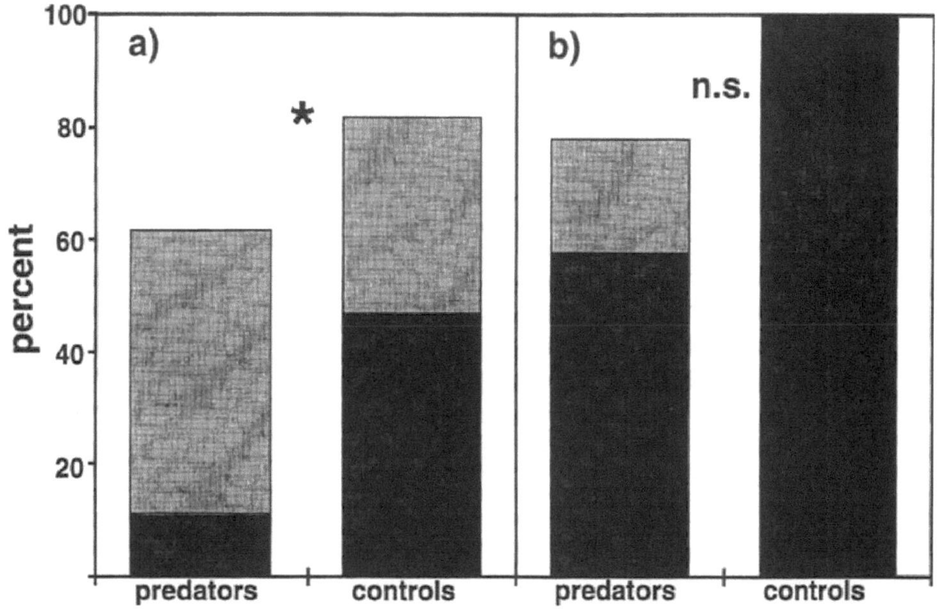

Figure 3. (a) Percentage of juveniles remaining as residents (dark shading) or emigrating (light shading) from experimental trays without (=controls) and with predators; $n=1150$ potential juveniles each in $N=3$ controls and $N=3$ predator treatments. (b) Percentage of females remaining as residents (dark shading) or emigrating (light shading) from experimental trays without and with predators; $n=18$ females each in $N=3$ controls and $N=3$ predator treatments.

Parenting mothers provide protection from epibenthic predators

Extended parental care provides some juvenile amphipods *L. pinguis* with the opportunity to grow to remarkable sizes in the burrow of their mother. The largest juveniles commonly found in females' burrows are between 6 mm and 10 mm in size. At these sizes the juveniles start to build their own burrows, but they are still susceptible to epibenthic predators: the escape of all sizes of juveniles in the experimental predator additions indicates that even large juveniles (6–10 mm) do not gain sufficient protection from epibenthic predation in their own burrows.

Unexpectedly, not all adult females survived in the predator treatments and some emigrated. The high numbers of recruiting juveniles that built their own burrows in the trays may have caused repeated destruction of the upper parts of female burrows. As a consequence, large females frequently had to come to upper sediment layers to repair their burrows thereby exposing themselves to predation. High burrow repair activity of large females in the experiments might thus have resulted in their unexpected disappearance and emigration from predator treatments. Nevertheless, large females are less affected by predator addition than their offspring, indicating that they are relatively safe from epibenthic predators in their deep burrows. These large females can provide a safe habitat for their offspring in their deep burrows, but many juveniles leave their mothers' burrows at a size still susceptible to epibenthic predation. With increasing size of the juveniles, overcrowding and increasing competition in their mother's burrows (there can be more than 100 juveniles in one female's burrow: Thiel et al., 1997) might cause some juveniles to emigrate, resulting in relatively high numbers of emigrating small juveniles (4–6 mm) in both, control and predator treatments (Figure 4). Before overcrowding effects occur, juveniles are much better off staying with their mothers than going out on their own. Extended parental care by the female (providing a protected burrow, maintaining and irrigating the burrow) enables small juveniles to survive in a habitat with high predation pressure (see e.g. Figure 4b).

Escape reaction of small juveniles during predator encounters

Juvenile *Leptocheirus pinguis* spend their first weeks in the protected burrow of their mother. After leaving

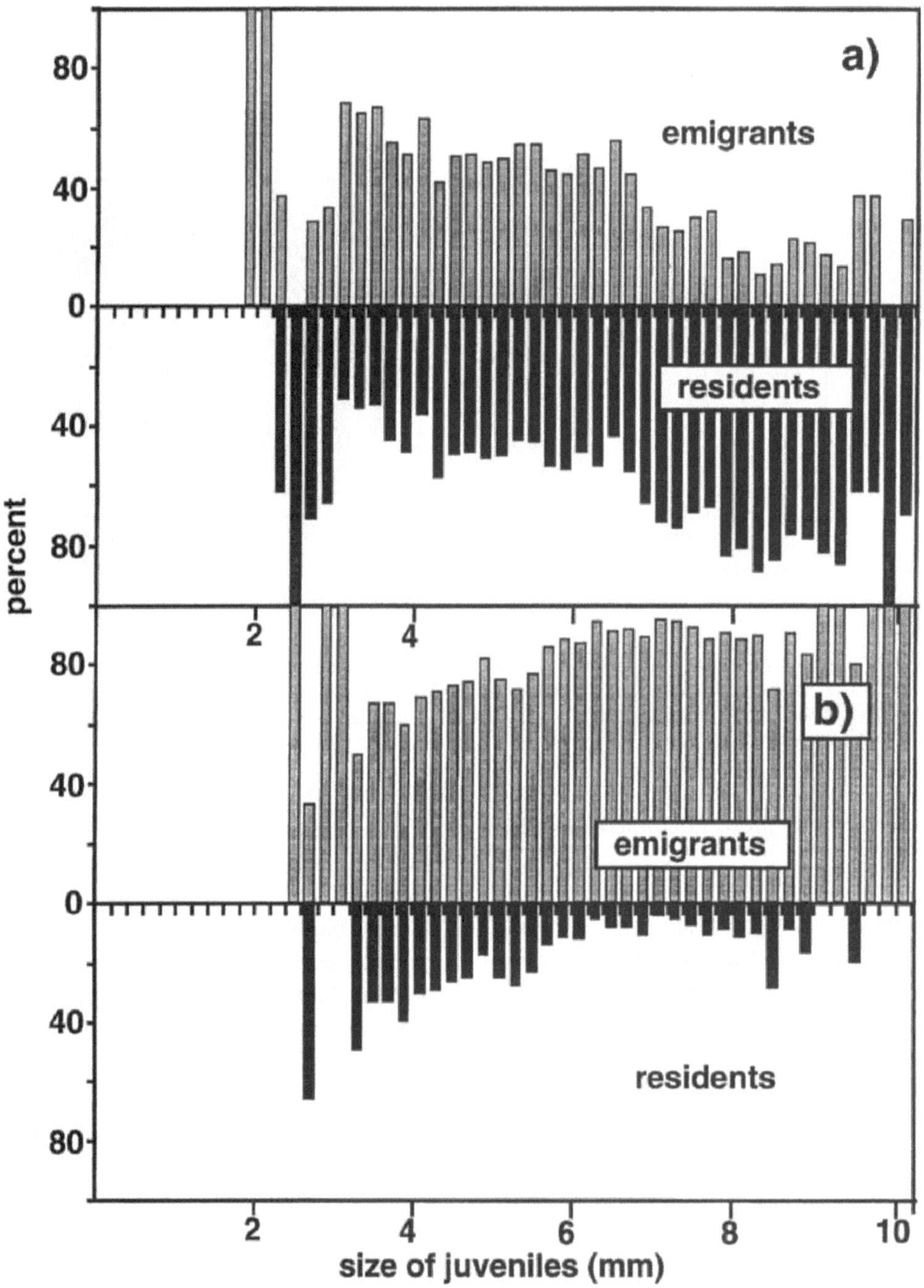

Figure 4. (a) Percentage of juveniles in each size class remaining as residents (dark shading) or emigrating (light shading) from experimental trays without predators; all juveniles recovered from $N=3$ controls were pooled. (b) Percentage of juveniles in each size class remaining as residents (dark shading) or emigrating (light shading) from experimental trays with predators; all juveniles recovered from N = 3 predator treatments were pooled.

their mothers' burrows, the juveniles establish their own burrows in the adult habitat (see large juveniles of 6–10 mm size in Figure 4a). These juveniles are still too small to build burrows deep enough to be well protected from predator encounters. They escape from their shallow burrows in response to immediate predator contacts (see Figure 4b: a high proportion of large juveniles emigrates). This strategy is relatively effective in

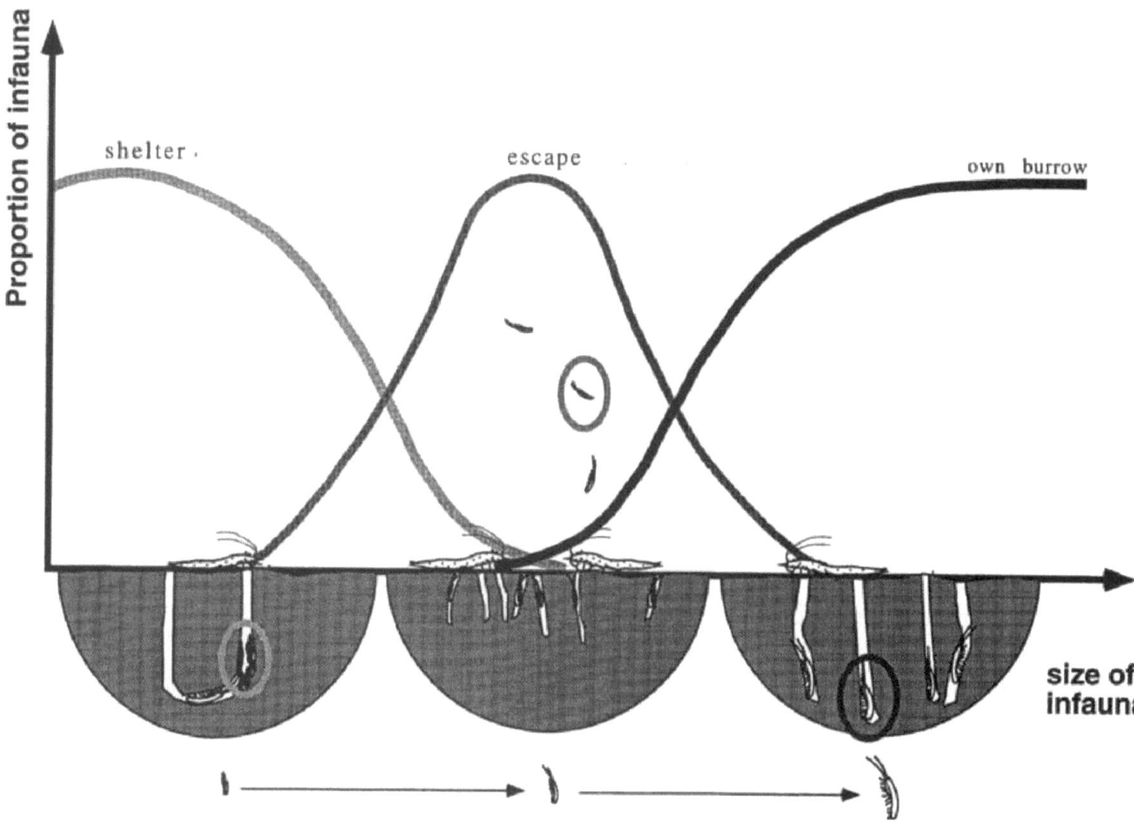

Figure 5. Predator avoidance strategies for small infauna of marine soft-bottoms in relation to their respective body sizes: high proportions of small infauna are expected to seek shelter in macrofauna burrows, medium-sized infauna is expected to avoid predators via escape reactions and large infauna is increasingly safe from epibenthic predation in their own burrows; all three predator avoidance strategies have been reported for respective infauna sizes (for references see text).

avoiding fatal predator encounters. The escape reaction in response to predators has been described for soft-bottom infauna (see e.g. Ambrose, 1984; Rönn et al., 1988; Thiel & Reise, 1993). Many escape reactions were observed in predator addition experiments and it can be assumed that they occur frequently in immediate response to predators. At least one study (Thiel & Reise, 1993) indicates that escaping amphipods only move short distances to get out of the immediate range of predators. Thus, these escape reactions can be considered as short-term as well as short-range response reactions to predators. Escapees seek a predator-free refuge within their general habitat. It has been discussed that high abundances of small benthic organisms in nocturnal plankton might be caused by escape reactions to benthic predators at night (e.g. Ambrose, 1984; Armonies, 1989, 1994). Thus, the escape reaction might be a very common predator avoidance strategy for soft-bottom fauna. This assumption, however, requires further experimental investigation.

Parental protection for juveniles or early larval stages in marine soft-bottom fauna has been described for small bivalve species (Gallardo, 1993), some polychaete species (Bartels-Hardege & Zeek, 1990) and some amphipod species (this study; Thiel et al., 1997). The duration of extended parental care is most likely controlled by limited space and increasingly insufficient resource supply for the growing offspring. The size of juveniles when leaving their parents will determine their survival chances as recruits. Large recruits are expected to have better survival chances, but if not yet big enough to build deep burrows, they can avoid fatal predator encounters via the escape reaction.

Protection strategy in marine soft-bottoms

Marine soft-bottoms do not provide many shelters, and in shallow waters the surface sediments are frequently reworked by epibenthic predators. Small organisms are very susceptible to epibenthic predation, as they often cannot achieve effective protection from predators. Cryptic habitats are relatively rare on marine soft-bottoms, and there are few reports of soft-bottom infauna containing untasty substances (Kem, 1985; Giray, pers. comm.). Small organisms (meiofauna and small macrofauna) can be found in high numbers in the burrows of larger macrofauna (e.g. Ockelmann & Muus, 1978; Ó Foighil & Gibson, 1984; Reise, 1985; Wetzel et al., 1995; Dittmann, 1996), seemingly undisturbed by the burrow founder (Figure 5). With increasing size of its guests, the burrow founder should pay more attention to them, as they might consume substantial amounts of oxygen, food or space in its burrow. Thus, with increasing size one should expect that shelter seeking organisms are kept out of the burrows of larger macrofauna. These organisms have to build their own burrows or tubes, but due to their relatively small size, their burrows will only reach a limited depths. The most effective strategy for these organisms is to stay in their burrows as long as they are undisturbed but to escape into the water column at predator encounters (Figure 5). With increasing size we expect to see fewer escape reactions, as larger organisms are able to build deep burrows safe from epibenthic predators (Figure 5).

It becomes evident that the protection strategy is not a risk-free undertaking for small infauna of marine soft-bottoms. However, it allows these organisms to inhabit areas with high predator abundance but otherwise optimal conditions (e.g. permanent water coverage and optimal food supply). Seeking shelter with larger organisms might be inhibited when resources for shelter-seekers or shelter-providers (food, oxygen) are limited. In general, small infauna engaged in the protection strategy are able to avoid risky migrations to predator-free areas, and they do not have to adjust their activity schedule to the presence or activity of predators.

Acknowledgments

Support for this study was received in form a graduate fellowship from the Center for Marine Studies at the University of Maine and grants from the Association of Graduate Students at the University of Maine. My participation at the 31st EMBS was supported by grants from the Association of Graduate Students, The Alumni Association and the School of Marine Sciences at the University of Maine. Ian Voparil and two anonymous reviewers provided important comments. My very special thanks extends to Svetlana and Andrey for their hospitality during my stay in St. Petersburg.

References

Ambrose, W. G., Jr., 1984. Increased emigration of the amphipod *Rhepoxynius abronius* (Barnard) and the polychaete *Nephtys caeca* (Fabricius) in the presence of invertebrate predators. J. exp. mar. Biol. Ecol. 80: 67–75.

Armonies, W., 1989. Meiofaunal emergence from intertidal sediment measured in the field: significant contribution to nocturnal planktonic biomass in shallow waters. Helgoländer Meeresunters 43: 29–43.

Armonies, W., 1994. Drifting meio- and macrobenthic invertebrates on tidal flats in Königshafen: a review. Helgoländer Meeresunters 48: 299–320.

Bartels-Hardege, H. D. & E. Zeek, 1990. Reproductive behavior of *Nereis diversicolor* (Annelida: Polychaeta). Mar. Biol. 106: 409–412.

Beukema, J. J. & J. de Vlas, 1989. Tidal-current transport of thread-drifting postlarval juveniles of the bivalve *Macoma balthica* from the Wadden Sea to the North Sea. Mar. Ecol. Prog. Ser. 52: 193–200.

Dittmann, S., 1996. Effects of macrobenthic burrows on infaunal communities in tropical tidal flats. Mar. Ecol. Prog. Ser. 134: 119–130.

Gallardo, C. S., 1993. Reproductive habits and life cycle of the small clam *Kingiella chilenica* (Bivalvia: Cyamiidae) in an estuarine sand flat from the South of Chile. Mar. Biol. 115: 595–603.

Günther, C.-P., 1990. Distribution patterns of juvenile macrofauna on an intertidal sandflat: an approach to the variability of predator/prey interactions. In Barnes, M. & R. N. Gibson (eds), Trophic Relationships in the Marine Environment (Proc. 24th Europ. Mar. Biol. Symp.). Aberdeen Univ. Press, Aberdeen, 77–88.

Hay, M. E. & W. Fenical, 1996. Chemical ecology and marine biodiversity: insights and products from the sea. Oceanogr. 9: 10–20.

Jormalainen, V. & J. Tuomi, 1989. Sexual differences in habitat selection and activity of the colour polymorphic isopod *Idotea baltica*. Anim. Behav. 38: 576–585.

Kamermans, P. & H. J. Huitema, 1994. Shrimp (*Crangon crangon* L.) browsing upon siphon tips inhibits feeding and growth in the bivalve *Macoma balthica* (L.). J. exp. mar. Biol. Ecol. 175: 59–75.

Kem, W. R., 1985. Structure and action of nemertine toxins. Am. Zool. 25: 99–111.

Ockelmann, K. W. & K. Muus, 1978. The biology, ecology and behaviour of the bivalve *Mysella bidentata* (Montagu). Ophelia 17: 1–93.

Ó Foighil, D. & A. Gibson, 1984. The morphology, reproduction and ecology of the commensal bivalve *Scintillina bellerophon* spec. nov. (Galeommatacea). The Veliger 27: 72–80.

Ólafsson, E. B. & L.-E. Persson, 1986. The interaction between *Nereis diversicolor* O. F. Müller and *Corophium volutator* Pallas as a structuring force in a shallow brackish sediment. J. exp. mar. Biol. Ecol. 103: 103–117.

Ólafsson, E. B., C. H. Peterson & W. G. Ambrose, Jr., 1994. Does recruitment limitation structure populations and communities of macro-invertebrates in marine soft sediments: the relative significance of pre- and post-settlement processes. Oceanogr. Mar. Biol. Ann. Rev. 32: 65–109.

Reise, K., 1985. Tidal Flat Ecology. Springer, Berlin, 191 pp.
Rönn, C., E. Bonsdorff & W. Nelson, 1988. Predation as a mechanism of interference within infauna in shallow water soft-bottoms; examples with an infauna predator, *Nereis diversicolor* O. F. Müller. J. exp. mar. Biol. Ecol. 116: 143–157.
Shillaker, R. O. & P. G. Moore, 1987. The biology of brooding in the amphipods *Lembos websteri* Bate and *Corophium bonnellii* Milne Edwards. J. exp. mar. Biol. Ecol. 110: 113–132.
Thiel, M. & K. Reise, 1993. Interaction of nemertines and their prey on tidal flats. Neth. J. Sea Res. 31: 163–172.
Thiel, M., 1997. Extended parental care in a high food environment – 'Babies don't go in the mud'. In Hawkins, L. E., Hutchinson, S., Jensen, A. C., Willian, J. A. & M. Sheader (eds), Responses of Marine Organisms to their Environment: Proceedings of the 30th European Marine Biology Symposium, Southampton, September 1995. Southampton Oceanography Centre, Southampton, in press.
Thiel, M., S. Sampson & L. Watling, 1997. Extended parental care in two endobenthic amphipods. J. Nat. Hist. 31: 713–725.
Vlas, J. de, 1985. Secondary production by siphon regeneration in a tidal flat population of *Macoma balthica*. Neth. J. Sea Res. 19: 147–164.
Wetzel, M. A., P. Jensen, O. Giere, 1995. Oxygen/sulfide regime and nematode fauna associated with *Arenicola marina* burrows: new insights in the thiobios case. Mar. Biol. 124: 301–312.

Hydrobiologia **355:** 21–28, 1997.
A. D. Naumov, H. Hummel, A. A. Sukhotin & J. S. Ryland (eds), Interactions and Adaptation Strategies of Marine Organisms.

Abundance, feeding behaviour and nematocysts of scyphopolyps (Cnidaria) and nematocysts in their predator, the nudibranch *Coryphella verrucosa* (Mollusca)

Carina Östman
Animal Development and Genetics, Uppsala University, Norbyvägen 18 A, S-752 36 Uppsala, Sweden

Key words: Scyphopolyp, *Aurelia*, Feeding behaviour, Nematocyst, Nudibranch, Cnidosac

Abstract

The abundance of *Aurelia* and *Cyanea* scyphopolyps on *Laminaria saccharina* was higher in sheltered, shallow areas compared with more exposed or deep ones. Liberated planulae probably were not transported far away from stranded and trapped jellyfish and settled nearby, preferably in patches on the downward side of the *Laminaria* thallus.

The principal prey for the scyphopolyps seems to be small copepods and the cladocerans *Podon* sp. and *Evadne* sp., which are abundant in surface waters during the summer. Temporarily abundant planktonic organisms, e.g., *Sagitta setosa*, *Pleurobrachia pileus* and hydromedusae might, also be important prey.

Harpacticids, halacarideans and *Corophium* sp., whose natural habitat is on *L. saccharina*, were not captured by the scyphopolyps. Scyphopolyps cultured in running sea water rich in detritus and phytoplankton fill their enteron with organic substrates, particularly diatoms.

A new category of heterotrichous microbasic rhopaloid nematocysts was identified in the scyphopolyps. These rhopaloids were earlier included within the euryteles and were not considered to be separate nematocysts. They are distinctive from the euryteles due to the two swellings on their discharged shaft.

The absence or presence of the nudibranch *Coryphella verrucosa* on laminarian thalli possibly has an effect on the number of scyphopolyps, as this nudibranch consumes numerous scyphopolyps. Isorhizas and the new category of rhopaloid nematocysts, identical to those present in the *Aurelia* polyps, occurred in the cnidosacs of examined *C. verrucosa*. The proportion of rhopaloid nematocysts compared with a-isorhizas was noticeably higher in *C. verrucosa* than in scyphopolyps. The nudibranch may selectively store rhopaloids.

Introduction

Hernroth & Gröndahl (1983, 1985) have investigated the biology of *Aurelia aurita* (Linnaeus, 1758) from the west coast of Sweden. At the end of summer, after the *Aurelia* medusae have released their larvae, scyphopolyps can occur abundantly on hard substrates or algae in shallow waters. The polyps were most frequently found growing upside-down on the downward side of the attached substrates (Brewer, 1976a, b). The majority of scyphopolyps release their ephyrae in late autumn, but some polyps live in the benthos throughout the winter releasing their ephyras in spring. The nudibranch *Coryphella verrucosa* is a major predator of the scyphopolyp (Hernroth & Gröndahl 1983). However, *C. verrucosa* is most abundant from late autumn to early spring. Later in the summer only few individuals are found in shallow waters in the Gullmar Fjord.

Scyphopolyps are generally assumed to be carnivores and capture zooplanktonic prey by means of their nematocysts. Recent examination of the feeding behaviour of cnidarians revealed that very small organisms, including bacteria, diatoms, dinoflagellates and tintinnids, could also be captured (Cornelius & Östman, 1987; Gili & Hughes, 1995). Unidentifiable particulate organic matter often made up the bulk of the enteron contents (Coma et al., 1994; Simkina, 1980).

The aim of this paper is to study the feeding behaviour of the scyphopolyps, to examine which plankton organisms they prefer and to verify the described upside-down settlement of the planulae (Brewer, 1976a, b). A brief description of the nematocysts identified in the scyphopolyps and in planulae is presented. The cerata of the nudibranch *C. verrucosa* are examined regarding their stored nematocysts. A more detailed description of the *Aurelia* nematocysts studied by light and scanning electron microscopy will be presented elsewhere.

Materials and methods

Scyphopolyps from the Gullmar Fjord (58°15′ N, 11°28′ E) on the Swedish west coast were collected from July to August 1990 and in August 1996. In the field no attempt was made to separate *Aurelia* and *Cyanea* polyps because, when young, their gross morphologies are similar and they cannot be distinguished with confidence by their nematocysts. The polyps were most abundant on the downward side of *Laminaria saccharina*, depth 0.3 to 3 m, but they were also present on *Fucus serratus, Mytilus edulis, Laminaria digitata* and on rocks and stones down to a depth of 10 m. Some polyps were also found on the fine, brown, annual filamentous algae, which are abundant in the Gullmar Fjord during August. The abundance of scyphopolyps growing on *L. saccharina* was investigated in sheltered creeks and on the exposed sides of Blåbergsholmen and Råttholmen, which are small islands at the mouth of the Gullmar Fjord.

The abundance of scyphopolyps on *L. saccharina* was estimated. Generally the polyps occurred in patches on the *Laminaria* thallus. Large patches were divided into 10 cm^2 areas and the numbers of polyps in the areas were counted. The number of polyps in large patches with densely occurring polyps could be estimated by multiplying the number of 10 cm^2 areas within the patch by the numbers of counted scyphopolyps in one representative 10 cm^2 area.

Scyphopolyps developed from liberated planulae of *Aurelia* medusae and scyphopolyps collected from the sea were cultured in small beakers under running sea water (28‰) in the laboratory. Sea water containing concentrated plankton was poured into the beakers with the polyps, or plankton organisms were pipetted towards the polyps. Plankton was collected by hauling a net close over the areas where the scyphopolyps were growing in the sea. Day and night plankton were also collected from other areas in the fjord down to 20 m depth.

The feeding behaviour of the polyps was studied under a stereo microscope equipped with a video camera. Some polyps were starved for a week before the feeding experiments, in order to study differences in feeding behaviour and prey selection between hungry and newly feed polyps.

Laminaria thalli in the studied areas were examined for the presence of *Coryphella verrucosa*.

Squash preparations for light microscopic (LM) investigations were made of fresh nematocysts from scyphopolyps, planulae and from the cerata of *C. verrucosa*. Nematocysts were examined and photographed with interference-contrast optics.

Nematocyst types present were identified according to the classification of Weill (1934) and Mariscal (1974) modified by Calder (1974), Östman et al. (1995) and Östman & Hydman (in press), with terminology according to Watson & Wood (1989).

Results

Habitat and abundance of scyphopolyps

From July to August 1990 the density of scyphopolyps was high on *Laminaria saccharina* growing in shallow and sheltered areas. Up to several thousand polyps were estimated to grow on a single *L. saccharina*. The planulae settled in patches, preferably on the downward side of the *Laminaria* thalli and in the small incisions, which were evenly scattered over the thallus. More than 100 polyps were counted in some 10 cm^2 areas. Often polyps grew on the major part of the downward side of the *L. saccharina*, which was not overgrown by other organisms. No polyps grew, however, on the oldest, partly worn out parts or on the most basal, youngest parts of the *Laminaria* thallus. However, they grew more basely on the thallus than hydroids, bryozoans and epiphytic algae. Few or no polyps were found on the upward side of the *L. saccharina*, which was often completely overgrown with algae, bryozoans and hydrozoans.

Scyphopolyps also grew on old basal parts of *Fucus serratus*. Their density seemed generally to be much lower than on *Laminaria*. A maximum of 46 polyps was found in a 10 cm^2 area.

L. saccharina growing on the more exposed localities or growing at a depth of 3 to 8 m were less covered with scyphopolyps, than laminarians growing in shallow, sheltered waters and the polyps were mostly present on the downward side of the algae.

During July–August, 1996, scyphopolyps were less numerous or were not found in the areas examined in 1990.

Coryphella verrucosa was not observed in the studied areas during the investigations performed in 1990 and 1996. Two specimens of *C. verrucosa* were, however, found on *Laminaria digitata* from the inner part of the Gullmar Fjord during July–August, 1990.

Planulae larvae and their settlement

Planulae placed into small beakers settled within 1–2 days. If small pieces of algae or detritus were present in the beaker they preferred to settle on them; otherwise they preferred to settle around the edge of the beaker or hanging upside-down on the surface water. Generally, they settled in patches. Planulae often swam in circles around already attached scyphopolyps before they settled. Some scyphopolyps ate planulae. One to two days after settlement tentacle buds grew out, and three days after planula release, some scyphopolyps had 8 tentacles. The larger polyps had up to 27 tentacles.

Feeding behaviour of the scyphopolyps

Cultured polyps starved for a week
These polyps had white, thin columns and their tentacles were long and extended when undisturbed. Polyps eagerly captured and ate plankton. Trapped prey was entangled by the contracted tentacles and was moved towards the mouth and engulfed. The polyps captured and ate prey until their columns bulged. Some polyps captured more prey than they could eat and continued until their oral discs and tentacles were completely covered with prey.

Evadne sp., *Podon* sp., small copepods, and *Sagitta setosa*, were captured when present. Even larger copepods such as *Calanus finmarchius* and *Anomalocera patersoni* were preyed upon. Other planktonic organisms eaten by the polyps were: hydromedusae, *Pleurobrachia pileus*, polychaetes and polychaete larvae of different species, crab zoeas, shrimp and fish larvae.

Some captured prey were several times longer than the scyphopolyps. Large prey might be partly dissolved outside the polyp. Enzymes of the polyp seem to be transported with the body fluid of the prey. In newly captured *Sagitta* circulation of body fluids was clearly seen, but eventually, the body of the *Sagitta* grew less transparent. Two scyphopolyps, simultaneously eating from the head and tail of one *Sagitta* were studied for 2,5 hours. The middle part of the *Sagitta* between the two feeding polyps lost its transparency and grew more opaque during this time. Finally, the middle of the *Sagitta* slowly dissolved. A starved scyphopolyp could engulf an entire *Sagitta* within 18 minutes.

Other polyps had simultaneously one or two *Sagitta setosa*, one hydromedusa or one *Pleurobrachia pileus* and several copepods on their tentacles. One large scyphopolyp with 23 tentacles which had caught two specimens of *Sagitta*, one specimen of *Pleurobrachia* and several copepods was studied. Within half an hour one *Sagitta* and the *Pleurobrachia* were engulfed, but complete engulfment of the second *Sagitta* took 4 hours 40 min. Meanwhile the polyp had engulfed the earlier caught copepods plus a few additional copepods. After the *Sagitta* was engulfed, the polyp showed only vague feeding reaction for several hours.

Newly collected scyphopolyps growing on Laminaria
These polyps were not as voracious eaters as were the starved ones. A single large polyp could eat 4–5 copepods before it ceased to capture a new prey. Thereafter the tentacles of the polyps remained contracted for 2–6 minutes before the tentacles were extended and ready to capture another prey.

Scyphopolyps kept in sea water with particulate organic material or phytoplankton
Polyps not fed with zooplankton but kept in slowly running sea water rich with organic particles engulfed the sedimentary materials. Their columns were thick and brownish and their tentacles were not extended in a capturing position. When fed with plankton, they showed no or vague feeding reaction. Materials ejected from the mouth were examined under light microscopy (LM). Diatoms, algae cells and discharged nematocysts were identified along with unidentified materials. The tentacles of the polyp and the incisions in between were heavily ciliated with long cilia. Ciliate transport of small particles could be followed under LM.

When squashed and investigated under LM, polyps cultured in water containing phytoplankton and small pieces of algae materials were observed to have green material including diatoms in their enteron. Particulate material coloured the epithelium of the enteron.

Large pieces of green tissue, presumably from *Zostera marina* were seen being engulfed.

Other observations

None of the captured large plankton prey was immediately paralysed. Copepods and other crustaceans were struggling furiously for several minutes after being caught. They were even seen moving in the enteron of the polyps after they had been engulfed. *Sagitta* was, however, immobilised faster and was motionless within a few minutes after being captured.

Noctiluca miliaris, undetermined ciliates, turbellarians, nematodes, halacarideans, harpacticoids, *Jassa* sp. and *Iarea* sp. were not observed to be caught and eaten.

Identified nematocysts

Two types of homotrichous isorhizous haploneme nematocysts and one type of heterotrichous microbasic rhopaloid heteronemes were identified in the scyphopolyp and planulae of *Aurelia aurita* (Figures 1–5). The different nematocysts are characterised and classified as similar nematocysts identified in *Cyanea* (Östman & Hydman, in press).

Homotrichous isorhizas haplonemes

1. Small a-isorhiza with ovate capsule. Inverted tubule densely regularly and horizontally coiled (Figure 2). Capsule length 4.0–7.0 μm, width 2.5–4.0 μm (corresponding to a-atrichs (Weill, 1934) or a-isorhizas (Calder, 1971, 1974)).

2. *Narrow and broad A-isorhizas with narrow-ovate to broad-ovate capsule.* Inverted tubule irregularly horizontally, densely coiled (Figures 3, 5). Capsule length 4.5–11.5 μm, width 2.0–6.0 μm. (corresponding to alpha isorhizas of Calder (1974), polyspiras of Spangenberg (1965) and to small A-isorhizas of Östman & Hydman (in press)).

Heterotrichous microbasic rhopaloid heteronemes

1. Small to medium rhopaloids with sub-spherical capsules. The shaft was visible as a thick, distinct rod inside capsule (Figures 3, 5). When discharged two swellings were visible on the shaft (Figure 1). Capsule length 5.5–11.0 μm, width 3.5–8.5 μm. (a new nematocyst category for *Aurelia*, earlier included in the euryteles, Calder (1977)).

All nematocyst types listed were found in the planulae and in the scyphopolyps as well as in the medusae of *Aurelia aurita*. The rhopaloids of the planulae and scyphopolyp were generally 1 to 2 μm smaller than rhopaloids of the medusae (Figure 5). An additional nematocyst category, the heterotrichous microbasic euryteles and one isorhiza with sub-spherical capsules were also present in the *Aurelia* medusae.

Nematocysts of the Aurelia polyps

The tentacle nematocysts. Nematocysts were most numerous on the tentacles and particularly at the tips. Most tentacular nematocysts were a-isorhizas (Figures 1, 2). Due to the long cnidocils of the a-isorhizas, reaching up to 23 μm, the tentacle tips had a spiny appearance (Figure 4). In some scyphistomae the a-isorhizas were almost the only nematocysts on the tentacles, but generally some rhopaloids were present at the tip or scattered along the tentacles (Figure 2). In other polyps the rhopaloids were more abundantly represented, specially at base of tentacles. The cnidocils of the rhopaloids were 4–5 μm long. In squash preparations many rhopaloids were discharged and released from the tissue (Figure 1).

The column nematocysts. The column bore loosely scattered rhopaloids and a-isorhizas between the richly occurring gland cells. Only a few narrow A-isorhizas and broad A-isorhizas were present. The broad A-isorhiza were more numerous in small 1–3 days old scyphistomae than in older ones. Rhopaloids dominated at the oral disc.

Nematocysts of Aurelia planulae

Small a-isorhizas dominated, followed by rhopaloids (Figure 3). The density of broad A-isorhizas was, however, larger in planulae than in polyps or medusae (Figures 2, 3), and they were most common in the anterior and the posterior end of the larvae.

Nematocysts of the nudibranch Coryphella verrucosa

The nematocysts in the tips of the cerata of *C. verrucosa* were densely packed in small cnidosacs containing between 20–50 nematocysts, with each end of the cerata having up to ten or more cnidosacs. Nematocysts on the long side of the cerata were not seen surrounded by any sac or membrane. One cerata could contain several hundred nematocysts. Rhopaloids were particularly numerous and clearly outnumbered the a-isorhizas. Only a few narrow A-isorhizas were identified.

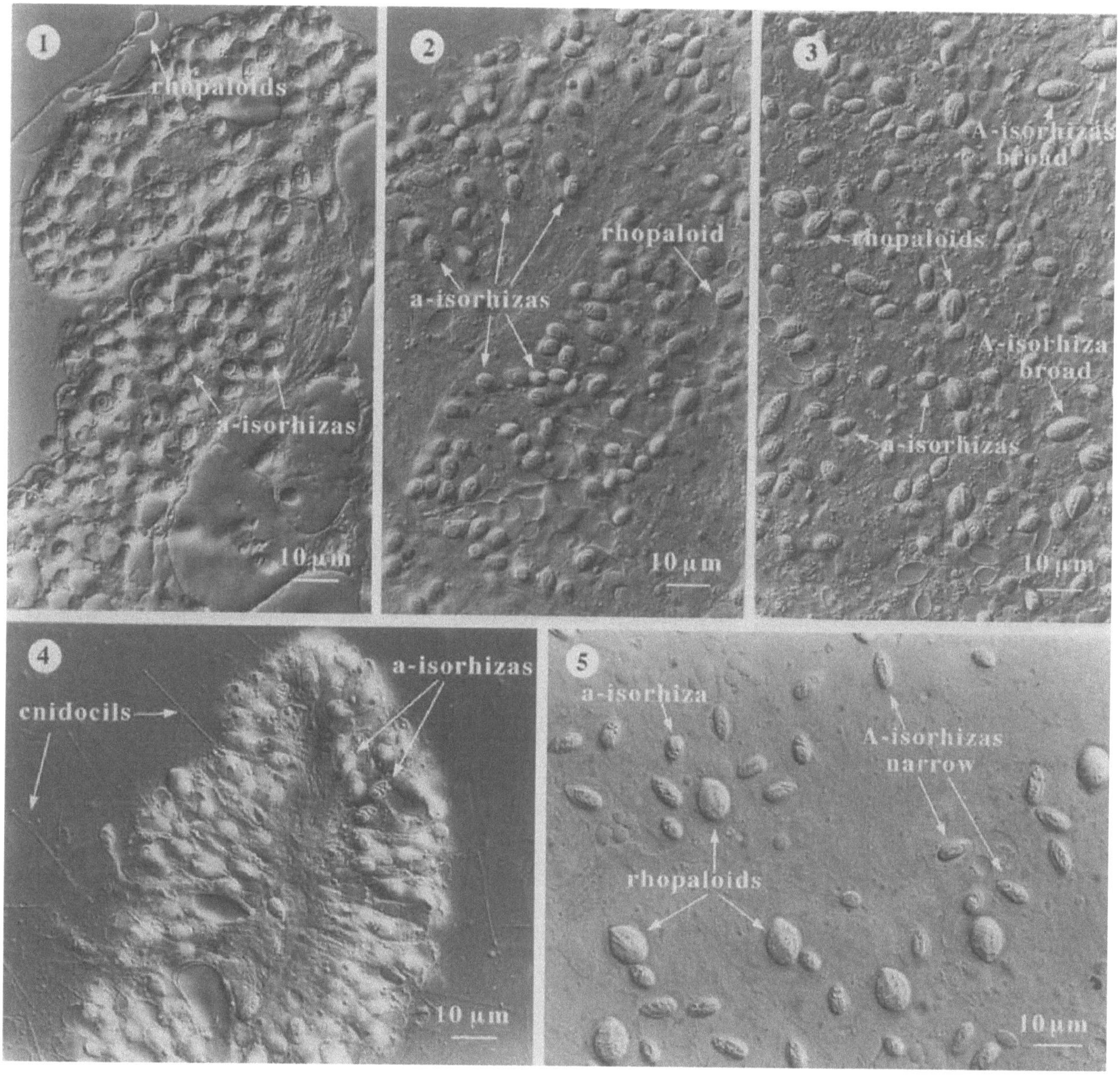

Figure 1. Light micrographs of nematocysts. (1, 2) Tentacles of *Aurelia* polyps. Note the high abundance of a-isorhizas compared with the few rhopaloids. (3) An *Aurelia* planula showing several a-isorhizas, broad A-isorhizas and rhopaloids. (4) Tip of tentacle from an *Aurelia* polyp showing the long cnidocils of the a-isorhizas. Most cnidocils are detached from the tentacle. (5) Base of oral tentacles of an *Aurelia* medusa showing several a-isorhizas, narrow A-isorhizas and rhopaloids. Note the slightly larger rhopaloid size compared with that of the planulae and polyps.

Discussion

Abundance of scyphopolyps

The large abundance of *Aurelia* medusae in the Gullmar Fjord during some weeks in June 1990, and the absence of *Coryphella verrucosa* in the later part of the summer may be correlated with the high abundance of scyphopolyps in August, the same year. In accordance with Brewer (1976a, b) the scyphopolyps were predominantly growing upside-down on the undersurface of the substrate. In laboratory experiments (Brewer, 1976a) planula larvae settled in small cavities or incisions on the substrates. Under natural conditions the planulae behaved in a similar way by settling in the small incisions on the *Laminaria* thallus. The numbers of polyps on *Laminaria saccharina* in sheltered, shallow areas were strikingly high compared with those

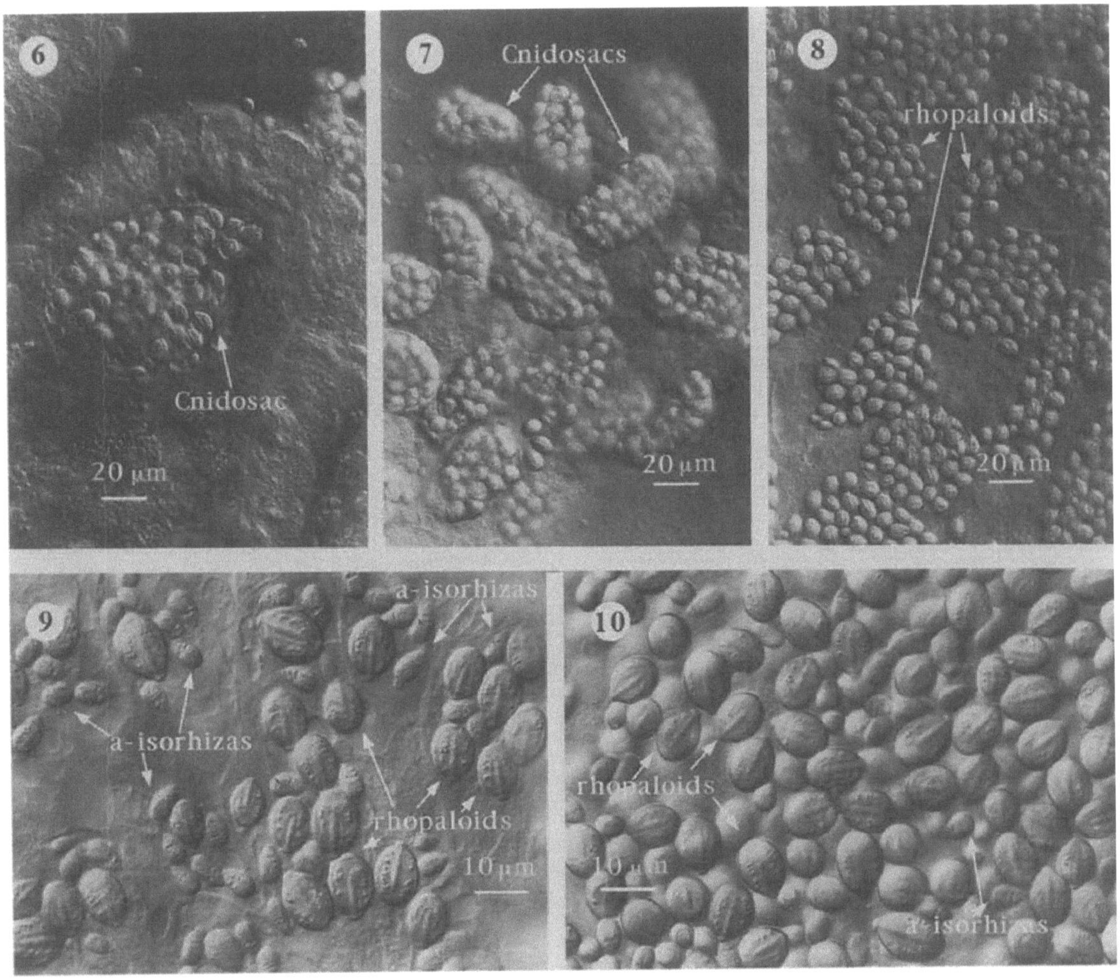

Figure 2. Nematocysts from the tip of cerata of *Coryphella verrucosa.* (6, 7) Slightly squashed tips showing cnidosacs with nematocysts. (8, 9, 10) Note the dominance of rhopaloids compared with the number of a-isorhizas.

of the more exposed areas. Up to several thousand scyphopolyps were estimated to be present on a single *Laminaria* thallus. Mature jellyfish, when trapped and stranded, may release their planulae. Most of the planulae settled probably nearby the trapped jellyfish. Newly released planulae descend to the bottom and to the lowest level in the area (Brewer, 1976a). According to the present laboratory studies planulae could settle within a day, and during such a short time planulae being close to the bottom in sheltered environments were probably not transported far away with water movements.

Due to the unusually cold spring and summer of 1996 on the Swedish west coast, the yearly *Aurelia* bloom was delayed with around a month and did not appear until late July and early August. Scyphopolyps were not found in the studied areas before the *Aurelia* bloom, and thereafter, in middle and late August, their abundance was less than in 1990. This might exemplify the influence of water temperature on a population.

Feeding, prey and diet

The main prey for the scyphopolyps appear to be small copepods and other small Crustaceans, which occur abundantly in the surface water during the summer. *Sagitta, Pleurobrachia* and hydromedusae seemed to be easy for the polyps to catch. They did not attempt to resist capture as energetically as crustacean plankton. Larger copepods and other crustaceans often succeeded in escaping after touching the tentacles of the polyps, although many of them were later found dead.

In agreement with other studies on feeding behaviour and diet of Cnidarians (see Introduction), diatoms and unidentified particulate organic matter were found in the enteron of the scyphopolyp. It can be questioned, however, whether stored energy expended in losing a nematocyst would be regained by the capture of a single minute prey organism. Discharged and detached nematocysts might pass with the prey organism into the enteron of the schypopolyp, however, where they could be digested along with the prey, and thereby regain some of the energy. Particulate, organic material that slowly sediments on the tentacles might be trapped by cilia movements without loss of any nematocysts. Ancestral Cnidarians from Prekambrium were probably consumers of bacteria and protists (Barnes et al., 1993) before the larger zooplankton organisms invaded the water column.

The feeding reaction of the polyp was probably stimulated when the enteron was empty. The reaction decreased in magnitude and finally ceased as the enteron was filled with prey or organic substrate.

Organisms such as the turbellarians, harpacticoids and halacarideans, which have their natural habitat on *Laminaria*, were never seen to be captured, even if they frequently touched the tentacles of the polyps. They might not trigger nematocyst discharge, or they may have evolved immunity against the scyphozoan toxin.

Nematocysts and Corryphella verrucosa

The function of the different nematocysts in the scyphopolyps remains to be studied. Similar nematocyst types in *Cyanea* medusae have been investigated with aid of scanning electron microscopy (Östman & Hydman, in press). The a-isorhizas and the rhopaloids were found to function by penetration, whereas the long tubes of large sized A-isorhizas were seen to entangle the prey.

C. verrucosa may actively store rhopaloids, since the proportion of rhopaloids compared with a-isorhizas was strikingly higher in the examined nudibranch than in the scyphopolyps (Figures 1, 2, 8–10). In the polyps the a-isorhizas outnumbered the rhopaloids several times. Or perhaps the nudibranchs had eaten *Aurelia* medusae since rhopaloids are abundant on their exumbrella. But the isorhizas with sub-spherical capsules, which also occur abundantly on their exumbrella, should have then been found in *Coryphella*. Further investigation is required to fully understand the relationship between *C. verrucosa* and *Aurelia*.

Acknowledgments

I thank Lars Johan Hansson and Prof. Rutger Rosenberg for stimulating advice and valuable criticism of the manuscript. The investigation was carried out at the Kristineberg Marine Research Station, Fiskebäckskil, on the west coast of Sweden. I thank the administrating director and staff for their hospitality and for providing facilities. Financial support was obtained from the Faculty of Science at Uppsala University.

References

Barnes, R. S. K., P. Calow & P. J. W. Olive. 1993. The Invertebrates: A New Synthesis. Blackwell Science, Oxford, 487 pp.

Brewer, R. H., 1976a. Larval settling behaviour in *Cyanea capillata* (Cnidaria: Scyphozoa). Biol. Bull. 150: 183–199.

Brewer, R. H., 1976b. Some microenvironmental influences on attachment behaviour of the planula of *Cyanea capillata* (Cnidaria: Scyphozoa). In Mackie, G. O. (ed.), Coelenterate Ecology and Behavior. Plenum Press, New York and London: 347–354.

Calder, D. R., 1971. Nematocysts of polyps of *Aurelia, Chrysaora,* and *Cyanea*, and their utility in identification. Trans. am. Microscop. Soc. 90: 269–274.

Calder, D. R., 1974. Nematocysts of the coronate scyphomedusa, *Linuche unguiculata*, with a brief re-examination of scyphozoan nematocyst classification. Short Papers and notes 15: 170–173.

Calder, D. R., 1977. Nematocysts of the ephyra stages of *Aurelia, Chrysaora, Cyanea*, and *Rhopilema* (Cnidaria, Scyphozoa). Trans. am. Microscop. Soc. 96: 13–19.

Coma, R., J. M. Gili, M. Zabala & T. Riera, 1994. Feeding and prey capture cycles in the aposymbiotic gorgonian *Paramuricea clavata*. Mar. Ecol. Progr. Ser. 115: 257–270.

Cornelius, P. F. S & C. Östman, 1987. Redescription of *Laomedea exigua* M. Sars, a hydroid new to Scandinavia, with comments on its nematocysts, life cycle and feeding movements. Zool. Scripta 16: 1–8.

Gili, J. M. & R. G. Hughes, 1995. Ecology of benthic hydroids. Oceanogr. Mar. Biol. annu. Rev. 33: 351–422.

Hernroth, L. & F. Gröndahl, 1983. On the biology of *Aurelia aurita* (L.) 1. The release and growth of *Aurelia aurita* (L.) ephyrae in the Gullmar Fjord, western Sweden, 1982–83. Ophelia 22: 189–199.

Hernroth, L. & F. Gröndahl, 1985. On the biology of *Aurelia aurita* (L.) 3. Predation by *Coryphella verrucosa* (Gastropoda, Opistobranchia), a major factor regulating the development of *Aurelia* populations in the Gullmar Fjord, western Sweden. Ophelia 24: 37–45.

Mariscal, R. N., 1974. Nematocysts. In Muscatine, L. & H. M. Lenhoff (eds), Coelenterate Biology. Academic Press, New York: 129–178.

Östman, C., A. Aquirre, M. Myrdal, P. Nyvall, J. Lindström & M. Björklund, 1995. Nematocysts in *Tubularia larynx* (Cnidaria, Hydrozoa) from Scandinavia and the northern coast of Spain. Sci. Mar. 59: 165–179.

Östman, C. & J. Hydman, 1997. Nematocysts in Scandinavian *Cyanea* medusae (Scyphozoa, Cnidaria). Sci. Mar.

Simkina, R. G., 1980. A quantitative feeding study of the colonies of *Pergonimus megas* (Hydroida, Bouigainvillidae). Zool. Zh. 59: 500–506.

Spangenberg, D. B., 1965. A study of strobilation in *Aurelia* under controlled condition. J. exp. Zool. 160: 1–9.

Watson, G. M. & R. L. Wood, 1989. Colloquium on Terminology. In Hessinger, D. A. & H. M. Lenhoff (eds), The Biology of Nematocysts. London: Academic Press, San Diego: 21–23.

Weill, R., 1934. Contribution à l' étude des cnidaires et de leurs nématocystes. I, II. Trav. Stn. zool. Wimereux 10 and 11: 1–701.

Hydrobiologia **355:** 29–39, 1997.

A. D. Naumov, H. Hummel, A. A. Sukhotin & J. S. Ryland (eds), Interactions and Adaptation Strategies of Marine Organisms.

The importance of intraspecific competition in a *Littorina littorea* population in the Wadden Sea

Christiane Fenske
Ernst-Moritz-Arndt-University, Zoological Institute, J.S.-Bach-Str. 11–12, D-17489 Greifswald, Germany (e-mail: fenske@rz.uni.greifswald.de)

Key words: Littorina littorea, food preferences, reaction to increased density, intraspecific competition

Abstract

Two field experiments were carried out to test whether effects of intraspecific competition in a *Littorina littorea* population can be detected in a short-term investigation. Different size classes of *L. littorea* showed no significant difference in preferences when offered four kinds of either possible food or substrata (*Fucus vesiculosus, Ulva lactuca, Carcinus maenas*, brick). Large and medium winkles preferred *Fucus vesiculosus*, followed by *Ulva lactuca*. Dead shore crabs (*Carcinus maenas*) were the least preferred objects for all size classes. On the first day of the experiment bricks were more attractive to small littorines than to larger ones. Considering all four days, the same ranking occurred for all size classes: *Fucus vesiculosus* > *Ulva lactuca* > brick > *Carcinus maenas*. The reaction of juveniles to increased densities was examined using an *in situ* caging experiment on a mussel bed. Mesh size of the cages allowed adult densities to be increased while juveniles could escape by passing through the meshes. However, there was no significant emigration of small winkles even from cages with 10 to 20 times natural density of large individuals. Of greater importance was the original number of winkles at the site. The available resources on the mussel beds appear to be sufficient to maintain a high population density. Intraspecific competition does not seem to play a major role in this *L. littorea*-population.

Introduction

Common periwinkles (*Littorina littorea* Linné, 1758) are very abundant in Königshafen bay, a shallow area at the Northern tip of Sylt island, North Sea. They prefer hard substrata, supplied in this area by mussel beds (*Mytilus edulis*). On these, *L. littorea* (> 5 mm shell height) appears in high densities (1300 m^{-2}). When juveniles (1–5 mm shell height) are included, densities of up to 1900 m^{-2} are possible. What are the factors controlling the population size?

Investigations of a number of physical factors indicate that *L. littorea* is able to withstand a wide range of harsh conditions. For example, *L. littorea* can tolerate extremes of cold and frost (Ziegelmeier, 1964; Ibing & Theede, 1975; Dörjes, 1980; Murphy & Johnson, 1980), high temperatures (Gowanloch & Hayes, 1926; Hayes, 1929; Fraenkel, 1960), different salinities (Colgan, 1910; Mayes, 1962) and desiccation of several weeks (Colgan, 1910; for a comprehensive review of littorine ecology see McQuaid, 1996).

There is also evidence to suggest that *L. littorea*-populations are not limited by either interspecific competition (Barnes, 1986; Frid & James, 1988) or parasitism (Werding, 1969; Dethlefs, 1995). Predation endangers only the smaller individuals (<11 mm) of the population (Scherer & Reise, 1981). In the case of birds it has been shown that *Littorina* does not represent a major food source (Dernedde, 1992; Hertzler, 1995). This leaves intraspecific competition as a possible regulating mechanism for this *L. littorea*-population.

In slow moving species, such as snails, there is no clear distinction between food and habitat. Hard surfaces, such as rocks or mussel shells, can be used for attachment but also as a food source. Littorines are known to feed on bacteria and microalgae, especially diatoms (Petraitis, 1983; Imrie et al., 1990). With their radulae they graze on the biofilms that develop on hard substrates. Concerning microbial food there is no intraspecific difference (Hylleberg & Tang Christensen, 1978). However, they also eat macroalgal spores, germlings and adult plants (Wilhelmsen & Reise, 1994). Furthermore it has been reported that *L. littorea* can be carnivorous (Hayes, 1929; Schäfer, 1950). Feeding preferences of littorines have been mainly investigated in the laboratory (Barkmann, 1956; Bakker, 1959; Watson & Norton, 1985; Wilhelmsen & Reise, 1994), often using algal extracts (Lubchenco, 1978; Imrie et al., 1989).

The purpose of this short-term study was twofold: firstly to investigate the requirements of different size classes (representing age groups) for the two major resources 'food' and 'space' and secondly to examine the periwinkles' interaction; whether there is an optimal density and if they attract or repel each other at high densities.

Materials and methods

Study area

Investigations were carried out in Königshafen, a sheltered bay at the northern tip of Sylt island, Germany (55 °02′ N, 8 °26′ E) (Figure 1). The bay comprises an area of 4.816 km^2 of intertidal flats and forms part of the northern Wadden Sea (North Sea). Tides are semi-diurnal with an amplitude of 1.8 m. The substrate consists of sand (90.0%), mud (9.0%) and mussel (*Mytilus edulis*) beds (1.0%). Salinity is approximately 30‰ and the mean annual water temperature about 9 °C, with a summer average of 15 °C and a winter average of 4 °C. More details about Königshafen are given in Wohlenberg (1937), Reise (1985) and Reise et al. (1994).

Regular counts

During the time of the experiments regular counts were carried out in four typical *L. littorea* habitats, each with five replicates: on mussel (*Mytilus edulis*) beds, on mussel clumps (= small aggregations of *M. edulis*, covered with *Fucus vesiculosus*), on a mud/sand flat and in a sea grass meadow, formed by *Zostera noltii*. To be able to count all *L. littorea* in all 20 marked areas during one low tide, squares of 1/16 m^2 were searched for winkles in the mussel habitats, and 1 m^2 in the other habitats.

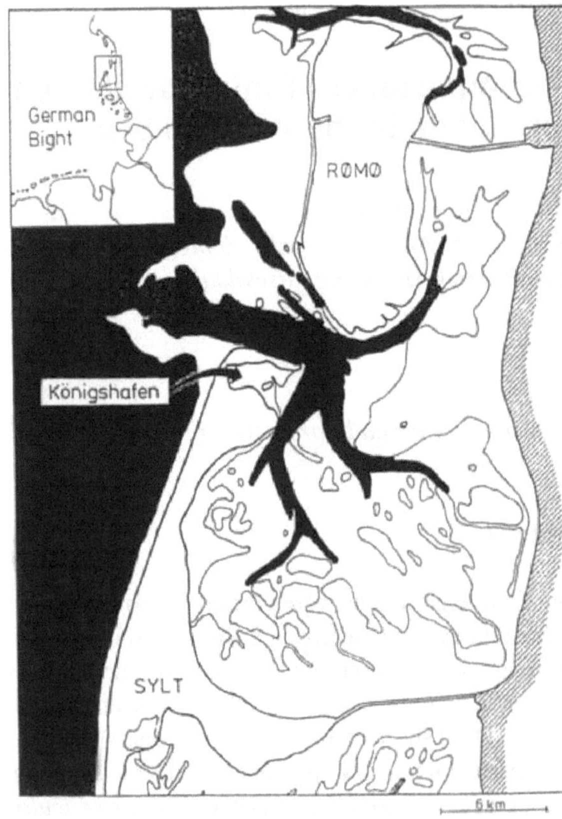

Figure 1. Study site: Königshafen, a shallow bay at the Northern tip of Sylt island (Wadden Sea). Intertidal flats: white area between thick and thin line; tidal inlets and North Sea >5 m depth: dark shading; subtidal flats: between thin line and dark shading; mainland coastline: hatched diagonally (from Reise & Gätje, 1994).

Food and substratum choice experiment

Three size classes of *Littorina littorea*, corresponding to different age groups, were tested for their food and substratum preferences. The size classes were defined as follows:

small:	6–13 mm shell height, corresponding to one year old winkles
medium:	14–19 mm shell height, corresponding to two years old winkles
large:	>19 mm shell height, corresponding to winkles three years and older

(cf. Moore, 1937; Smith & Newell, 1955; Williams, 1964; Guyomarc'h-Cousin, 1975).

The periwinkles were collected one day before the experiment from the intertidal population in Königshafen/Sylt and marked with pink nail varnish on the apex of the shell. The experiment used 30 winkles per size class and was replicated six times for each size class, giving a total of 180 per size class and 540 winkles altogether.

The four substrata offered were:

(1) a brick (well scrubbed to remove the microbial and microalgal cover) (2) *Ulva lactuca* (Chlorophyceae) (3) *Fucus vesiculosus* (Phaeophyceae) (4) a dead *Carcinus maenas* (Crustacea, Decapoda)

These substrata were also collected one day prior to the experiment and stored at 5 °C. At low tide on the day of the experiment, the four substrata of each set were arranged on the tidal flat in a semicircle so that each animal was equidistant from each of the substrata. 30 marked winkles of each size class were placed at a distance of 50 cm from the substrata. The arrangement of the four substrata was varied for each replicate. Likewise, the six replicates of each size class were alternated with the replicates of the other size classes.

Algae and *Carcinus* were fixed with metal pegs. Only entire fresh algae of similar mass were used, to exclude exudates from damaged plants as a source of attraction. The winkles were released inshore of the substrata at low tide. The direction of their movement was noted at the beginning of the experiment (20/7/1994), subsequently all winkles on the substrata were counted after one tidal cycle of 12 hours and then again after three and seven tidal cycles (21/7, 22/7 and 24/7/1994). Unmarked winkles, that had arrived from the surrounding area, were counted in addition to marked experimental animals.

An analysis of variance (ANOVA) was carried out to test the substrate preference of winkles of different size classes.

Cage experiment with increased densities of adult winkles

To test if young winkles would leave areas with a high number of adult winkles, a cage experiment was conducted on a bed of the mussel *Mytilus edulis*. Twenty cylindrical cages were set up (25 cm high, 20 cm ⌀, area included ≈ 314 cm^2). These were made from wire netting with hexagonal meshes (maximum extension of 15 mm), which allowed only small winkles (≤ 17 mm shell height) to pass through. Three to four steel bars (50 cm long, 6 mm ⌀) were woven through the meshes to stabilise the form of the cage and to anchor the cage into the mussel bed. The bars were pushed approximately 5–8 cm into the substratum. The cages were fixed carefully to the mussel bed to prevent winkles from passing underneath.

The upper end of the cage was fixed with a plastic ring (20 cm ⌀). 10 cm of the wire netting was bent outwards over the plastic ring to form a brim. The most exterior meshes were opened with pliers to make a sharp-edged rim. This hinders winkles both from escaping and entering from the outside. The mesh size used prevented winkles from crawling on the netting. Smaller meshes would allow large winkles to crawl over the netting foiling a real enclosure or exclosure effect. Using open cages (not covered on top) had two advantages: photosynthesis was not inhibited and the cages were less likely to be swept away because of their lower resistance to currents. The distance between cages was 50 cm.

In this experiment four size classes of periwinkles were used:

juvenile	(1–5 mm shell height)
small	(6–13 mm shell height)
medium	(14–17 mm shell height)
large	(>17 mm shell height)

After putting up the cages all *L. littorea* in them were counted. The average density of winkles >17 mm within this mussel bed was two per cage. Then the number of large winkles was altered. Different numbers of winkles >17 mm (too large to pass through the meshes) were placed in the cages. Four cages were left without large winkles, the remainder had multiples of two per cage, i.e. 2, 4, 6....30. The final cage had 40 large winkles, that is a 20 fold increase over normal density. On other mussel beds densities are often higher than at the experimental site, therefore the increase is not beyond natural variation.

Physical conditions within cages were standardised by establishing them at the same tidal height, ensuring that none were at the edge of the mussel bed and that all were at the same level which was determined by the number of mussel layers. Treatments (i.e. density of large winkles) were assigned randomly to the cages. The cage experiment began on August 5th, 1994 and was run for one month. After manipulation of the number of large winkles, all winkles in the cages were counted after the subsequent tide (6/8), and then on the following dates: 7/8, 9/8, 11/8, 16/8, 21/8, 25/8, 4/9.

Results

Regular counts

Although *L. littorea* occurs in different habitats, the size classes are not distributed equally (Figure 2). Only on mussel beds, the optimum habitat, close to the low water line, all size classes are represented. Juvenile and small winkles do not occur on the mud/sand flat whereas no medium or large winkle is found in the *Zostera*-meadow. This indicates a habitat segregation. The relative abundances of different size classes were more or less constant throughout July and August (Figure 3). There was no crash or sudden increase in number of small, medium, or large winkles. Juveniles began to settle from the plankton at the end of July. Their number increased most during the second week of August.

Food and substratum choice experiment

There was a clear preference for *Fucus vesiculosus* among medium and large winkles (Figure 4). The second favoured substratum was the green algae *Ulva lactuca*. Only few individuals (one large, one medium and no small winkle) went to the dead shore crab, therefore this substratum was excluded from the analysis of variance to avoid transformation of all data.

The number of winkles was significantly affected by substratum (ANOVA, $F=9.37$, df=2, $p=0.0004$), but not size class (ANOVA, $F=1.41$, df=2, $p=0.2547$). For the small size class the brick and *Fucus* seemed to be equally attractive: totals of 24 and 23 respectively were found at these objects in all six replicates after one tide. Including all four days in the analysis, there is again a significant difference between the objects (one-way-ANOVA with repeated measurements, $F=16.635$, df=2, $p=0.00002$) but no significant difference between the size classes (one-way-ANOVA with repeated measurements, $F=2.678$, df=2, $p=0.869$). The ranking was *Fucus* > *Ulva* > brick> *Carcinus*.

Besides the marked experimental winkles, unmarked ones were also found on the substrata. Taking these into account as well and considering all four days, there is a consistent pattern of preference (*Fucus* > *Ulva* > brick > *Carcinus*), though this could not be tested because of the lack of independence. The percentage of each size class (marked and unmarked winkles) per substratum is shown in Figure 5. The number of marked and unmarked winkles at the objects throughout the experiment can be seen in Table 1. Particularly high numbers were achieved on the second day.

Table 1. Sum of marked and unmarked winkles of three size classes, found at all four objects in the food and substratum experiment.

Winkle size	21/7/1994	22/7/1994	24/7/1994
Large	142	372	396
Medium	129	464	115
Small	78	115	88

In contrast to the large and medium winkles, the small winkles (6–13 mm shell height) proved to be very active. They started to crawl shortly after they had been placed on the ground, although not always in the direction of the experimental substrata. The recovery rate of marked winkles (i.e. number of found winkles at the substrata) after one tide was about 48%, 39% and 32% for the large, medium and small winkles respectively. After seven tides there were about the same percentages of each size class at the substrata, ranging between 10.6% (large winkles) and 12.2% (medium winkles).

Increased densities of adult winkles

Increasing the densities of large periwinkles (>17 mm) in cages on a mussel bed did not lead to obvious emigration of small individuals ($\leq$ 17mm) although they could move through the meshes of the cages. More important in determining the number of small winkles in a cage during and at the end of the experiment seemed to be the original number of winkles of all sizes at the cage site. There is a significant correlation between the contribution of each cage to the total of winkles $\leq$ 17 mm before and after the experiment ('after' here means average of all counts): Spearman rank correlation coefficient $r_S = 0.558$ (20 df; $p<0.05$).

The number of winkles $\leq$ 17 mm fluctuated strongly in each cage, which means that if winkles were small enough to pass through the meshes they were not hindered by the wire netting but moved freely on the mussel bed. The number of small winkles found in a cage did not vary according to density of large winkles: Spearman rank correlation: $r_S = -0.2388$ (20 df, $p>0.05$).

The rank orders compared were:

(a) the number of large winkles (>17 mm shell height) per cage (0, 2, 4, 6...40) and

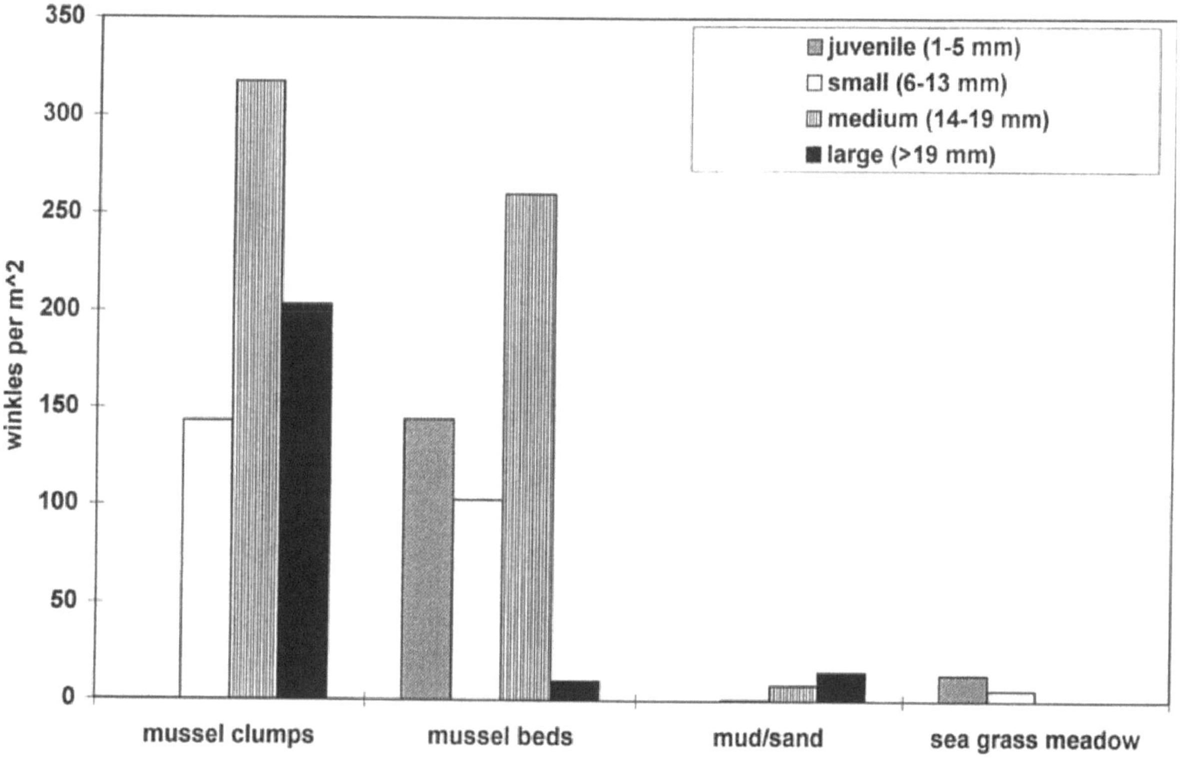

Figure 2. Number of winkles per square meter in four different habitats. Mean of five replicates, average of six counts (July–August 1994).

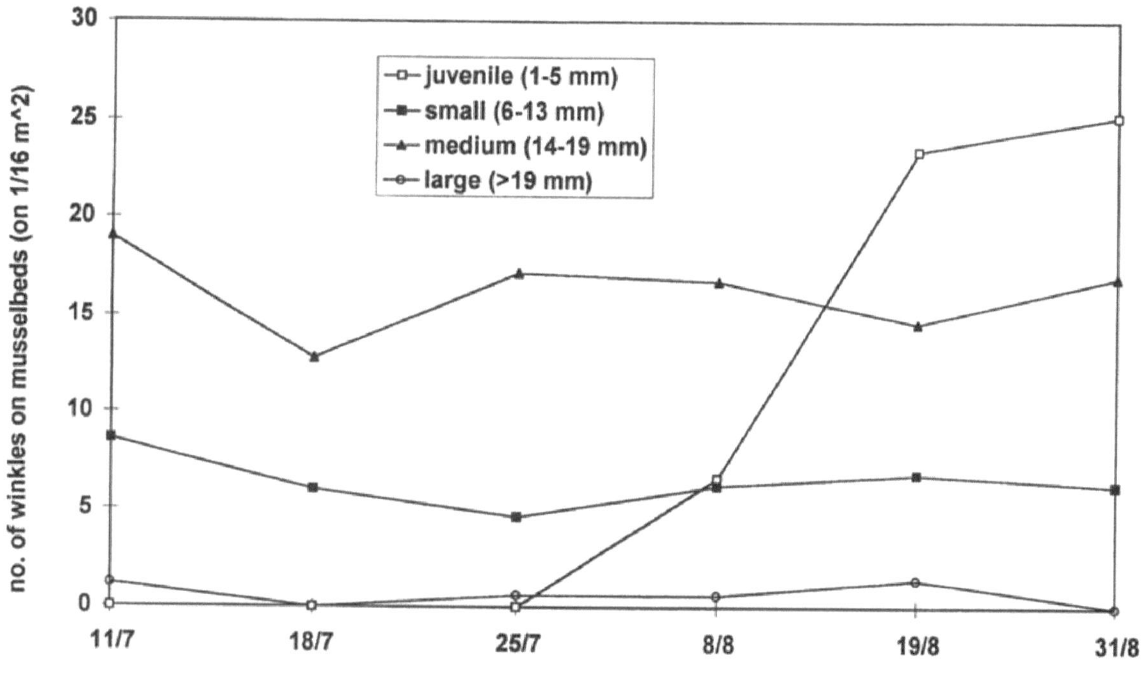

Figure 3. Number of winkles on mussel beds (mean of five replicates) during July and August 1994.

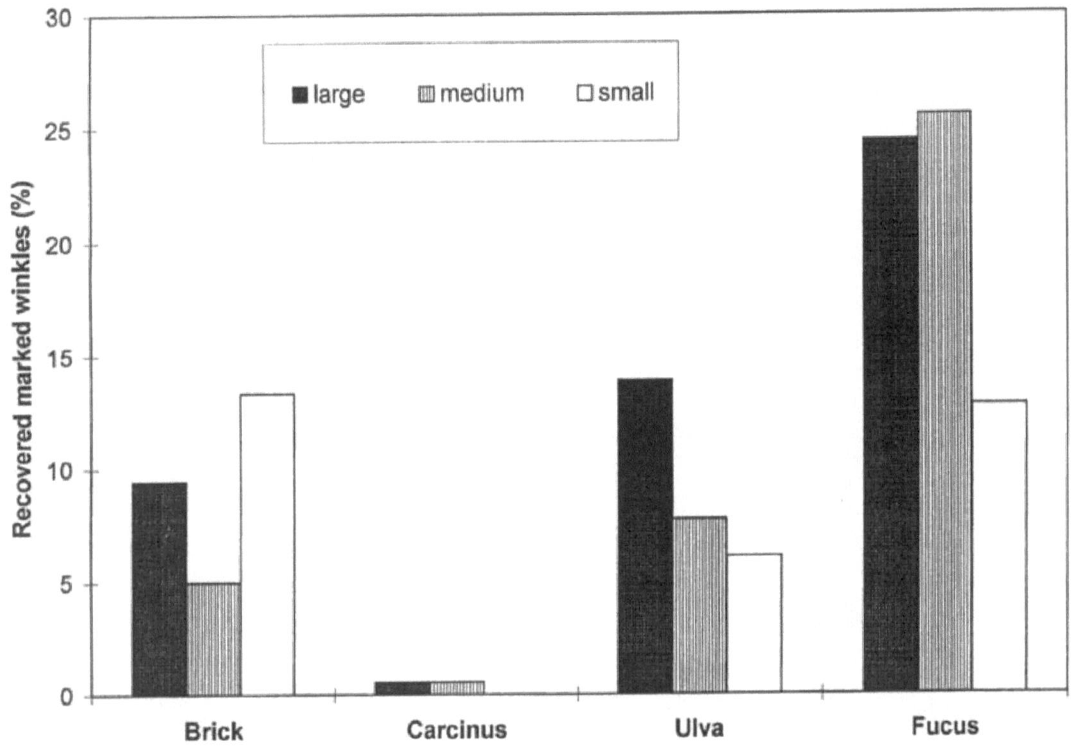

Figure 4. Percentage of marked winkles collected after the first tidal cycle at each substrate.

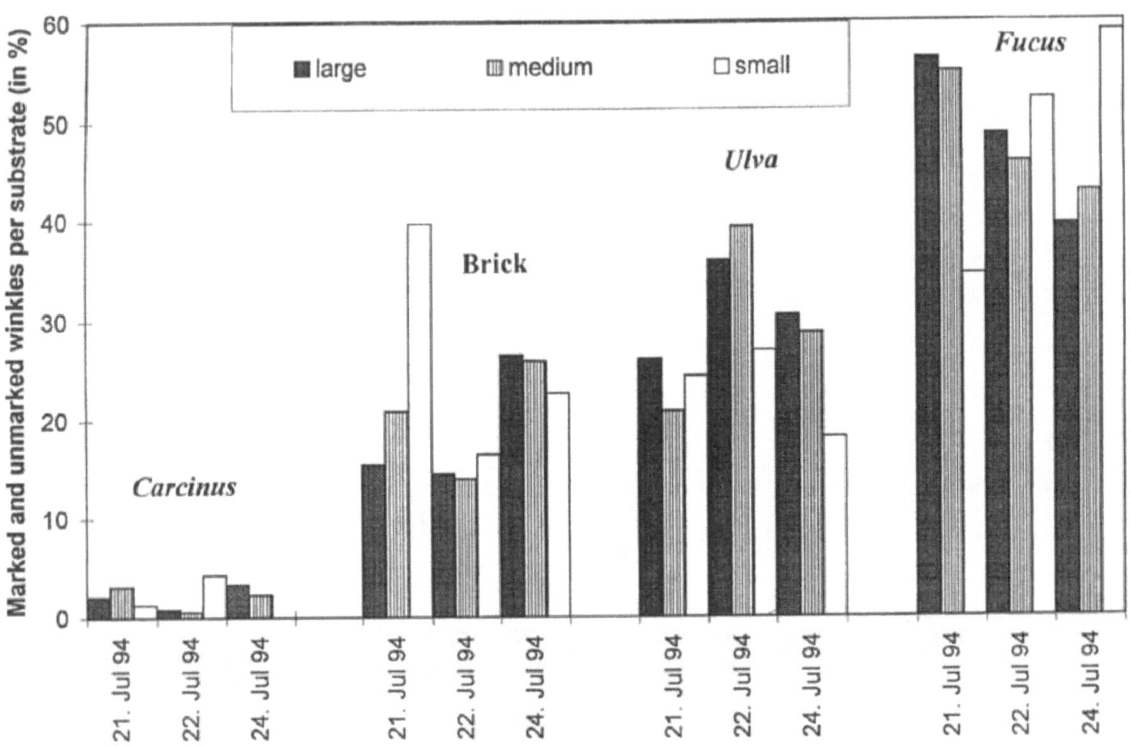

Figure 5. Sum of marked and unmarked winkles: percentage found per substrate per day.

(b) the percentage of total number of small winkles (≤ 17 mm) present in each cage using an average of seven counts.

This comparison is more useful than just comparing the first and the last day of the experiment because it includes the development during the whole month of the experiment. The change in number of winkles ≤ 17 mm in all cages is shown in Figure 6. There is no correlation between the number of large winkles and the change in number of small winkles.

The Spearman rank correlation was also applied to compare the number of winkles >17 mm with those of 6–17 mm shell height. Neither of the comparisons was significant ($p>0.1$) (before experiment vs last day/ second last day/third last day of experiment, and before experiment vs. average of all counts during experiment). There was no correlation between the number of winkles >17 mm and those of 6–17 mm shell height.

The number of juvenile (1–5 mm), small (6–13 mm) and medium winkles (14–17 mm) found in the cages varied greatly. Largest variations were observed in the medium size class as seen by calculating the relative variation coefficient: $V_r = s\ x^{-1}\ n^{-0.5}$, with V_r = relative variation coefficient, s = standard deviation, x = mean and n = amount of counts. The results were: V_r (juveniles) = 3.62%; V_r (6–13 mm) = 6.26%; V_r (14–17 mm) = 22.25%. The large variations are probably due to the low number of winkles of the medium size class. Maximum numbers per day were 74 medium, 102 small, and 561 juvenile winkles in all cages.

Discussion

Food and habitat

Littorina littorea was presented with different sorts of food which occur naturally in its habitat. The brick, being scrubbed, offered a surface for attachment only; *Ulva* and *Fucus* represented green and brown algae respectively which may have had epiphytes. The dead *Carcinus maenas* offered the opportunity for scavenging as well as a surface which was certainly covered with bacteria and microalgae.

Littorines of the medium and large size class clearly preferred *Fucus vesiculosus*. This stands in contrast to the results achieved by Lubchenco (1978) and Norton et al. (1990) who reported that *L. littorea* is less attracted by *Fucus* as compared to ephemeral green algae. However, when offered different species of *Fucus*, *L. littorea* prefers *Fucus vesiculosus* (Barker & Chapman, 1990). The first choice of small winkles led to similar numbers at the bricks and on *Fucus* (totals of 24 and 23 resp.). But they soon left the bricks again and on the subsequent days *Fucus* and *Ulva* were as attractive to them as to the larger size classes (Figure 5). Using natural units of algae, it is possible that there was slightly more space available on *Fucus* than on *Ulva* because of the ramified structure of the former. Still, the differences between the numbers of winkles at these objects were significant. Had there been a clear preference for *Ulva*, winkles would not have chosen *Fucus* so frequently. Even if the periwinkles in my experiment did not always feed directly on *Fucus* but only on its microalgal epiphytes, it was conspicuous how often littorines went to this object. The more so as they had not been starved for weeks as in Watson & Norton's (1985) experiment but had only been collected one day prior to the experiment. As far as could be observed, trail following did not occur frequently and the results do not suggest that this phenomenon was important. The odoriferous *Carcinus* was hardly ever chosen by the winkles as a food source in the experiment although I often observed many *L. littorea* on dead shore crabs outside the experiment.

Sensing

In laboratory experiments littorines are able to recognise plant exudates from a distance and move towards favoured foods (Norton et al., 1990). Barkmann (1956) examined *L. obtusata* in aquariums and found that winkles were even attracted to algae from a distance of 150 cm but only if a single species of alga was presented. Offering several species at once did not lead to a clear food preference by the winkles. There is a lack of experiments to test sensory capacities in the natural habitat. It remains unclear to what extent *L. littorea* makes use of its olfactory and optical senses. Lubchenco (1978) and Watson & Norton (1985) found that little or no orientation towards food species takes place. As littorines are opportunistic with regard to their food preferences, it may not be essential to have sensory capabilities as elaborate as in a food specialist.

The recovery rate of marked winkles in the choice experiment after one tide was about 48%, 39% and 32% for the large, medium and small winkles respectively. If the animals had made a selective choice, recovery rates (sum of recovered marked winkles at all objects) should have been above 50% as there were no other suitable substrata in the study area which might

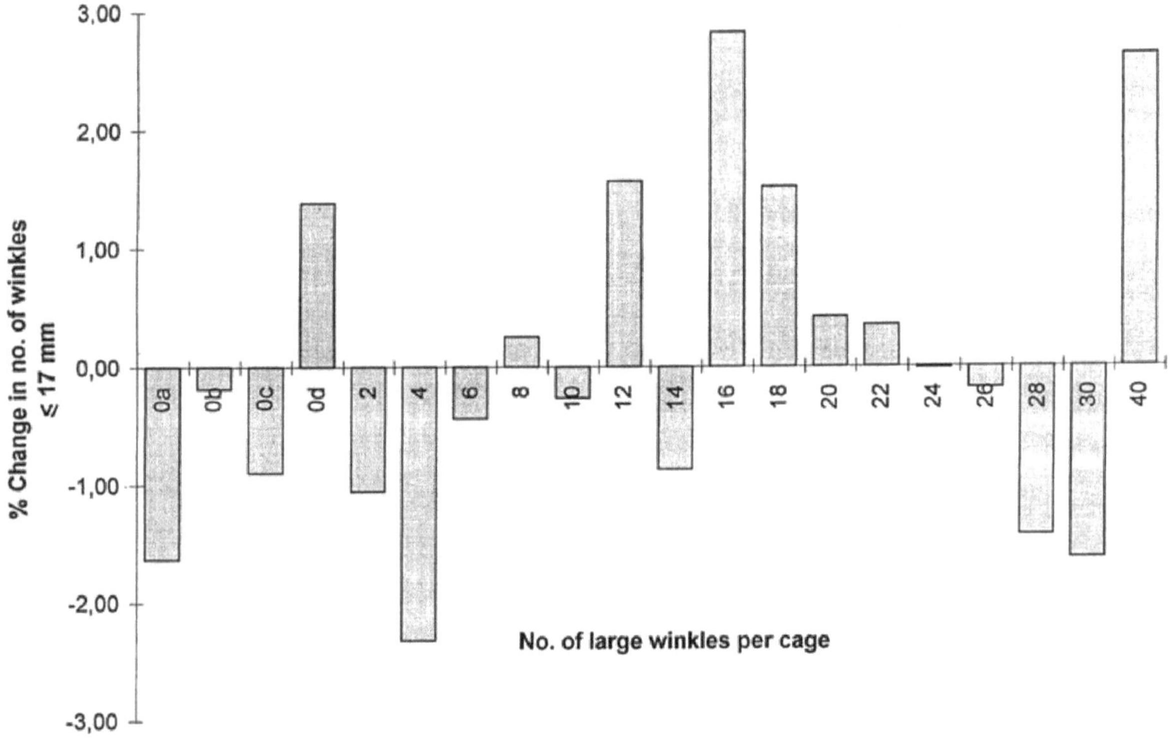

Figure 6. Percentual change in number of winkles ≤17 mm per cage, comparing the contribution of each cage to the total of winkles ≤17 mm before the experimental manipulation and afterwards (average of seven counts).

have been attractive to the littorines. The results for the first day therefore suggest that the winkles arrived by chance at the food and substrata offered. When the number of winkles at the different objects on the second and fourth days of the experiment were included with the non-marked winkles there were clear preferences for all size classes (*Fucus* > *Ulva* > brick > *Carcinus*) (Figure 5). This indicates that different age classes have to cope with resource overlap. To compensate for this, habitat segregation may play an important role in reducing the competition between the age classes (Figure 2).

Increased densities

To test for intraspecific competition in populations of marine molluscs other authors have used cages in which all individuals were enclosed, i.e. the mesh size was such that the animals could not escape (Underwood, 1978; Creese & Underwood, 1982; Sherell, 1981; Branch & Branch, 1981; Fletcher & Creese, 1985; Janke, 1990). The effects of increased densities were then measured as reduced growth rates, reduced body weight, modified feeding behaviour or higher mortality.

The cages used in this experiment allowed winkles ≤ 17 mm to crawl through the mesh. Other species in the eulitoral (such as the lug worm *Arenicola marina* and the cockle *Cerastoderma edule*) have been shown to compensate for high densities by emigrating and/or by reducing their growth rates (Pardo, 1984; Reise, 1985).

In this study, small *L. littorea* did not respond systematically to increased densities of large periwinkles. Juveniles always formed a high proportion of the total number of winkles in the cages. On account of their small size (1–5 mm shell height) they can use those little crevices and niches on a mussel bed that cannot be reached and grazed by the adults (Geller, 1991). This may explain why the juveniles were not pushed out by a large number of adult winkles. But even the small and medium size classes (6–13 and 14–17 mm shell height), that certainly have already the same demands as the large ones, did not emigrate systematically.

Food resources on the mussel bed

Increased densities of adult winkles ($\geq$ 3 years) did not push out juvenile or young winkles (1–2 yrs) systematically. Could it be that there was still enough food?

The mussel bed on which the cage experiment was conducted did not contain macroalgae. Therefore, the main food source available was microalgae. The area available for grazing within a mussel bed is, of course, much greater than the area enclosed by a cage. The surface area for attachment of microalgae and grazing can be estimated from the number of shells within a cage and the surface area of a single shell. This in turn can be calculated from its length and width using the equation for an ellipse. Even assuming only half the theoretical surface was available for diatom attachment, this produces a density of littorines per unit surface area far below the densities which Castenholz (1961) suggests are able to keep rocks free of microalgae. Castenholz worked with another species (*L. scutulata*) and at another locality (Oregon, U.S.). Although a direct comparison is difficult, the results indicate that food is unlikely to have been limiting in the present study. The large resources on the mussel bed consisting of microalgae and sedimenting particles may be the reason why juvenile, small and medium winkles were not systematically pushed out of cages with high densities of large winkles.

Importance of competition

Intraspecific competition occurs mainly in populations at high densities (Ayala, 1970). However, mortality, one possible effect of competition, is surprisingly low in experiments in which densities are manipulated (Branch & Branch, 1981; Peterson, 1982; Fletcher & Creese, 1985). Growth rates are directly proportional to the available amount of food and are reduced at high densities (Sherell, 1981; Stiven & Kuenzler, 1979). However, the response depends also on the density of the population from which the experimental animals are taken. If they come from a high density population, a further increase in density has no effect, e.g. on the body mass (Branch & Branch, 1981). The *L. littorea* in the investigations presented here were taken from such a high density-population.

Moreover, it is essential to distinguish between the intensity and the importance of competition (Welden & Slauson, 1986). The intensity corresponds to the process, the importance to the result of competition, i.e. the effect competition has on the offspring, mortality etc. Even a low intensity of competition can be very important for an organism if this is the only factor affecting fitness. On the other hand, strong competition can be of little importance if the organism is mainly influenced by other factors such as abiotic stress, predation or disturbances.

Competition often plays an important role on exposed shores, whereas predation is the major force on sheltered shores (Menge, 1976; Reise, 1977, 1978). Furthermore, the complexity of trophic organisation also determines the importance of competition: the more complex the organisation the more stable is the association (Hutchinson, 1959) and the more important is competition (Menge & Sutherland, 1976).

The sheltered Wadden Sea habitats and the wide range of food sources of *L. littorea*, mainly consisting of algae, suggest that competition is likely to have low importance.

Conclusion

(1) A food and substratum choice experiment did not show significant differences between size/age groups of *Littorina littorea*.

(2) High densities of large individuals on a mussel bed did not lead to a systematic emigration of small winkles.

(3) Apparently there are no mechanisms of intraspecific competition that hinder small winkles from settling or staying where there are already a large number of adults.

It is possible that the investigated *L. littorea* population is mainly limited by larval supply and predation on larval and juvenile stages.

Acknowledgments

I would especially like to thank Prof. K. Reise for his valuable advice and support during all phases of the work. His wide experience was a great help for the experiments and the interpretation. Drs A. Albrecht, D. Schories and W. Armonies likewise provided me with useful information and patient assistance. I am also grateful to Prof. C. D. McQuaid and Dr C. Halbert for critical remarks, questions, and encouragement. This work was initiated during the course of a M.Sc. thesis, degree in biology, done at the Biologische Anstalt Helgoland, Wattenmeerstation List/Sylt, Germany.

References

Ayala, F. J., 1970. Competition, coexistance and evolution. In Hecht, M. K. & W. C. Steere (eds), Essays in Evolution and Genetics in Honor of Theodosius Dobzhansky. Amsterdam: 121–158.

Bakker, K., 1959. Feeding habits and zonation in some intertidal snails. Arch. Néerl. Zool. 13: 230–257.

Barker, K. M. & A. R. O. Chapman, 1990. Feeding preferences of periwinkles among four species of *Fucus*. Mar. Biol. 106: 113–118.

Barkmann, J. J., 1956. On the distribution and ecology of *Littorina obtusata* (L.) and its subspecific units. Arch. Néerl. Zool. 11: 22–86.

Barnes, R. S. K., 1986. Daily activity rhythms in the intertidal gastropod *Hydrobia ulvae* (Pennant). Estuar. coast. Shelf Sci. 22: 325–334.

Branch, G. M. & M. L. Branch, 1981. Experimental analysis of intraspecific competition in an intertidal gastropod, *Littorina unifasciata*. Aust. J. mar. Freshwat. Res. 32: 573–589.

Castenholz, R. W. 1961. The effect of grazing on marine littoral diatom populations. Ecology 42: 783–794.

Colgan, N., 1910. Notes on the adaptability of certain littoral mollusca. Irish Nat. 19: 127–133.

Creese, R. G. & A. J. Underwood, 1982. Analysis of inter- and intraspecific competition amongst limpets with different methods of feeding. Oecologia (Berlin) 53: 337–346.

Dernedde, T., 1992. Untersuchungen zur Ernährung der Möwen im Königshafen auf Sylt. Master thesis, University of Kiel, Germany, 92 pp.

Dethlefs, B., 1995. Reproduktion der Strandschnecke *Littorina littorea* (L.) im Wattenmeer bei Sylt. Master thesis, University of Hamburg, Germany, 68 pp.

Dörjes, J., 1980. Auswirkungen des kalten Winters 1978/79 auf das marine Makrobenthos. Natur und Museum 110: 109–115.

Fletcher, W. J. & R. G. Creese, 1985. Competitive interactions between co-occurring herbivorous gastropods. Mar. Biol. 86: 183–191.

Fraenkel, G., 1960. Lethal high temperatures for three marine invertebrates: *Limulus polyphemus, Littorina littorea* and *Pagurus longicarpus*. Oikos 11: 171–182.

Frid, C. L. J. & R. James, 1988. Interactions between two species of saltmarsh gastropods, *Hydrobia ulvae* and *Littorina littorea*. Mar. Ecol. Prog. Ser. 43: 173–179.

Geller, J. B., 1991. Gastropod grazers and algal colonization on a rocky shore in northern California: The importance of the body size of grazers. J. exp. mar. Biol. Ecol. 150: 1–17.

Gowanloch, J. N. & F. R. Hayes, 1926. The physical factors, behaviour and intertidal life of *Littorina*. Contr. Canad. Biol. Fish. 3: 134–165.

Guyomarc'h-Cousin, C., 1975. Etude de la croissance d'un gastéropode prosobranche gonochorique: *Littorina littorea* L. Cah. Biol. Mar. 16: 483–494.

Hayes, F. R., 1929. Development, growth and behaviour of *Littorina*. Contr. Canad. Biol. Fish. 4: 415–430.

Hertzler, I., 1995. Nahrungsökologische Bedeutung von Miesmuschelbänken für Vögel (Laro-Limikolen) im Nordfriesischen Wattenmeer. Master thesis, University of Göttingen, Germany, 119 pp.

Hutchinson, G. E., 1959. Hommage to Santa Rosalia, or why are there so many kinds of animals. Am. Nat. 93: 145–149.

Hylleberg, J. & J. Tang Christensen, 1978. Factors affecting the intra-specific competition and size distribution of the periwinkle *Littorina littorea* (L.). Natura jütl. 20: 193–202.

Ibing, J. & H. Theede, 1975. Zur Gefrierresistenz litoraler Mollusken von der deutschen Nordseeküste. Kieler Meeresforschungen 31: 44–48.

Imrie, D. W., S. J. Hawkins & C. R. McCrohan, 1989. The olfactory-gustatory basis of food preference in the herbivorous prosobranch, *Littorina littorea* (Linnaeus). J. Moll. Stud. 55: 217–225.

Janke, K., 1990. Biological interactions and their role in community structure in the rocky intertidal of Helgoland (German Bight, North Sea). Helgoländer Meeresunters. 44: 219–263.

Lubchenco, J., 1978. Plant species diversity in a marine intertidal community: importance of herbivore food preference and algal competitive abilities. Am. Nat. 112: 23–29.

Mayes, P. A., 1962. Comparative investigations of the euryhaline character of *Littorina* and the possible relationship to intertidal zonation. Nature 195: 1269–1270.

McQuaid, C. D., 1996. Biology of the gastropod family Littorinidae. II Role in the ecology of intertidal and shallow marine ecosystems. Oceanography and Marine Biology: an Annual Review 34: 263–302.

Menge, B. A., 1976. Organization of the New England rocky intertidal community: role of predation, competition, and environmental heterogeneity. Ecol. Monogr. 46: 35–93.

Menge, B. A. & J. P. Sutherland, 1976. Species diversity gradients: synthesis of the roles of predation, competition and temporal heterogeneity. Am. Nat. 110: 351–369.

Moore, H. B., 1937. The biology of *Littorina littorea*. Part I. Growth of the shell and tissues, spawning, length of life and mortality. J. mar. biol. Ass. U.K. 21: 721–742.

Murphy, D. J. & L. C. Johnson, 1980. Physical and temporal factors influencing the freezing tolerance of the marine snail *Littorina littorea* (L.). Biol. Bull. 158: 220–232.

Norton, T. A., S. J. Hawkins, N. L. Manley, G. A. Williams & D. C. Watson, 1990. Scraping a living: a review of littorinid grazing. Hydrobiologia 193: 117–138.

Pardo, L. P., 1984. Dichteabhängige Wanderungen in einer Population von *Arenicola marina* (L.) im Gezeitenbereich der Nordsee. Master thesis, University of Bochum, Germany, 57 pp.

Peterson, C. H., 1982. The importance of intra- and interspecific competition in the population biology of two infaunal suspension-feeding bivalves, *Protothaca staminea* and *Chione undellata*. Ecol. Monogr. 52: 437–475.

Petraitis, P. S., 1983. Grazing patterns of the periwinkle and their effect on sessile intertidal organisms. Ecology 64: 522–533.

Reise, K., 1977. Predator exclusion experiments in an intertidal mud flat. Helgoländer wiss. Meeresunters. 30: 263–271.

Reise, K., 1978. Experiments on epibenthic predation in the Wadden Sea. Helgoländer wiss. Meeresunters. 31: 55–101.

Reise, K., 1985. Tidal Flat Ecology. Springer, Berlin, 191 pp.

Reise, K. & C. Gätje 1994. Königshafen: The natural history of an intertidal bay in the Wadden Sea – An introduction. Helgoländer Meeresunters. 48: 141–143.

Reise, K., E. Herre & M. Sturm, 1994. Biomass and abundance of macrofauna in intertidal sediments of Königshafen in the northern Wadden Sea. Helgoländer Meeresunters 48: 201–215.

Schäfer, W., 1950. Über Nahrung und Wanderung im Biotop bei der Strandschnecke *Littorina littorea*. Arch. Moll. 79: 1–8.

Scherer, B. & K. Reise, 1981. Significant predation on micro- and macrobenthos by the crab *Carcinus maenas*, L. in the Wadden Sea. Kieler Meeresforsch., Sonderh. 5: 490–500.

Sherell, R. M., 1981. Intraspecific competition in the periwinkle, *Littorina littorea*. Biol. Bull. 161: 131.

Smith, J. E. & G. E. Newell, 1955. The dynamics of the zonation of the common periwinkle (*Littorina littorea* (L.)) on a stony beach. J. anim. Ecol. 24: 35–56.

Stiven, A. E. & E. J. Kuenzler, 1979. The response of two salt-marsh molluscs, *Littorina irrorata* and *Geukensia demissa*, to field manipulations of density and *Spartina* litter. Ecol. Monogr. 49: 151–171.

Underwood, A. J., 1978. An experimental evaluation of competition between three species of intertidal prosobranch gastropods. Oecologia (Berlin) 33: 185–202.

Watson, D. C. & T. A. Norton, 1985. Dietary preferences of the common periwinkle *Littorina littorea* (L.). J. exp. mar. Biol. Ecol. 88: 193–211.

Welden, C. W. & W. L. Slauson, 1986. The intensity of competition versus its importance: an overlooked distinction and some implications. The Quarterly Review of Biology 61: 23–44.

Werding, B., 1969. Morphologie, Entwicklung und Ökologie digener Trematoden-Larven der Strandschnecke *Littorina littorea*. Mar. Biol. 3: 306–333.

Wilhelmsen, U. & K. Reise, 1994. Grazing on green algae by the periwinkle *Littorina littorea* in the Wadden Sea. Helgoländer Meeresunters. 48: 233–242.

Williams, E. E., 1964. The growth and distribution af *Littorina littorea* (L.) on a rocky shore in Wales. J. anim. Ecol. 33: 413–432.

Wohlenberg, E., 1937. Die Wattenmeer-Lebensgemeinschaften im Königshafen von Sylt. Helgoländer wiss. Meeresunters 1: 1–92.

Ziegelmeier, E., 1964. Einwirkungen des kalten Winters 1962/63 auf das Makrobenthos im Ostteil der Deutschen Bucht. Helgoländer wiss. Meeresunters 10: 267–282.

Hydrobiologia **355**: 41–47, 1997.
A. D. Naumov, H. Hummel, A. A. Sukhotin & J. S. Ryland (eds), Interactions and Adaptation Strategies of Marine Organisms.

Occurrence of epifauna on the periwinkle, *Littorina littorea* (L.), and interactions with the polychaete *Polydora ciliata* (Johnston)

G. F. Warner
School of Animal and Microbial Sciences, The University of Reading, Whiteknights, P.O. Box 228, Reading RG6 2AJ, U.K.

Key words: Littorina littorea, epibiota, *Polydora ciliata*, intertidal zonation, interactions

Abstract

A population of *Littorina littorea* on the north shore of Southampton Water, UK, has been examined for the occurrence of epifauna. Epifaunal cover was not affected by shore level below MTL but was greatest on larger shells, which were commoner on the lower shore close to MLWS. Epifaunal components included barnacles, serpulids and bryozoans. The polychaete *Polydora ciliata* occurred most frequently on the lower shore where it excavates galleries in periwinkle shells. Heaviest infestations were found in larger shells. *P. ciliata* settles preferentially in or beside irregularities on the shell surface. Settlement of barnacles close to the shell aperture, where new shell growth takes place, produces shell distortion which creates sites for settlement of *P. ciliata*. The angle between serpulid tubes and the shell surface also attracts *P. ciliata*. It is suggested that colonisation by barnacles and serpulids facilitates settlement by *P. ciliata*, and that the latter decreases survival of the host by progressively destroying the shell.

Introduction

The shells of gastropods and of some bivalve molluscs are well known to attract the settlement of a variety of epibiota such as algae and sessile animals. Such encrustations have a range of effects on their hosts such as affording camouflage, increasing water resistance and repelling or attracting predators (Wahl, 1989; Wahl & Hay, 1995). In the case of the common periwinkle, *Littorina littorea* (L.), the occurrence of epibiota has been observed by a number of authors (Smith & Newell, 1955; Orrhage, 1969; Wahl & Sönnichsen, 1992). Common periwinkles occur over a wide range of North Atlantic intertidal habitats, often at population densities in excess of 100 m^{-2}; they feed by grazing algae from the surrounding substratum (Fretter & Graham, 1994). Some populations are relatively unfouled by epibiota, but Wahl & Sönnichsen (1992) were unable to detect antifouling adaptations in this species other than grazing on each others shells at high population densities. Some effects on the periwinkle by its epibiota have been investigated by Wahl (1996) who found that hydrodynamic drag on fouled snails was much higher than on unfouled specimens, and that in steady flow the growth rate of fouled snails was reduced in comparison with unfouled snails, indicating an energy cost of the epibiota.

The spionid polychaete *Polydora ciliata* (Johnston) is a common member of the epibiotic community colonising *L. littorea* shells (Smith & Newell; 1955; Orrhage, 1969; Wahl & Sönnichsen, 1992). This small worm lives in a mud and mucus tube on a variety of benthic substrata, but on calcareous substrata such as mollusc shell it bores into the shell, creating extensive galleries (Dorsett, 1961). This habit of *Polydora* spp is well known in mussels and oysters, in both of which the presence of the worms in the shell is associated with poor condition of the host (Kent, 1979; Ambariyanto & Seed, 1991; Wargo & Ford, 1993). Another effect of *Polydora* spp. is to weaken the shell structure so that it is more easily cracked by predators (Ambariyanto & Seed, 1991; Buckley & Ebersole, 1994).

In this study, I have addressed the following questions. Does the height at which the periwinkle population occurs on the shore affect the cover by epifauna? Does the size of individual periwinkles affect their epifaunal cover? And does the sessile calcareous epifauna affect the co-occurrence of *P. ciliata*? The answers to these questions have implications for the life history and habitat strategies of this fairly long-lived gastropod (ca 5 years, but can live to 20: Fretter & Graham, 1994), which can migrate up and down the shore and is susceptible to many different stresses including trematode infections (Williams & Ellis, 1975; Huxham et al., 1993).

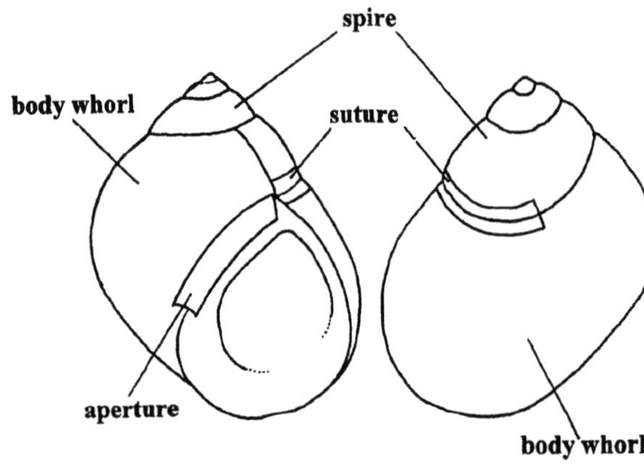

Figure 1. Drawings of *L. littorea* shells showing the four positions at which *P. ciliata* holes were counted.

Methods

Southampton Water is an elongated, sheltered inlet on the south coast of England which receives the input from several rivers. The surrounding land is urbanised and industrial, and there is considerable shipping activity. However, the tidal range is up to 5 m and the salinity is normally about 30‰. At Netley, on the north shore of Southampton Water, southeast of the docks, wide mud and gravel flats are exposed at low spring tide. From high to low water the distance is about 300 m.

Three sampling sites were established on a part of this shore where the substratum was gravelly (see below) and *Littorina littorea* was common. The sites were at about Mean Low Water Springs (0.7 m Above Chart Datum), 80 m up-shore (1.4 m ACD) and a further 80 m up-shore (2.7 m ACD). These levels will be referred to as lower, middle and upper respectively, despite the fact that the upper site was at about Mean Tide Level. The substrate at the upper site consisted of gravel and small stones on a firm sand and gravel base, the middle site was muddy sand with gravel, stones and shells, the lower site was sandy mud with stones and shells. At each site, approximately monthly collections of between 30 and 80 *L. littorea* were made within 50 × 50 cm quadrats, giving estimates of population density. On return to the laboratory, *L. littorea* were killed by boiling (ca 30 s), the flesh was removed and the shells dried. The maximum length of the shells was then measured with callipers and the shells were examined for the presence of epifauna. Sessile epifaunal percent cover was estimated by eye as 0%, <10%, 10–30%, 30–50% and >50%. The sessile epifauna were then scraped off and the number of burrows (holes) made by *Polydora ciliata* counted, and their positions on the shell recorded. Four shell positions were used: a strip 1 mm either side of the suture between the spire and the body whorl, the spire itself, the body whorl, and a 2 mm wide strip bordering the aperture (Figure 1). The surface areas of these four regions were calculated from measurements from several shell sizes to obtain the relative areas of each region. The data presented here are taken from 8 samples collected between June 1995 and March 1996.

Results

The population structure was different at the three levels. It was usually bimodal (Figure 2), with the largest proportion of smaller periwinkles at the middle site, where population density was highest (96–152 m^{-2}). The largest snails occurred at the lower site where population density varied from 31 to 78 m^{-2}. At the upper site, population density was affected by winter migration downshore and varied from 7 m^{-2} in December to 90 m^{-2} in June. Figure 2 shows typical frequency distributions.

The epifauna on the winkle shells consisted mainly of barnacles, with serpulids and bryozoans being commonly found, especially at the lower site. The commonest barnacle species was *Elminius modestus* Darwin, with *Semibalanus balanoides* (L.) occurring occasionally and *Balanus crenatus* Bruguière occurring rarely and at the lower site only. The serpulids included *Pomatoceros lamarcki* (Quatrefages), *Pomatoceros triqueter* (L.) and *Hydroides* spp. The bryozoans were *Conopeum reticulum* (L.) and *Cryptosula palliasiana* (Moll). *Polydora ciliata* was common, especially at the lower site, and individual periwinkles were found containing more than 100 burrows. The encrusting brown alga *Ralfsia verrucosa* Aresch. and

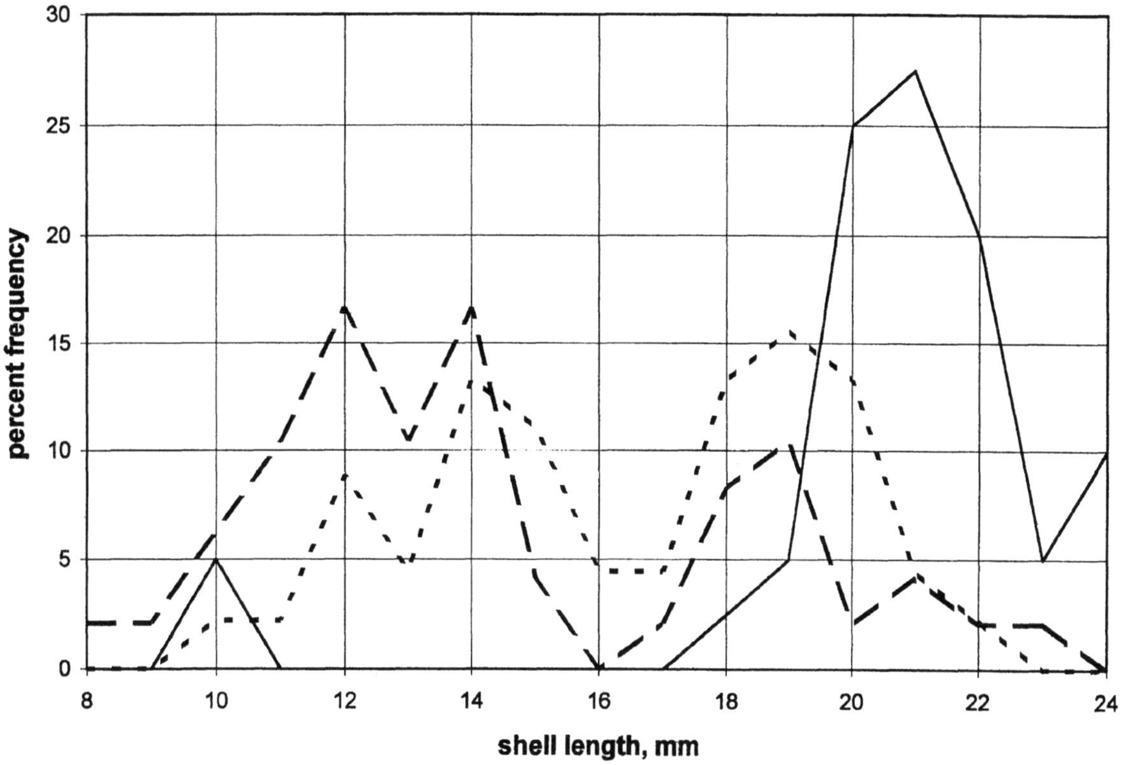

Figure 2. An example of the population structure of *L. littorea* at three levels on the shore. Dotted line = upper site ($n = 45$), dashed line = middle site ($n = 48$), solid line = lower site ($n = 40$). These samples were collected in June 1995.

the red alga *Chondrus crispus* (L.) commonly occurred on *L. littorea* shells, but were not considered in this study.

Smaller shells have less epifauna (Figure 3), and this was found to be the case at all levels of the shore. Since there were more smaller winkles at the upper and middle sites than at the lower site (Figure 2), the whole population showed a significantly lower epifaunal cover at the upper and middle levels than at the lower level (Table 1). However, at a shell length of 20 mm, there was little difference in epifaunal cover due to tidal level, and this applied to all of the larger shell lengths (Table 1). Equivalent relationships were shown for *P. ciliata* (Table 1, Figure 4). As with the other epifauna, *P. ciliata* is more frequent in larger shells (Figure 4). However, the distribution is much more strongly affected by shore level, with the larger, 20 mm, shells having significantly fewer holes at higher shore levels (Table 1).

Many more holes of *P. ciliata* were counted on the body whorl than on the other three possible positions (Table 2). T-tests showed that, despite the large standard deviations, all differences between mean counts are significant ($p<0.001$) except that between suture and spire. However, the four positions are very different in surface area: the aperture position had the smallest area, with the suture, spire and body whorl being larger in the ratio 1:1.2:8.4:30 respectively. When the numbers of holes are expressed in terms of relative area (Table 2) it can be seen that holes were most frequent at the suture and aperture. Distortions of the shells were frequently observed at the suture and sometimes at the aperture, caused by shell growth being impeded by barnacles. Distortions at the suture take the form of a raised lip formed by the edge of the body whorl where it meets the spire, and they persist as small cavities after the barnacle which caused them has died. Distortions caused by barnacles at the aperture leave similar cavities between the smooth nacreous layer of the aperture and the rough surface of the body whorl. These, however, are eventually covered by further shell growth. It was observed that distortions on winkles at the lower site were usually colonised by *P. ciliata*. For instance, on 10 September 95, 56 winkles were collected from the lower site, 38 were >20 mm long and carried epifauna; 17 of these had at least one shell distortion (16 at the suture and 3 at the aperture), all were colonised by *P. ciliata* and all distortions were

Table 1. Occurrence of sessile epifauna and *P. ciliata* on all winkles, and on 20 mm long winkles, collected between June 1995 and March 1996 at three shore levels. Percentages are given in parentheses. Probabilities (χ^2 test) are given in italics.

	Shore levels			
	Upper	Middle	Lower	
Total winkles	208	597	359	
Total + epifauna	46 (22)	173 (29)	251 (70)	*p<0.0001*
Total + *Polydora*	8 (4)	36 (6)	240 (67)	*p<0.0001*
Total 20 mm winkles	31	34	53	
20 mm + epifauna	17 (55)	24 (71)	33 (62)	*ns*
20 mm + *Polydora*	5 (16)	9 (26)	34 (64)	*p<0.01*

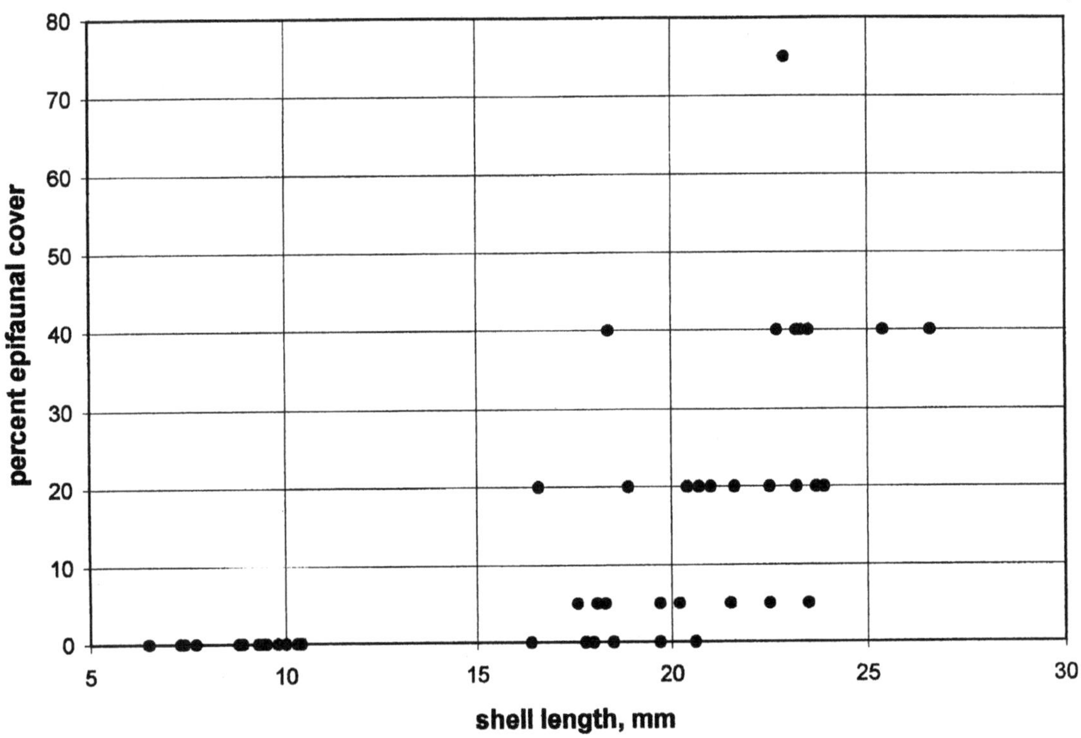

Figure 3. An example of the relationship between *L. littorea* shell length and percent epifaunal cover. This sample ($n=49$) was collected in March 1996 at the lower site. Percent cover has been arbitrarily assigned to the middle of the range estimated (i.e. 10–30% = 20%).

Table 2. The mean numbers of *P. ciliata* holes per infected shell (±SD) in 4 regions of shells; relative surface areas of shell regions; and proportions of holes per unit area in the 4 regions. Means are calculated from all infected periwinkles collected at the lower site between 6/95 and 3/96 (n = 238). Proportions were obtained by dividing mean numbers of holes by relative areas and adjusting the smallest proportion to 1.

	Shell regions			
	Suture	Spire	Body whorl	Aperture
Number of holes	3.9±4.65	4.5±5.95	10.5±15.5	1.8±2.64
Relative areas	1.2	8.4	30	1
Proportions per unit area	9.6	1.59	1.0	5.29

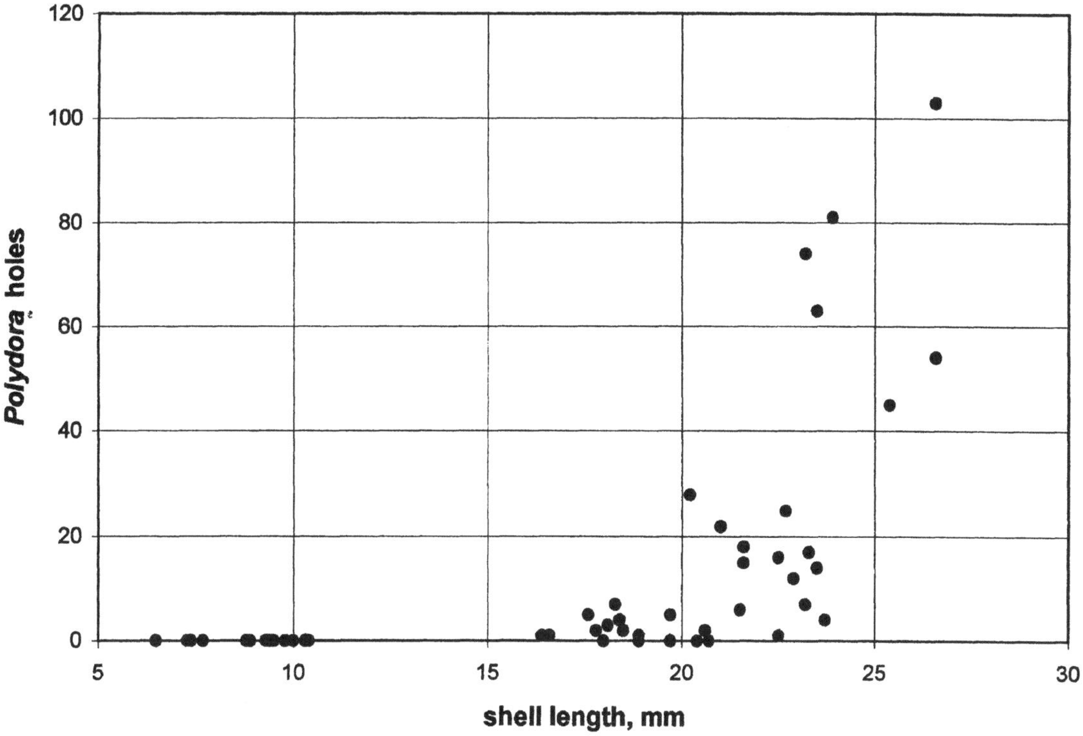

Figure 4. An example of the relationship between *L. littorea* shell length and the number of *P. ciliata* holes. This sample ($n = 49$) was collected in March 1996 at the lower site.

colonised; the mean number of holes per shell was 31 (SD = 21.6, range 1-93). In contrast, of the remaining 21 winkles with epifauna but without distortions, only 13 were colonised by *P. ciliata* and the mean number of holes per colonised shell was only 7 (SD = 6.9, range 1-27). These differences are significant (*t*-test, $p<0.001$). Another place on the shells where *P. ciliata* holes were often observed was along the edge of serpulid tubes where the tube made an angle with the surface of the shell.

It is clear that, for a given size of shell, a high percent cover of epifauna is normally associated with high numbers of *P. ciliata* holes (Table 3). Despite the large differences between mean numbers of *P. ciliata* in shells of different epifaunal cover, not all of the differences are significant (Table 3). This is because the variation in numbers of holes per shell (see standard deviations) is very large.

Discussion

Previous studies on winkle populations have also shown that smaller winkles tend to occur towards the middle and upper regions of the shore, and larger periwinkles tend to occur near to MLWS (Smith & Newell, 1955; Williams, 1964; Fish, 1972), but only Smith and Newell (1955) commented on the lack of epifauna on small (<10 mm) winkles. The relationship found here between epifaunal cover and the size of the host shell was clear (Figure 3) and applied at all tidal levels (Table 1). The most likely explanation is that cover is a function of shell age: the longer the shell has been available for settlement, the more likely it is to be colonised. There are, however, two additional possibilities. First, settling larvae may respond to the size of a potential settlement site (Jackson, 1977) and may avoid small ones. Second, some other features of smaller shells may inhibit settlement. There is little experimental evidence for the first proposition at the level of the sizes of periwinkle shells, but differences in settlement between species related to substratum size have been recorded, and have been linked to differences in competitiveness (Keough, 1984; Butler, 1991). The second possibility would appear to be ruled out by the findings of Wahl & Sönnichsen (1992) that *L. littorea* lack antifouling defences. However, I have observed that periwinkles at this site grow con-

Table 3. The mean number of *P. ciliata* holes (±SD) in winkle shells of 21 and 22 mm lengths collected from low tide level between 6/95 and 3/96 and related to the epifaunal cover on the shells; (n is given in parentheses). Different letters indicate significant differences between means within a size group ($p<0.05$) (t-tests).

	Epifaunal cover			
	0%	<10%	>10%	>30%
21 mm	0.38±0.74	3.41±4.56	14.95±15.06	26.87±22.43
	(8) a	(34) b	(22) c	(8) c
22 mm	1.50±3.21	12.50±14.77	18.20 ±22.27	29.30±18.44
	(6) a	(20) b	(27) bc	(14) c

tinuously until a length 14–17 mm and an age of 1–2 years (unpublished data). The shell laid down during this rapid growth is smooth with a thin periostracum. Growth is then checked, presumably during the first breeding season, and, when it resumes, a clear distinction can be seen between old and new shell. The old shell has lost the periostracum and has become progressively worn and pitted (this is a feature of shells at this site and may not occur elsewhere). A rough, uneven surface texture is well known to attract more settlement than does a smooth surface (e.g. barnacles: Crisp & Barnes, 1954), while periostracum has been recognised as having defensive functions in some molluscs (Kumar & Ayyakkannu, 1991; Harper & Skelton, 1993). I conclude therefore that the shells of smaller winkles are less colonised by epifauna than those of larger winkles partly because they have been exposed for less time, but also because they are less attractive, being smoother and having a greater proportion of the surface covered by periostracum. The population density at the middle site was often in excess of 100 m^{-2}, a level that should provide some antifouling by mutual grazing (Wahl & Sönnichsen, 1992); however snails at this site were not significantly less fouled than at the other two sites where population density was normally closer to 50 m^{-2}.

The results of this study indicate that on this sheltered shore the epifaunal cover on periwinkles is not affected by the range of tidal levels covered in this work (MTL-MLWS). Above MTL it is likely that epifaunal cover would decrease, as it does on the surrounding suitable substrata (pers. obs.), but on this shore periwinkles do not extend much above MTL. Smith & Newell (1955) studied a population which occurred mainly above MTL and found that the epifaunal cover (*Elminius modestus*) decreased above MTL in the same way on both winkles and on the surrounding substrata. They used this feature to support their contention that individual snails maintain their zonal position on the shore. Others, however, have observed a downward migration during the winter affecting particularly the upper shore snails (Williams & Ellis, 1975) and this also seems to occur at Southampton Water. Recognition of the zone of origin of Southampton periwinkles from their epifauna is not possible by percentage cover alone, but the epifaunal species composition alters down the shore, with a higher species diversity and more bryozoans and serpulids at the lower levels.

Turning to *Polydora ciliata*, this animal is strongly affected by tidal level in Southampton Water, occurring in large numbers close to MLWS and more rarely at higher levels. Ambariyanto & Seed (1991), however, found no relation between the tidal level and the occurrence of *P. ciliata* in mussel shells. *P. ciliata* was found below MTL by Dorsett (1961), and Smith & Newell (1955) noted that *L. littorea* shells from the upper shore contained few *P. ciliata* whereas those from the lower shore were 'always considerably excavated'.

As in the case of the other epifauna, *P. ciliata* was mainly found in larger periwinkle shells, probably for the same reasons as those given above for other epifauna. However, evidence is presented here in support of the hypothesis that the presence of sessile calcareous epifauna actually increases the settlement of *P. ciliata* by providing additional irregularities on the shell, and distortions at the suture and aperture which attract settling larvae. An alternative explanation of the data in Table 3 is that all epifauna, including *P. ciliata*, respond only to shell age and that periwinkle size, controlled for in Table 3, is a poor guide to age. However, the close association of *P. ciliata* with shell distortions suggests a more direct link with epifauna, as do the many observations of *P. ciliata* holes bordering serpulid tubes. Like many sessile animals, *P. ciliata* probably settles gregariously (Blake, 1969) and this habit is no doubt partly responsible for the very large variations found in numbers of holes per shell (Tables 2, 3) which were also noted in mussel shells by Ambariyanto & Seed (1991).

The association of *P. ciliata* with *L. littorea* is bad for the latter since it leads to the gradual destruction of the shell. This probably makes affected periwinkles more susceptible to predators such as the crab *Carcinus maenas* (L.) (an example of 'shared doom': Wahl & Hay, 1995). Williams (1964) suggested that since the largest periwinkles are normally to be found on the lower shore, this habitat may be optimal (longer time for feeding: faster growth). At Southampton Water, however, this supposed advantage must be weighed against the risk of shell destruction by *P. ciliata*. From a life history point of view, migration to the lower shore may not be an optimal strategy.

References

Ambariyanto & R. Seed, 1991. The infestation of *Mytilus edulis* Linnaeus by *Polydora ciliata* (Johnston) in the Conway Estuary, North Wales. J. moll. Stud. 57: 413–424.

Blake, J. A., 1969. Reproduction and larval development of *Polydora* from northern New England (Polychaeta: Spionidae). Ophelia 7: 1–63.

Buckley, W. J. & J. P. Ebersole, 1994. Symbiotic organisms increase the vulnerability of a hermit crab to predation. J. exp. mar. Biol. Ecol. 182: 49–64.

Butler, A. J., 1991. Effect of patch size on communities of sessile invertebrates in Gulf St Vincent, South Australia. J. exp. mar Biol. Ecol. 153: 255–280.

Crisp, D. J. & H. Barnes, 1954. The orientation and distribution of barnacles at settlement with particular reference to surface contour. J. anim. Ecol. 23: 142–162.

Dorsett, D. A., 1961. The behaviour of *Polydora ciliata* (Johnst.). Tube-building and burrowing. J. mar. biol. Ass. UK 41: 577–590.

Fish, J. D., 1972. The breeding cycle and growth of open coast and estuarine populations of *Littorina littorea*. J. mar. biol. Ass. UK 52: 1011–1019.

Fretter, V. & A. Graham, 1994. British prosobranch molluscs, 2nd edition. Ray Soc. London, 820 pp.

Harper, E. M. & P. W. Skelton, 1993. A defensive value of the thickened periostracum in the Mytiloidea. Veliger 36: 36–42.

Huxham, M., D. Raffaelli & A. Pike, 1993. The influence of *Cryptocotyle lingua* (Digenea: Platyhelminthes) infections on the survival and fecundity of *Littorina littorea* (Gastropoda: Prosobranchia); an ecological approach. J. exp. mar. Biol. Ecol. 168: 223–238.

Jackson, J. B. C., 1977. Habitat area, colonization, and development of epibenthic community structure. In Keegan, B. F., P. O'Céidigh & P. J. S. Boaden (eds), Biology of Benthic Organisms. Pergamon Press, Oxford: 349–358.

Kent, R. M. L., 1979. The influence of heavy infestations of *Polydora ciliata* on the flesh content of *Mytilus edulis*. J. mar. biol. Ass. UK 59: 289–297.

Keough, M. J., 1984. Effects of patch size on the abundance of sessile marine invertebrates. Ecology 65: 423–437.

Kumar, S. A. & K. Ayyakkannu, 1991. Periostracum of the gastropod *Hemifusus pugilinus*: natural inhibitor of boring and encrusting organisms. J. mar. biol. Ass. India 33: 374–378.

Orrhage, L., 1969. On the shell growth of *Littorina littorea* (Linné) (Prosobranchiata Gastropoda) and the occurrence of *Polydora ciliata* (Johnston) (Polychaeta Sedentaria). Zool. Bidr. Upps. 38: 137–153.

Smith, J. E. & G. E. Newell, 1955. The dynamics of the zonation of the common periwinkle *Littorina littorea* (L.) on a stony beach. J. anim. Ecol. 24: 35–56.

Wahl, M., 1989. Marine epibiosis. I. Fouling and antifouling: some basic aspects. Mar. Ecol. Prog. Ser. 58: 175–189.

Wahl, M., 1996. Fouled snails in flow: potential of epibionts on *Littorina littorea* to increase drag and reduce snail growth rates. Mar. Ecol. Prog. Ser. 138: 157–168.

Wahl, M. & M. E. Hay, 1995. Associational resistance and shared doom. Oecologia 102: 329–340.

Wahl, M. & H. Sönnichsen, 1992. Marine epibiosis. IV. The periwinkle *Littorina littorea* lacks typical antifouling defences – why are some populations so little fouled? Mar. Ecol. Prog. Ser. 88: 225–235.

Wargo, R. N. & S. E. Ford, 1993. The effect of shell infestation by *Polydora* sp. and infection by *Haplosporidium nelsoni* (MSX) on the tissue condition of oysters, *Crassostrea virginica*. Estuaries 16: 229–234.

Williams, E. E., 1964. The growth and distribution of *Littorina littorea* (L.) on a rocky shore in Wales. J. anim. Ecol. 33: 413–432.

Williams, I. C. & C. Ellis, 1975. Movements of the common periwinkle *Littorina littorea* (L.), on the Yorkshire coast in winter and the influence of infection with larval Digenea. J. exp. mar. Biol. Ecol. 17: 47–58.

Hydrobiologia **355:** 49–59, 1997.
A. D. Naumov, H. Hummel, A. A. Sukhotin & J. S. Ryland (eds), Interactions and Adaptation Strategies of Marine Organisms.

Effects of epibiosis on consumer–prey interactions

M. Wahl[1], M. E. Hay[2] & P. Enderlein[1]
[1]*Zool. Instit., Univ. Kiel, D-24098 Kiel, Germany (e-mail: mwahl@zoologie.uni-kiel.de)*
[2]*Univ. of North Carolina at Chapel Hill, Instit. Mar. Sci., Morehead City, NC 28557, U.S.A.*

Key words: indirect effects, interaction modification, epibiosis, predation, herbivores, carnivores

Abstract

In many benthic communities predators play a crucial role in the population dynamics of their prey. Surface characteristics of the prey are important for recognition and handling by the predator. Because the establishment of an epibiotic assemblage on the surface of a basibiont species creates a new interface between the epibiotized organism and its environment, we hypothesised that epibiosis should have an impact on consumer–prey interactions. In separate investigations, we assessed how epibionts on macroalgae affected the susceptibility of the latter to herbivory by the urchin *Arbacia punctulata* and how epibionts on the blue mussel *Mytilus edulis* affected its susceptibility to predation by the shore crab *Carcinus maenas*.

Some epibionts strongly affected consumer feeding behavior. When epibionts were more attractive than their host, consumer pressure increased. When epibionts were less attractive than their host or when they were repellent, consumer pressure decreased. In systems that are controlled from the top-down, epibiosis can strongly influence community dynamics. For the *Carcinus/Mytilus* system that we studied, the *insitu* distribution of epibionts on mussels reflected the epibiosis-determined preferences of the predator. Both direct and indirect effects are involved in determining these epibiont-prey–consumer interactions.

Introduction

In succession theory, highest diversity occurs at an intermediate stage in a community's progression from low-diversity pioneer stage to a low-diversity climax stage. This theoretical development of a community is governed by recruitment, subsequent interference competition for resources, and exclusion of some species by consumers or stressful physical conditions (Connell & Slatyer, 1977; Sousa, 1979). If left undisturbed, the elimination of inferior competitors or poor colonists leads to a low-diversity community dominated by a few species. In nature, however, this hypothetical climax is rarely reached. Physical (oxygen-deficiency, ice, storms...) and biological disturbances (predation, parasitism...) eliminate individuals or species and throw the community, or patches within the community, back to an earlier point in succession (e.g. Menge & Sutherland, 1987; Lampert & Sommer, 1993). Amplitude, frequency and regularity of disturbance determine its impact on the community (e.g. Sommer, 1995). In many marine communities, consumers (both herbivores and carnivores) constitute a major source of disturbance.

Consumer-prey interactions are necessarily mediated by the potential prey's body surface. Whether consumers forage using optical, olfactory or gustatory and tactile cues, the signals they perceive are a property of, or transmitted by, the prey's surface: shape, colour, scent, texture, consistency and so on. Also, many anti-consumer defenses are characteristics of the prey's surface: mimicry, camouflage, spines, mucus, toxins. The prey's body surface may thus be expected to play a primary role in its interactions with consumers.

Epibiosis, the colonisation of a living surface by sessile animals or plants, can substantially change the basibiont's surface properties. In fact, following epibiotic colonisation the basibiont/water interface is replaced by a basibiont/epibiont(s)/water interface. Successful consumption of a prey item occurs via four serial phases: (1) encounter→(2) recognition→(3) capture/handling→(4) consumption (after Lampert & Sommer, 1993). Interference in any of these phases impairs consumption. The presence of epibionts on a prey's surface may interfere with phases 2 (recognition) or 3 (capture/handling), and probably not 4 (consumption), because most predators detect unpalatability during handling.

We hypothesised that consumer/prey interactions may be affected by epibiosis. The questions we asked were:

– Does epibiosis influence consumer pressure by modifying the preference behaviour of consumers?

– If so, do different epibiont species shift consumer/prey interactions in different directions (increase vs decrease of predation pressure)?

In this investigation, we studied the effects of epibiosis on both herbivory (sea urchin *Arbacia punctulata* grazing on macroalgae) and carnivory (the shore crab *Carcinus maenas* preying on the mussel *Mytilus edulis*).

Material and methods

Herbivore–prey interactions

Details on methodology and an expanded presentation or data are available in Wahl & Hay (1995), but a short overview is provided below. Most prey algae were collected between 0.5 and 2 m below MLW from jetties and pilings around Morehead City and Beaufort (32° 42′ N, 76° 41′ W), North Carolina, USA. Only the brown alga *Zonaria* was taken 41 km offshore from Wilmington, North Carolina, at a depth of 28 m. Host seaweeds used in this study were the green alga *Codium fragile,* the brown algae *Sargassum filipendula* and *Zonaria tournefortii* and the red algae *Gigartina acicularis*, *Gracilaria tikvahiae* and *Agardhiella subulata.* Epibiont species found on one or more of these host species in sufficient abundance to be used in this investigation were the brown alga *Ectocarpus* sp. (on *Sargassum, Codium, Gracilaria,* and *Agardhiella*), the red algae *Polysiphonia* sp. (on *Sargassum, Codium, Gigartina, Gracilaria,* and *Agardhiella*) and *Audouinella* sp. (on *Sargassum,* and *Codium*), the bryozoans *Bugula neritina* (on *Sargassum*) and *Membranipora membranacea* (on *Zonaria*), and eggs of the gastropod *Anachis floridana* (on *Sargassum*). After their first introduction, all organisms will be referred to by their generic names.

All feeding assays were run with the omnivorous sea urchin, *Arbacia punctulata*, which is the most abundant inshore sea urchin and one of the most ecologically important herbivores in this area. Urchins were collected at Radio Island jetty near Beaufort, NC, and kept in shallow flow-through tanks. Between assays, urchins were fed on a variety of available algae, mostly *Gracilaria, Enteromorpha* sp., *Ulva* sp., *Gigartina,* and *Codium.*

To determine how epibionts and hosts affected each other's probability of being (a) found and (b) eaten by this generalist grazer, we employed two experimental approaches. First, we tried to determine the position of each host or epibiont on *Arbacia*'s feeding preference hierarchy. Secondly, we determined whether a host's susceptibility to urchin grazing changed if it was epiphytized. Because movement to a prey (= 'choice') and consumption of that prey were significantly correlated (Spearman $r=0.615$, $p=0.009$; see Wahl & Hay [1995]), only choice data will be presented here.

All experiments were run in four 2.3 m × 0.65 m × 0.15 m flow-through tanks. In each tank 10 urchins were fenced individually in plastic mesh cylinders (d = 30 cm, h = 20 cm, 1 cm mesh width). These 40 urchins were regarded as independent replicates. Urchins were always offered a two-way choice with the 2 prey items initially being positioned 20 cm from the urchin and separated from each other by 1 urchin diameter (5–10 cm). 'Choice' was determined by checking each replicate at about 30 minute intervals for the first 8–12 h of the experiment and recording which food was first contacted by the urchin's oral field.

To predict how a host's susceptibility to urchin grazing would change as a function of being fouled, the theoretical change in rank (ΔR) of a fouled host was computed as $\Delta R = R_h - (R_e + R_h)/2$ where R_h = the rank of the clean host and R_e = the rank of the epiphyte on the urchin's preference hierarchy (see Wahl & Hay, 1995).

Carnivore-prey interactions

All test organisms (the shore crab *Carcinus maenas*, as well as fouled and clean mussels *Mytilus edulis*) were collected by SCUBA from shallow near-shore habitats around Kiel (Western Baltic, 54° 22′ N, 10° 09′ E). Between experiments, crabs were kept in a 2 m^3 tank that was connected to a closed, recirculating system (volume: 5 m^3, salinity: 17–18‰, temperature: 15 °C.). They were fed with blue mussels collected from the same habitat. Mussels were kept in a separate tank that was connected to the same seawater system as the crabs. Only mussels smaller than 5 cm shell length fall into the prey spectrum of the crabs used in the predation preference experiments. Mussels used for these tests were fouled by barnacles *(Balanus improvisus)*, hydroids (*Laomedea flexuosa*), bryozoans (*Electra pilosa*), or filamentous algae. All but the dominant epibiont species of the epibiotic assemblage on a given mussel were scraped off in order to avoid mixture of cues.

To test for the possibility that fouled and unfouled mussel differ in some predation-related property unrelated to the presence of epibionts (e.g., physiology, value as a food, ...), the shells of fouled mussels were thoroughly cleaned (after having noted the composition of their epibiotic community) and predation on these cleaned mussels was compared with predation on similar sized mussels that had not initially supported epibionts.

Experiments started within 36 h after the mussels had been collected. The preference experiments were run in a 250 cm × 70 cm × 15 cm sized flow-through table that was subdivided by movable walls into 12 compartments (L: 30 cm, W: 30 cm, H: 15 cm). The bottom of each compartment was covered with sand because crabs behaved more naturally and were less frantic if they were allowed to burrow. All experiments were run as two-way preference tests, where (in each replicated compartment) one crab was randomly offered two prey items. The crabs chose either between a clean and a fouled mussel, or between a clean mussel and an originally fouled, but then cleaned mussel. We positioned the mussels in the middle of each compartment and separated them by at least the distance of one crab-width. Care was taken to match mussel size in each replicate.

Feeding preference (= actual consumption of the first of the two mussels offered) was recorded. Crabs that had not fed after 12 h were replaced. Crabs and mussels were removed from the compartments as soon as one of the prey items had been consumed. They were only used once per test series. In order to distinguish between the effects of visual and olfactory cues in determining prey choice by the crabs, a different experimental set-up was used. In these experiments, successful feeding was precluded by the experimental set-up (see below), instead 'choice' as first contact between crab and prey was registered. To avoid disturbance of crab behaviour by a continuous presence of the experimenter, this survey had to be automated: each of the 2 prey items per compartment was placed in the centre of a circle (diameter 10 cm) of six photo sensors hidden in the sand. Signals of the sensors, activated by the shadow of a crab contacting a prey, together with activation time were registered by computer. To produce only visual signals, prey (clean or fouled mussel) were offered to crabs in closed glass vessels that allowed crabs to see but not smell or touch the prey. To produce only chemical signals, prey (clean or fouled mussel) were offered to crabs in black but finely perforated vessels that allowed crabs to smell but not see or touch the prey.

Natural distribution of Mussel-Epibiont-Associations

At one of our experimental sites (an abandoned harbour), both mussels and crabs were very abundant. While mussels grew on all hard substrata, some of these substrata were inaccessible to crabs. In a SCUBA survey we assessed crab abundance (as an indicator for predation pressure by crabs) and the distribution of differently fouled mussels (clean, with barnacles, hydrozoans...) on various man-made substrata (pilings, ladders, ropes, pipes). Subsequently, we tested whether the observed epibiotic patterns could be explained by our data showing how epibionts affected crab feeding preferences.

To evaluate treatment effects, we used contingency table analyses or Fisher's exact test (if appropriate due to small cell sizes) and, when appropriate, a paired U-test.

Results

Herbivore–prey interaction

The generalist sea urchin *Arbacia punctulata* exhibited strong preferences among the 12 prey species offered (Figure 1). Potential prey could therefore be ranked linearly on a preference gradient from least (rank 12) to most (rank 1) preferred food. In the majority of cases, urchin preference differed significantly ($p<0.05$) among the various prey species. Only at the extremes of the gradient did preferences became less distinct (group of least preferred: *Ectocarpus 1, Sargassum*, group of most preferred: *Anachis* eggs, *Agardhiella, Ectocarpus 2*). During this study, *Ectocarpus* sp. apparently changed food quality drastically between spring (*E.1*) and summer (*E.2*), probably due to physiological stress (Wahl & Hay, 1995; Cronin and Hay, 1996).

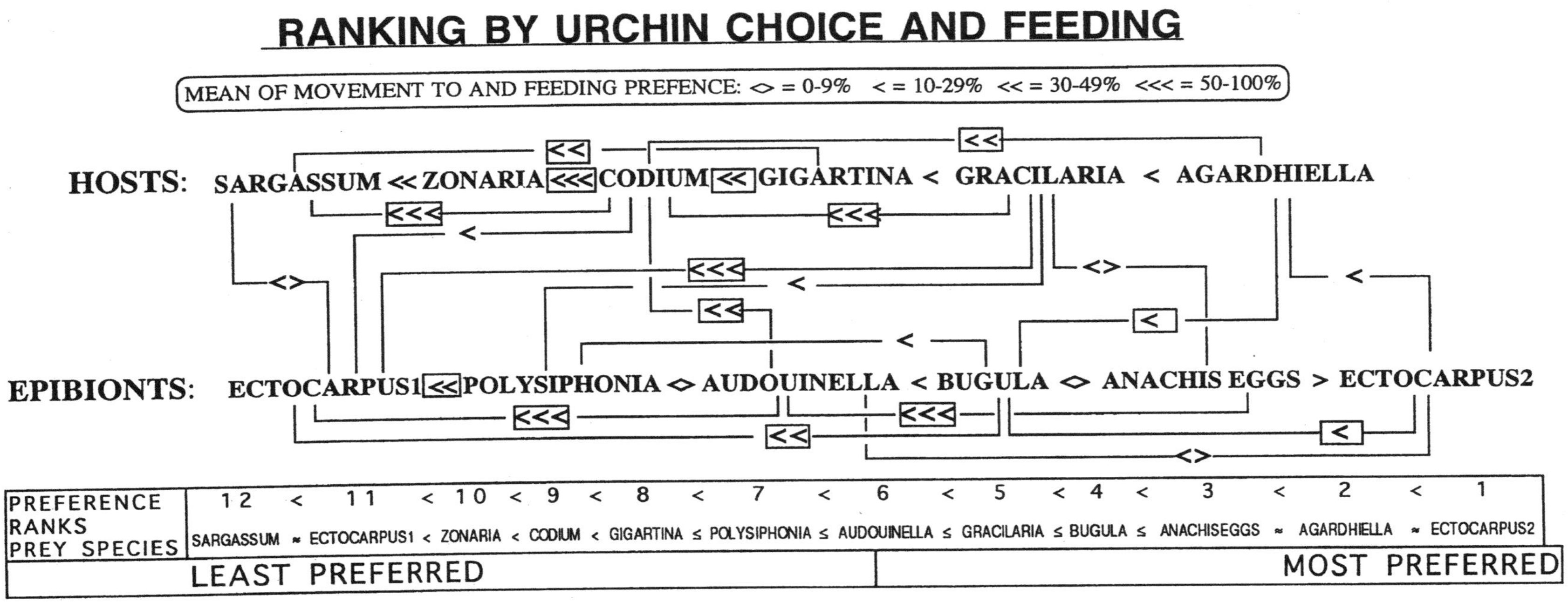

Figure 1. Ranking of epibionts and hosts by urchin choice. A key to the meaning of the various symbols is provided at the top of the figure. The final ranking of the 12 species is given at the bottom of the figure: 1 is the most (*Ectocarpus* 2) and 12 the least preferred species (*Sargassum*). Although, species grouped at the extremes of the preference gradient are almost indistinguishable, they were attributed individual ranks based on comparisons with common neighbours. Quantitative data used to construct this preference hierarchy are given in Wahl & Hay (1995). Framed in-equality signs stand for significant choice or feeding preference differences.

We predicted that the attractiveness or repulsiveness of a fouled alga would be determined by the relative food values of the epibionts and the basibiont based on the assumption that an epibiotic association combines characteristics of both epibionts and basibionts in approximately equal proportions. Surprisingly, this extreme simplification of calculating the preference rank of a fouled prey as the arithmetic mean of epibiont and basibiont preference ranks turned out to be a good predictor of how the urchins would react to the composite organism (Figure 2). Fouling by a more palatable epibiont enhanced consumption (shared doom), while colonization by less a preferred epibiont reduced consumption by the urchins (associational resistance). The greater the difference between epibiont and basibiont rank, the greater the epibiosis-induced change in how urchins treated the fouled host.

These effects of epibiosis can be strong enough to switch host algal position on the urchin's preference gradient, either increasing or decreasing an alga's susceptibility relative to other co-occurring seaweeds (Figure 3). The ecological consequences for community dynamics are discussed later, but could be far-reaching.

Carnivore-prey interactions

Epibionts on mussels strongly modified predator-prey interactions (Figure 4). After cleaning, mussels originally bearing different epibionts (barnacles, hydrozoans) did not differ ($p>0.43$) from unfouled mussels in their susceptibility to predation. Crabs consumed equal quantities of clean vs formerly hydrozoan-fouled mussels (15 vs 12, $p=0.43$) and of clean vs formerly barnacle-fouled mussels (15 vs 18, $p=0.47$). This shows that mussels without epibionts whether previously fouled or not did not differ in any property that produced a significant change in susceptibility to crab predation.

Fouling by filamentous algae did not affect the crab' predation on mussels (5 vs 5, $p=1$), but fouling by barnacles or hydrozoans strongly influenced crab predation on the mussel host (Figure 4). *Balanus improvisus* was highly attractive and increased its host's susceptibility to crab predation by a factor of three. When given the choice between clean and barnacle-fouled mussels, *Carcinus* consumed 5 clean vs 15 epibiotized prey ($p=0.0021$). In contrast, fouling by the hydroid *Laomedea flexuosa* had the opposite effect, and significantly lowered predation on host mussels ($p=0.0001$). Crabs consumed 20 clean mussels, but only 4 mussels fouled by hydrozoans. This represents a 5 fold decrease in susceptibility. Only 5 mussels of the size class 3–5 cm fouled by the bryozoan *Electra pilosa* could be found. This bryozoan-mussel association was taken 4 times as often as clean mussels, but, possibly due to the small number of replicates, this difference was not statistically significant ($p=0.11$).

When offering the crabs mussels fouled by barnacles vs. hydrozoans, *Carcinus* exhibited an even more dramatic difference in predation preference than when one of the prey alternatives was a clean mussel. Nine crabs consumed the mussel bearing barnacles, while only one crab fed on the mussel carrying hydrozoans on its shell. This difference in predation was highly significant ($p<0.001$).

Summarising, filamentous algae did not affect the host mussel's susceptibility to predation, barnacles increased susceptibility, and hydrozoans decreased susceptibility of the host mussel.

With unobstructed perception (all stimuli), 90% of the crabs preferred barnacle-fouled over hydrozoan-fouled mussels. For this most unequal pairing with regard to crab preference, we tried to determine whether the observed effects of epibionts on crab behaviour were caused by visual or olfactory properties of the epibiosis (Figure 5). When perceiving only visual cues (prey in glass vessels), the significant preference for barnacle-fouled mussels was lost. In fact, 8 out of 12 crabs first approached the (usually less preferred) hydrozoan-mussel association. When only olfactory stimuli were available to searching crabs (prey in perforated black vessels), 11 crabs first approached the hydrozoan-mussel association, whereas 13 crabs contacted the barnacle-mussel vessel in the first place. No preference was apparent ($p=0.58$). If one considers not only the first but all contacts crabs made with the offered prey, then the search seems even more 'cueless'. 88 (barnacle-mussel) versus 99 (hydrozoan-mussel) and 211 (barnacle-mussel) versus 208 (hydrozoan-mussel) contacts made in the visual and olfactory search, respectively.

Summarising, neither optical nor olfactory properties of the epibionts *per se* are responsible for the observed effects of epibiosis on crab predation.

On the hard substrata investigated, *Carcinus maenas* was unevenly distributed, reflecting their epibenthic life-style. On substrata that did not reach the bottom (ropes, ladders, pipes) no crabs were found. These objects were classified as 'inaccessible' (to crabs). On pilings, accessible to crabs, *Carcinus* abundance exhibited a decreasing gradient from the bottom to the water surface (Figure 6).

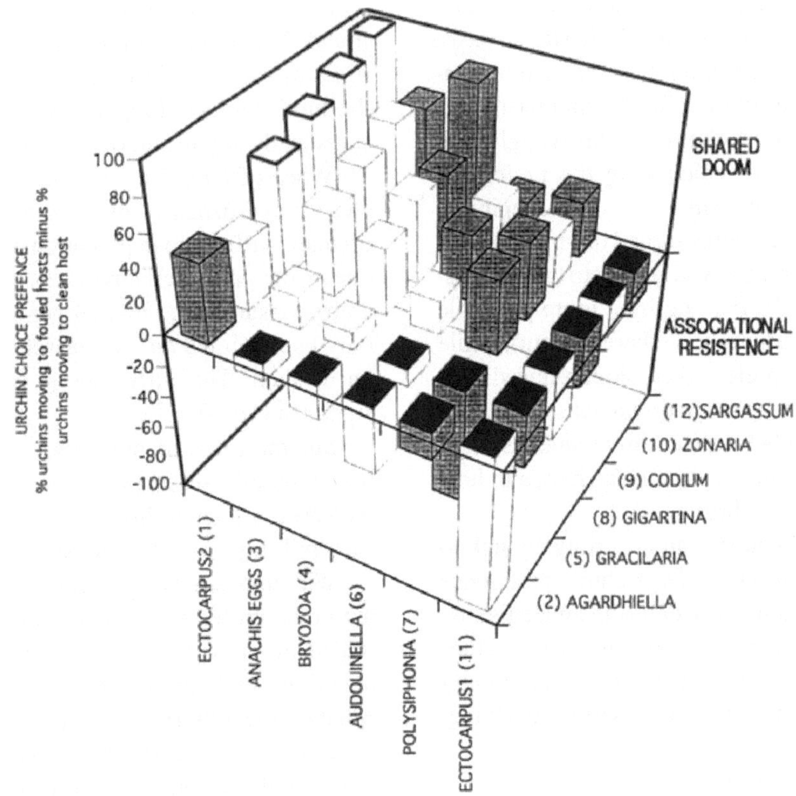

Figure 2. Predicted (white columns) and observed (shaded columns) changes in urchin choice of a prey species due to presence of epibionts. Positive values indicate shared doom effects; negative values indicate associational resistance effects. Columns with black bottoms represent negative changes. In this presentation shaded columns (= results) have replaced predicted (white) columns. Results match predictions well. For details of predictions see the text or Wahl & Hay (1995).

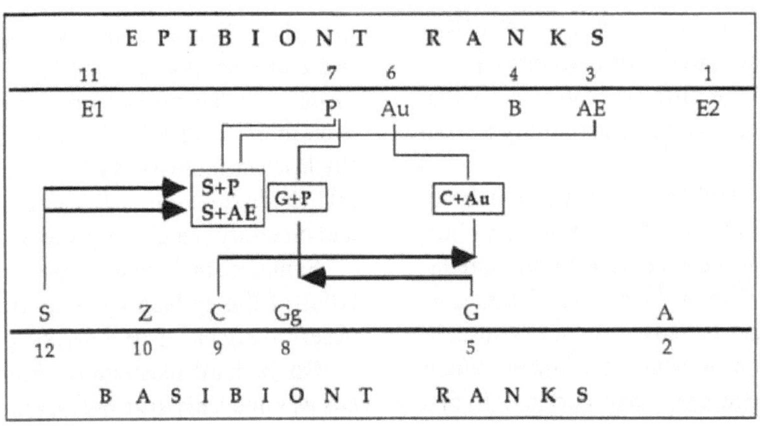

Figure 3. Rank shifts caused by epibiosis. Predator-prey interaction modification by epibionts may be strong enough to make prey species shift position on the urchin preference gradient. Thus, their susceptibility to grazing relative to neighboring prey species is altered. As an example, *Codium* when fouled by *Audouinella* surpasses the formarly preferred *Gigartina* in palatability and becomes equivalent to *Gracilaria* (formerly almost 6 ranks more attractive).

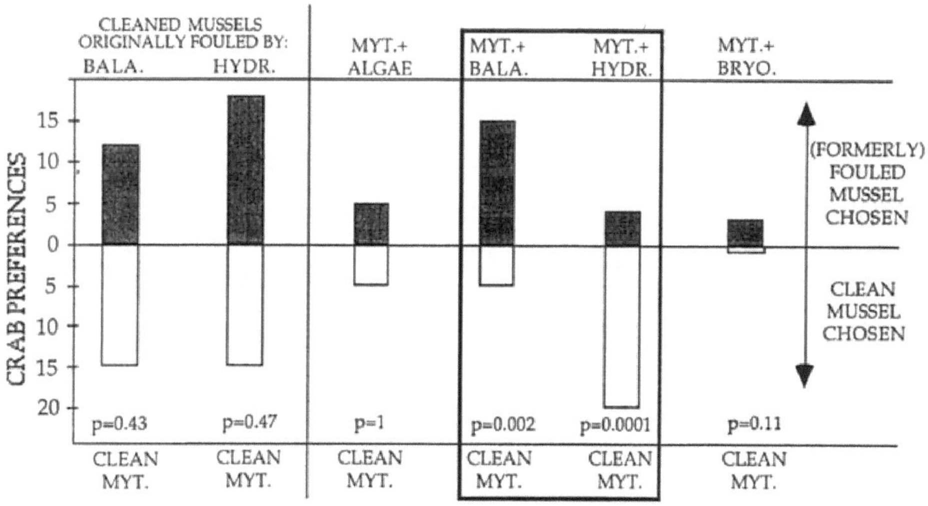

Figure 4. Crab feeding preferences when offered fouled, previously fouled, or clean mussels. *p*-values by Fisher's Exact test following a Contingency Table analysis. Abbreviations: Bala. = barnacle *B. improvisus*, Hydr. = hydroid *L. flexuosa*, Bryo. = bryozoan *E. pilosa*, Myt. = mussel *M. edulis*; y-axis: number of mussels consumed.

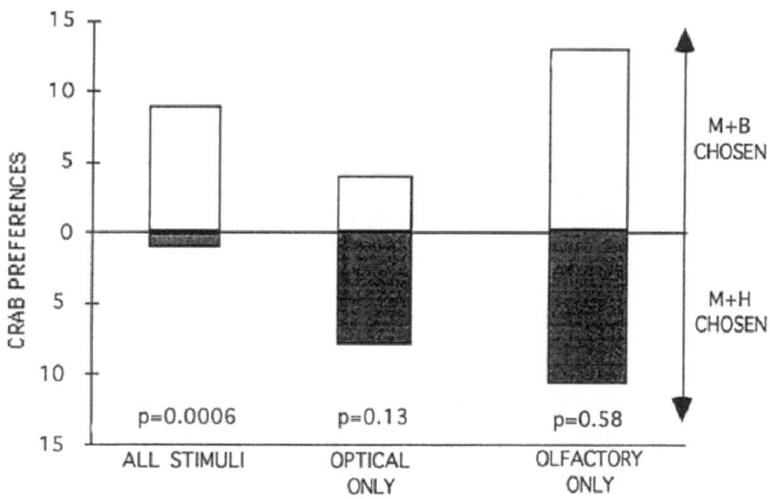

Figure 5. Preferences of crabs choosing between barnacle-fouled (M + B) and hydroid-fouled mussels (M + H) providing different stimuli. 'All stimuli' = visual + olfactory + gustatory/tactile, 'Optical only' = olfactory and gustatory/tactile stimuli suppressed, 'Olfactory only' = visual and gustatory/tactile stimuli suppressed; *p*-values: Fisher's Exact following a Contingency Table analysis; y-axis: number of mussels consumed.

Mussels occurred at similar abundances on all substrata and at all investigated depths. In contrast, the distribution of epibiotic communities on mussels varied strongly (Figure 6). On substrata without crabs, 70–80% of the mussels were fouled by barnacles, less than 20% of the mussels, on average, were clean, and epibiotic hydrozoans did not occur. On substrata with crabs, clean mussels dominated (45–65%), barnacles fouled 20–40% of the mussels, and abundance of hydrozoa fouled mussels increased with depth from 0% to 15–20%.

The most conspicuous result is that hydrozoan-mussel associations occurred exclusively in the presence of *Carcinus*. The relative abundance of the former correlates directly with crab abundance (Spearman Rank Correlation, rho = 0.7, $p = 0.006$).

The relative abundance of the barnacle-mussel association correlates inversely with the abundances of hydrozoan-mussel epibioses (rho = – 0.71, $p = 0.013$) and with *Carcinus* density (rho = – 0.74, $p = 0.01$).

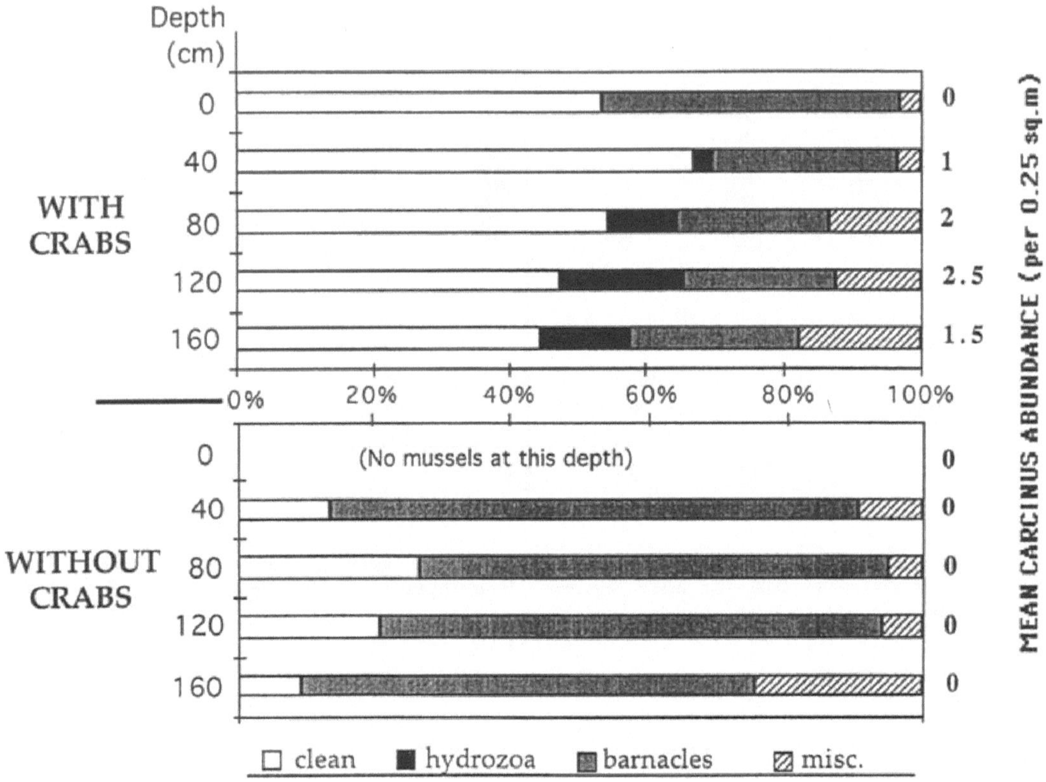

Figure 6. Proportional distribution (in % of population) of mussels fouled by different epibiont species in habitats inaccessible (lower part) and more or less accessible (upper part) to crabs. 'misc.' = mussels fouled by algae, bryozoans or sponges. Crab abundances on the right side of the graph.

Discussion

Epibiosis can drastically alter basibiont susceptibility to both herbivores and carnivores. Epibionts ranking higher on a preference hierarchy than their host will increase predation (shared doom); those ranking lower in attractiveness will decrease predation (associational resistance) on their host. Finally, the distribution of mussel/epibiont associations at the one location inventoried seems to reflect the epibiont – dependent differential mortality of mussels when predators may have been strongly affecting the community.

We are aware of very few reports on the role of epibionts in affecting predator-prey interactions. Feifarek (1987) described how an epibiotic sponge impeded starfish predation on a spiny oyster, Barkai & McQuaid (1988) found that whelks in shells overgrown by a bryozoan were resistant to lobster predation, and Gil-Turnes et al. (1989) reported that bacteria growing on the surface of shrimp embryos chemically defended the embryos from pathogenic fungi. Bach (1980) mentioned how normally immune corn may be moderately attacked by cucumber herbivores when grown intermixed with cucumber, Threlkeld et al. (1993) found that zooplankton may become more susceptible to predation when fouled, and epibiotic bryozoans have been shown to increase the attractiveness of kelp blades for grazing fishes (Bernstein & Jung, 1979) and echinoids (Ryland, 1976). In contrast, Prescott (1990) could not discover any effect of scallop epibionts on predation by seagulls or whelks.

The two interaction webs investigated here differed not only in the type of basibiont prey (algae and animals, resp.) but also in the diet breadth of the consumer species. *Arbacia punctulata* is an extreme generalist and it consumed all host and epibiont species when the alternative was a neutral non-food (agar agar, Wahl & Hay, 1995). Consequently, the urchins were not repelled by any epibiont. Rather the epibionts either enhanced or masked attractive signals of the basibiont. The facts that *Arbacia* is blind and that choice and consumption correlated so strongly imply that the urchins identify their preferred prey from a distance. The water-borne chemical signals responsible are presumably exudates of the alga and, if fouled, its epibionts. Because increasing the biomass of a prey did not notably affect urchin preference (Wahl & Hay, 1995), the species-specific chemical effects of the epibionts appeared to be responsible for the changes in foraging rather then these changes being caused by the fact that hosts with epibionts represent a larger resource than hosts alone.

The carnivorous generalist, *Carcinus maenas*, appeared to have more narrowly defined food choices than the urchin. Of the prey items offered, the crabs readily consumed *Balanus* and *Mytilus*, but did not eat the epibiotic alga or hydrozoan. *Balanus* (and probably the bryozoan *Electra*) enhanced predation on the mussels (shared doom), whereas the hydrozoan *Laomedea* reduced it (associational resistance). The red algal epibiont *Ceramium* did not affect predation rates.

For localisation and identification of a potential prey, *Carcinus maenas* may theoretically use cues on three informational levels: visual, olfactory, and gustatory/ tactile. Zimmer-Faust et al. (1995) have recently explored in detail the olfactory search behaviour of the blue crab *Callinectes sapidus*. They found that this crab species was capable of localising prey by binary comparison of chemical concentrations of cues inside and outside an odour plume originating from the prey. Our stimulus-experiments showed that in the case of *Carcinus maenas* neither visual nor olfactory information alone was sufficient to produce the clear preferential behaviour that the crab exhibited when all stimuli were offered. Either, only the simultaneous perception of visual and olfactory cues enables the crab to localise and identify preferred prey, or *Carcinus* also depends to a significant degree on gustatory/ tactile information for prey recognition. A prominent role in food search of the gustatory funnel-canal organs in the dactyl tips of *Carcinus* has been proposed by Schmidt & Gnatzy (1987).

In predator-controlled (top-down) communities, such as the Baltic mussel population we studied, our lab results on crab preferences would lead us to expect that *in situ* predation pressure should (in part) depend on the kind of epibionts a mussel bears. Differential survival should produce observable patterns of mussel-epibiont distribution. Indeed, the SCUBA inventory of *in situ* mussel-epibiont associations showed that under natural conditions, epibiont species were not distributed randomly between habitats characterized by different densities of crabs. The different mussel-epibiont associations in crab-controlled as compared to crab-free habitats reflected clearly the unequal susceptibilities to predation of these different associations.

On mussels not accessible to *Carcinus* predation, the barnacle *Balanus improvisus* was clearly the dominant epibiont, clean mussels were relatively rare, and hydrozoan epibionts did not occur. In contrast, when crabs were present, the proportion of barnacle-fouled mussels was much reduced, clean mussels dominated the population, and hydrozoan-fouled mussels (which are resistant to crab predation) correlated positively with crab abundance.

Thus, epibiotic communities differed sharply as a function of crab absence or presence. We believe, this may be explained by the effects of the different epibiont species on *Carcinus* feeding. *Carcinus* feeds preferentially on barnacles and on barnacle-fouled mussels. In habitats accessible to crabs, this preference is reflected by a reduced proportion of mussels bearing barnacles, accompanied by an increased proportion of clean (or rather cleaned?) mussels. The 'attractant/decoy scenario' of Atsatt & O'Dowd (1976) describes how species may benefit from the presence of more preferable species, the latter attracting a common predator away from the former. In an earlier paper (Wahl & Hay 1995) it was suggested that epibiosis might be too close an association for many large predators to distinguish between attractive epibiont and less palatable basibiont. Such an association should rather lead to shared doom. While this observation was true for urchins, it may not apply to the same extent to the more delicately discriminating crabs. In fact, during the experiments presented here, *Carcinus* several times only consumed highly attractive epibionts (e.g. barnacles) without feeding on the basibiotic mussel. It is conceivable, that a rich barnacle aufwuchs on mussels could sometimes satisfy the nutritional needs of *Carcinus* thus sparing the basibiotic mussels from predation. Such a situation would correspond to the 'attractant/decoy scenario'. Possibly, a portion of the clean mussel contingent had actually been cleaned of their barnacles by selectively feeding crabs.

The associational resistance effect of hydrozoan epibionts may be responsible for the observation, that mussel-hydrozoan associations increased in dominance with increasing predation pressure. The absence of epibiotic hydrozoa from mussels inaccessible to crabs could suggest that barnacles, released from crab predation, out-compete hydrozoans. An alternative explanation could be that in the absence of crabs hydrozoan-predators (e.g. nudibranchs) proliferate. Indeed, without predation, direct effects between different epibiotic species (e.g. competition) and between epibionts and hosts (e.g. Wahl *in press*) may gain greater importance.

Concluding, epibionts have the potential to dramatically increase or decrease the mortality of their hosts. Direct effects of epibiosis have frequently been described (reviewed in Wahl, 1989 and Wahl *in press*). The two investigations presented here reveal strong indirect effects that operate via the presence of a third species (the consumer in this case). These indirect effects occurred due to the presence of the epibionts and not because epibionts happen to colonize individual hosts that aleady differ in other characteristics that attract or deter consumers. Crabs were unable to discriminate between clean and cleaned mussels, which previously had been epibiotized by barnacles (attractive) or hydrozoans (repulsive), and urchins preyed with equal intensity on clean *Sargassum* and on cleaned *Sargassum* which previously had carried the attractive epibiont *Bugula neritina* (Wahl & Hay, 1995).

The epibionts affected their hosts' mortality *via* influencing predator behaviour by enhancing or masking chemical cues or (in the case of the hydrozoan-crab interaction) by repelling the contacting predator. Because predation pressure changed as a consequence of altered behaviour and not an abundance increase of the predator, this effect would qualify as an 'interaction modification', as opposed to an 'interaction chain' with serial direct effects (Wootton, 1993).

We are aware of only one study which described similar types of indirect effects. Schmitt (1987) investigated the predation of various species (lobster, octopus, whelk) on gastropod and bivalve prey. He found that by the addition of a prey species, more predators were attracted and predation pressure increased on all prey species. These interactions favored spatial segregation of the two prey species, a phenomenon he called 'apparent competition' (see also Menge, 1995 for a comprehensive treatment of the subject). Our shared doom scenario falls in the same category of indirect effects. For instance, epibiotic barnacles increase predation on fouled mussels. Probably as a consequence of this, in nature, barnacles and mussels co-occurred less often (epibiotically) in the presence than in the absence of predators. On the other hand, associational resistance does not seem to fit any of Menge's types of indirect effects. It has, however, been described repeatedly from terrestrial (e.g. Root, 1973; Atsatt & O'Dowd, 1976; Bach, 1980) and marine environments (Hay, 1986; Littler et al., 1986; Pfister & Hay, 1988). As in some of these earlier studies, associations providing hosts with protection (hydrozoan-mussel) were favoured by the presence of predators.

Up to now we have worked only with generalist predators. It could be rewarding to assess the effects of epibionts on predators that are specialized on particular host species and that therefore may not have the physiological or behavioral flexibility to switch to an alternative food when their prey has become less attractive or unrecognisable due to epibiosis.

Because epibiosis creates a new interface between a host and its environment it may be expected that many other interactions, for example with physical factors, with conspecifics, with parasites, etc., may also be modified. The direct and indirect effects of epibiosis on community dynamics are doubtlessly worth investigating more thoroughly.

References

Atsatt, P. R. & D. J. O'Dowd, 1976. Plant defense guilds. Science 193: 24–29.

Bach, C. E., 1980. Effects of plant diversity and time of colonization on a herbivore–plant interaction. Oecologia 44: 319–326.

Barkai, A. & C. McQuaid, 1988. Predator-prey role reversal in a marine benthic ecosystem. Science 242: 62–64.

Bernstein, B. B. & N. Jung, 1979. Selective pressure and coevolution in a kelp canopy community in Southern California. Ecol. Monogr. 493: 335–355.

Connell, J. H. & R. O. Slatyer, 1977. Mechanisms of succession in natural communities and their role in community stability and organization. Am. Nat. 111: 1119–1144.

Cronin, G. & M. E. Hay, 1996. Susceptibility to herbivores depends on recent history of both the plant and animal. Ecology 77: 1531–1543.

Feifarek, B. P., 1987. Spines and epibionts as antipredator defenses in the thorny oyster *Spondylus americanus* Hermann. J. exp. mar. Biol. Ecol. 105: 39–56.

Gil-Turnes, M. S., M. E. Hay & W. Fenical, 1989. Symbiotic marine bacteria chemically defend crustacean embryos from a pathogenic fungus. Science 246: 116–118.

Hay, M. E., 1986. Associational plant defenses and the maintenance of species diversity: turning competitors into accomplices. Am. Nat. 128: 617–641.

Lampert, W. & U. Sommer, 1993. Limnoökologie. Georg Thieme Verlag Stuttgart, New York, 440 pp.

Littler, M. M, P. R. Taylor & D. S. Littler, 1986. Plant defense associations in the marine environment. Coral Reefs 5: 63–71.

Menge, B., 1995. Indirect effects in marine rocky intertidal interaction webs: patterns and importance. Ecol. Monogr. 65: 21–74.

Menge, B. & J. P. Sutherland, 1987. Community regulation: variation in disturbance, competition, and predation in relation to environmental stress and recruitment. Am. Nat. 130: 730–757.

Pfister, C. A. & M. E. Hay, 1988. Associational plant refuges: convergent patterns in marine and terrestrial communities result from differing mechanisms. Oecologia 77: 118–129.

Prescott, R. C., 1990. Sources of predatory mortality in the bay scallop *Argopecten irradiance* (Lamarck): interactions with seagrass and epibiotic coverage. J. exp. mar. Biol. Ecol. 144: 63–83.

Root, R. B., 1973. Organization of a plant-arthropod association in simple and divers habitats: the fauna of collards (*Brassica oleracea*). Ecol. Monogr. 431: 95–120.

Ryland, J. S., 1976. Physiology and ecology of marine bryozoans. Adv. mar. Biol. 14: 285–443.

Schmidt, M. & W. Gnatzy, 1987. Contact chemoreceptors on the walking legs of the shore crab, *Carcinus maenas*. Ann. N.Y. Acad. Sci. 1987: 589–590.

Schmitt, R. J., 1987. Indirect interactions between prey: apparent competition, predator aggregation, and habitat segregation. Ecology 68: 1887–1897.

Sommer, U., 1995. An experimental test of the Intermediate Disturbance Hypothesis using cultures of marine plankton. Limnol. Oceanogr. 40: 1271–1277.

Sousa, W. P., 1979. Experimental investigations of disturbance and ecological succesion in a rocky intertidal community. Ecol. Monogr. 49: 227–254.

Threlkeld, S. T., D. A. Chiavelli & R. L. Willey, 1993. The organization of zooplankton epibiont communities. Trends Ecol. Evol. 8: 317–321.

Wahl, M., 1989 . Marine epibiosis. I. Fouling and antifouling: some basic aspects. Mar. Ecol. Progr. Ser. 58: 175–189.

Wahl, M. (in press). Living attached: aufwuchs, fouling, epibiosis. In Nagabhushanam, R. & M. F. Thompson (eds), Fouling Organisms of the Indian Ocean: Biology and Control Technology. Oxford & IBH Publ. Co. Put Ltd, New Delhi: 31–83.

Wahl, M. & M. E. Hay, 1995. Associational resistance and shared doom: effects of epibiosis on herbivory. Oecologia 102: 329–340.

Wootton, J. T., 1993. Indirect effects and habitat use in an intertidal community: interaction chains and interaction modifications. Am. Nat. 141: 71–88.

Zimmer-Faust, R. K., C. M. Finelli, N. D. Pentcheff & D. S. Wethey, 1995. Odor plumes and animal navigation in turbulent water flow: a field study. Biol. Bull. 188: 111–116.

Hydrobiologia **355**: 61–70, 1997.
A. D. Naumov, H. Hummel, A. A. Sukhotin & J. S. Ryland (eds), Interactions and Adaptation Strategies of Marine Organisms.

Parasites on an intertidal *Corophium*-bed: factors determining the phenology of microphallid trematodes in the intermediate host populations of the mud-snail *Hydrobia ulvae* and the amphipod *Corophium volutator*

Kim N. Mouritsen, Tomas Jensen & K. Thomas Jensen
Department of Marine Ecology, Institute of Biological Sciences, University of Aarhus, Finlandsgade 14, DK-8200 Aarhus N, Denmark
(Phone: [+45] 8942 4386; Fax: [+45] 8942 4395; e-mail: KNM@bio.aau.dk)

Key words: Corophium, Hydrobia, intertidal mud flats, microphallid trematodes, phenology

Abstract

The phenology of microphallid trematodes within their intermediate host populations has been studied on an intertidal mud flat. The parasites use the mud snail *Hydrobia ulvae* and the infaunal amphipod *Corophium volutator* as first and secondary intermediate host, respectively. Migratory shorebirds act as final hosts. Our results show a general trend of decline in the density of infected intermediate hosts during both spring and autumn, which could mainly be ascribed to shorebird predation. During summer the density of both infected snails and infected amphipods increased considerably, with a culmination in June within the snail population (1000 infected m^{-2}) and in August within the amphipod population (40 000 infected m^{-2}). This time lag in parasite occurrence could be related to (1) the development time of larval trematodes within the snails, (2) higher ambient temperatures in late summer increasing parasite transmission between snails and amphipods during this period, and (3) a general increase in the *Corophium* population during late summer. From samples collected between 1990 and 1995 it is shown that microphallid trematodes occasionally may give rise to mass mortality in the amphipod population. The prerequisites for such an event are a high parasite prevalence within the first intermediate host population and unusually high ambient temperatures, facilitating parasite transmission to the secondary intermediate host, *C. volutator*.

Introduction

Considering a community of infaunal macro invertebrates at an intertidal mud flat, an array of reasons why a given species should show wide fluctuations in abundance can be given. Among the well-known factors the often harsh environmental conditions that prevail in tidal areas can be mentioned, e.g. prolonged and strong offshore winds causing desiccation, strong onshore winds causing substrate erosion, thermal and osmotic stress during low water and mechanical stress due to ice pack during winter at higher latitudes (Reise, 1985; Barnes & Hughes, 1988; Nybakken, 1993). However, also biotic factors may have profound influence on the population dynamics of intertidal softbottom organisms. For instance, well-defined periods of recruitment usually result in substantial increases in abundance of many benthic organisms, whereas seasonality in the often very significant predation pressure from fish, shorebirds and other epibenthic and infaunal predators may cause sudden declines in density of many infaunal organisms, including *C. volutator* (Reise, 1985; Posey, 1987; Peer et al., 1986; Ambrose, 1991; Wilson, 1991).

However, one biotic factor only very rarely considered in ecological studies of softbottom macrozoobenthic communities is parasitism. In the light of the growing body of evidence emphasizing parasites as more or less cryptic determinants of host population dynamics and even community structure (Dobson & Hudson, 1986; Anderson, 1991; Minchella & Scott, 1991; Holt, 1993), and given that most if not all soft-bottom macro invertebrates are hosts of one or usually more species of parasites (Kinne, 1980; Sousa, 1991), the lack of attention to the ecological consequences of parasitism is unfortunate.

An initial step in achieving more information about the parasites' effect on their host population is to obtain knowledge of the phenology of the parasites themselves in order to identify factors controlling their abundance. The present study aimed at describing the phenology of parasites in their intermediate host populations of the mud snail *Hydrobia ulvae* (Pennant) and the tube-dwelling amphipod *Corophium volutator* (Pallas). The life cycle of the microphallid digenean trematodes *Maritrema subdolum* Jägerskiöld, 1909 and *Microphallus claviformis* (Brandes, 1888) on intertidal mud flats is used as a model. Special reference is given to the parasites' influence on the population dynamics of *Corophium. Hydrobia* and *Corophium* are both widely distributed and abundant on intertidal mud flats, and hence, can be considered central macro-zoobenthic organisms in the softbottom intertidal ecosystem.

Materials and methods

The microphallid life cycle

The two species of microphallid trematodes have a similar but rather complex life cycle (Figure 1). The adult parasites reproduce sexually in the digestive tract of their final vertebrate hosts, being various shorebirds. Their eggs are expelled to the exterior through the droppings of the birds. Left on the sediment surface, some of these eggs may be eaten (probably accidentally) by the first intermediate host, *Hydrobia* spp. Within the gonads of the snails, the parasites reproduce asexually and develop into sporocysts that upon the right stimuli proliferate myriads of tail-carrying cercariae. The cercariae then seek out and penetrate the cuticle of the secondary intermediate amphipod host, in the Wadden Sea usually *Corophium* spp. Within the body cavity of the amphipods the now tail-less cercariae encyst as metacercariae that gradually develop into adult trematode worms. The life cycle is completed when the infected *Corophium* specimen is eaten by a suitable shorebird host.

The study site

As study site we selected a mid-tidal station on an approximately 80 ha large *Corophium*-bed at Højer tidal flat (approximately 1500 m intertidal zone) in the southern part of the Danish Wadden Sea (Figure 2). At low water, a *Corophium*-bed is often characterized by a mosaic of alternating tidal pools and emerged areas (Figure 3). The latter supports very high numbers of amphipods, whereas considerably fewer specimens occur in the pools. The mud snail *H. ulvae* can be found abundantly on both the emerged areas and in the pools. The *Corophium*-bed is very attractive to feeding waders, especially the Dunlin *Calidris alpina*, which in this area occur at average densities above 200 foraging birds per hectare during the autumn and spring migration periods (Mouritsen, 1994).

Collection and treatment of animals

At each sampling event, five to ten core samples (50 cm^2, 25 cm deep) were collected on the emerged areas of the *Corophium*-bed and sieved on location through a 500 μm mesh sieve. Retained animals were preserved in pH-buffered 4% formaldehyde. In the laboratory densities were calculated and the length of the amphipods was measured from rostrum to telson. Snails and amphipods were subsequently dissected and examined for the presence of microphallid trematodes. Parasites were identified in accordance with Deblock (1980), Lauckner (1987) and own unpublished experimental results regarding identification of metacercariae in the amphipods.

Measurements and dissection were carried out on animals from unbiased subsamples at each sampling occasion (*Hydrobia*: 33–220 specimens; *Corophium*: 102–150 specimens).

Analysis and presentation of data

Regarding the main phenological description (Figure 4), the data originate from seven sampling occasions in 1995. Statistical treatments of these data are restricted to 95% confidence limits (density of hosts, Figure 4B) and the Z-test (Snedecor & Cochran, 1989) when dealing with parasite prevalence (part of Figure 4A, see below). When evaluating trends in infection intensity (Figure 4C), the non-parametric Mann-Whitney test has been applied rather than 95% confidence limits, due to the overdispersed distribution of parasites within the amphipods. For convenience, ordi-

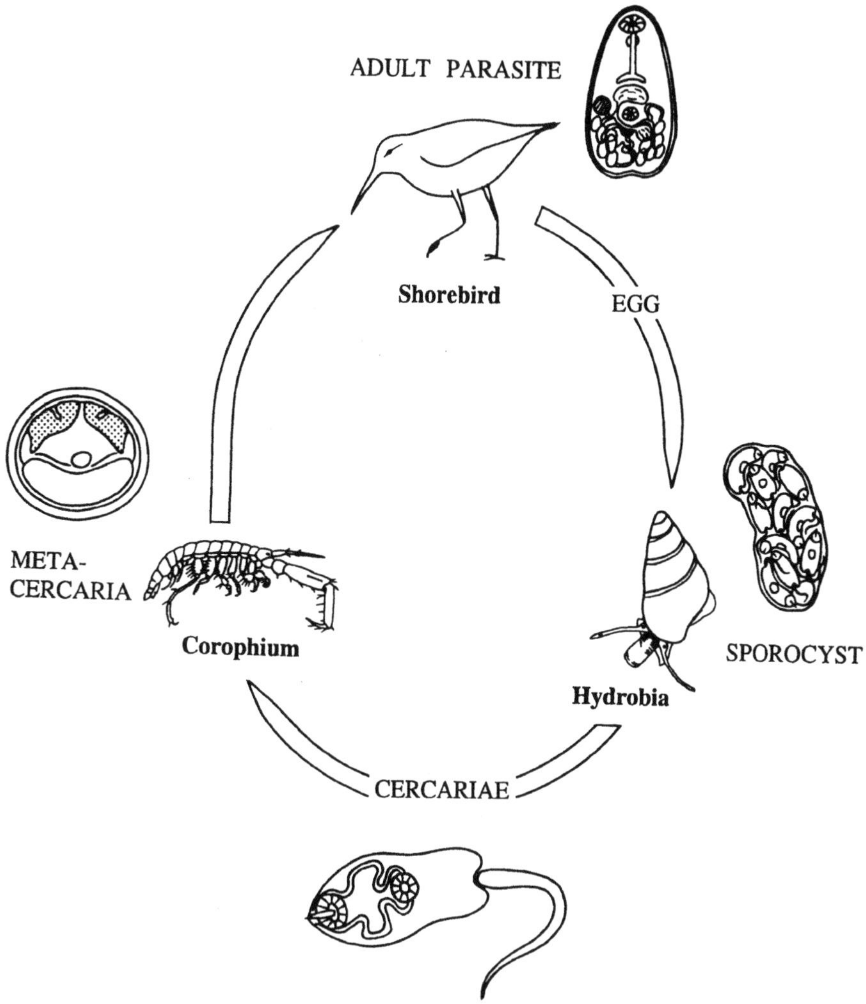

Figure 1. The life cycle of the microphallid trematodes *Maritrema subdolum* and *Microphallus claviformis*. See text for details.

nary standard error of the mean is nevertheless shown in the figure.

Figure 4A, showing the temporal development in density of infected hosts, is based on mean host density multiplied by trematode prevalence. Since the latter is obtained from a subsample, the statistical significance of the observed changes can not be tested formally. However, if both mean host density and trematode prevalence show similar and statistically significant trends, or when one of those parameters shows no significant trend whereas the other one does, it is reasonable to consider an observed change in the density of infected hosts to be significant. In Figure 4A such significant changes are indicated by an asterisk.

Indication of the main periods of bird presence at the study site is based on own bird counts in the area in concert with data presented by Laursen & Frikke (1984). Temperature data originate from a nearby meteorological station.

In Figure 5 the mean densities of hosts are shown without attached variation in order to simplify the figure. The statistical significance (using one-way ANOVA and Z-test) of any trends mentioned is, however, indicated in the figure legend.

All statistical tests are two-tailed, using the default 5% level of significance.

Results and discussion

The phenology

Considering the microphallid life cycle (Figure 1) it is possible to make at least four obvious predic-

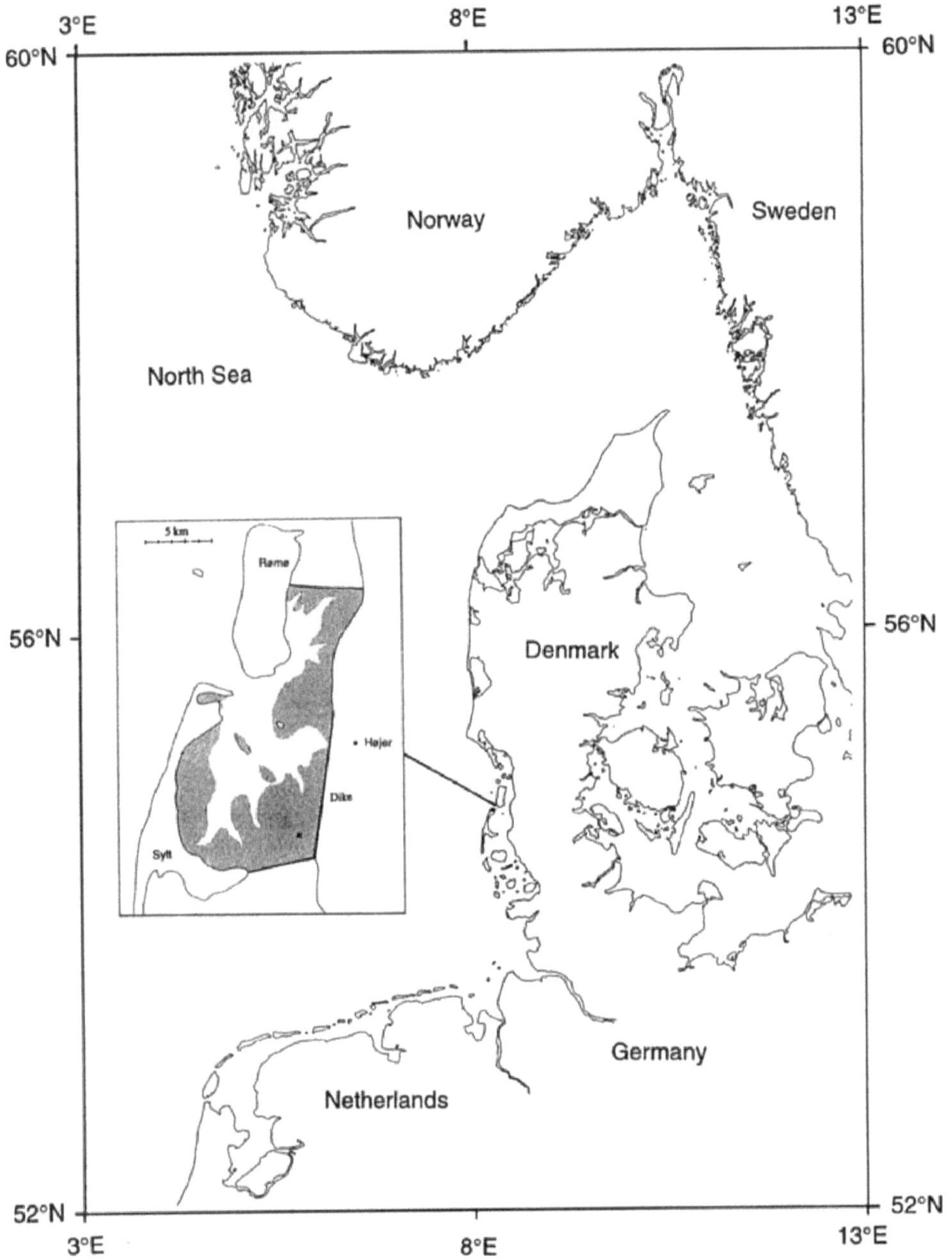

Figure 2. The location of the study site in the Danish Wadden Sea (■). The intertidal flats are shown in the Lister Dyb tidal area (between Rømø and Sylt) on the inserted map (shaded areas).

tions regarding the phenology of these parasites within their intermediate host populations. Firstly, the abundance of microphallids in the amphipods should depend on the density of infected *Hydrobia*, which in turn is dependent on former bird presence. Secondly, their occurrence in *Corophium* should lag behind their occurrence in *Hydrobia* because it may take about 1–1.5 month before these parasites reach maturity within the snail host (Jensen & Mouritsen, 1992). Thirdly, the occurrence of metacercariae within the amphipods should increase as ambient temperatures rise due to a positive association between temperature and cercarial emergence. In experiments, we showed that the number of shed *Maritrema*-cercariae increased by a factor of nine as the water temperature rose from 15 °C to 25 °C (Mouritsen & Jensen,

Figure 3. A view of a well-developed *Corophium*-bed at Højer tidal flat in the southern part of the Danish Wadden Sea.

1977; see also Ginetsinkaya, 1988). Fourthly, the size of the microphallid metapopulation within the amphipod hosts should also depend on the amphipod density itself. This should occur, not because of the usual expectation of density dependence due to e.g. malnutrition causing the infective stages to transmit more readily (e.g. Anderson, 1991 and references therein), but simply because more hosts may be able to absorb more of the released cercariae. The microphallid cercariae become exhausted after approximately six hours, and if they fail to find a suitable host within this period they will die (K. N. Mouritsen, unpubl. data).

The above predictions are fully confirmed by our results. During spring a decline in density of infected *Hydrobia* and *Corophium* is observed (Figure 4A), corresponding to a decrease in the density of these hosts in general (Figure 4B). These changes are most likely due to bird predation which is quite intensive during spring. The most numerous shorebird species at our study site, Dunlins, prey on both *Corophium* and *Hydrobia* (Worrall, 1984; Mouritsen, 1994), and it has previously been shown that shorebirds are capable of diminishing prey abundance, including *Corophium*, significantly (e.g. Wilson, 1991; Peer et al., 1986).

Following the spring migration period of the birds, a considerable and rapid increase in the density of microphallid infected snails to about 1000 specimens m^{-2} in early June is observed (Figure 4A). This increase should be seen as a consequence of the previous presence of birds giving rise to new infections via transmission of microphallid eggs from their droppings to the snails. In support of this interpretation no more new infections appear during the summer period in which migratory shorebirds are absent. Regarding the density of infected amphipods, a considerable peak of about 40 000 infected specimens m^{-2} can be observed in mid August. Hence, almost half of the entire *Corophium* population is infested with microphallid trematodes just before the shorebirds return during their autumn migration.

As predicted there is a time lag between the increase in density of infected snails and infected amphipods (Figure 4A). Three points should sufficiently explain this pattern. During spring and especially summer a significant increase in the abundance of amphipods in general is observed, which can be ascribed to two recruitment periods (Figure 4B). From April to May the proportion of *Corophium* individuals larger than three mm (the approximate size from which the amphipods can be sexed by the presence of oostegites) shows a decline from 89% to 38%. A similar trend is observed from June to August, where the proportion of the larger specimens is reduced from 50% to 28%. In conclusion, the population of amphipods increases during the summer, and more specimens will catch up the released microphallid cercariae. The larval trematodes do not

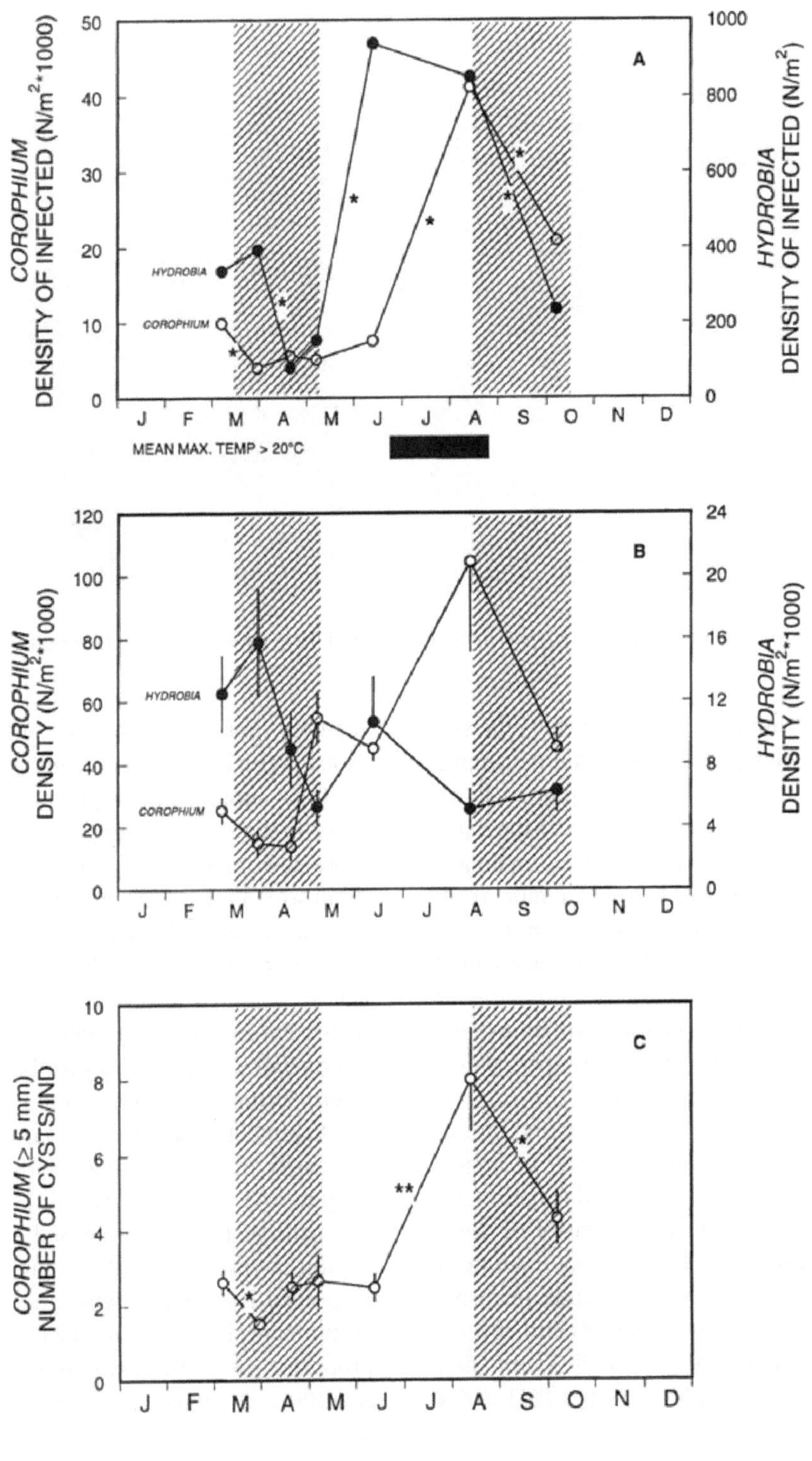

Figure 4. The phenology of hosts and parasites from spring to autumn 1995 on Højer tidal flat. Shaded areas denote the principal periods of wader presence. A: The seasonal changes in the density (numbers m^{-2}) of microphallid infected *Hydrobia ulvae* and *Corophium volutator*. The black bar below the x-axis shows the period in which the weekly mean of daily maximum temperature exceeded 20 °C ($*$ denotes changes that can be considered significant; see Materials and Methods). B: The population dynamics of the *Hydrobia* and *Corophium* host (mean density $m^{-2} \pm$ 95% confidence limits). C: Changes in the mean number ($\pm$ S.E.) of microphallid trematodes within infected amphipod individuals ($n = 19$–39 *Corophium* specimens). $*$ $p<0.05$ (Mann-Whitney test, Z>2.01), $**$ $p<0.001$ (Mann-Whitney test, Z=3.74).

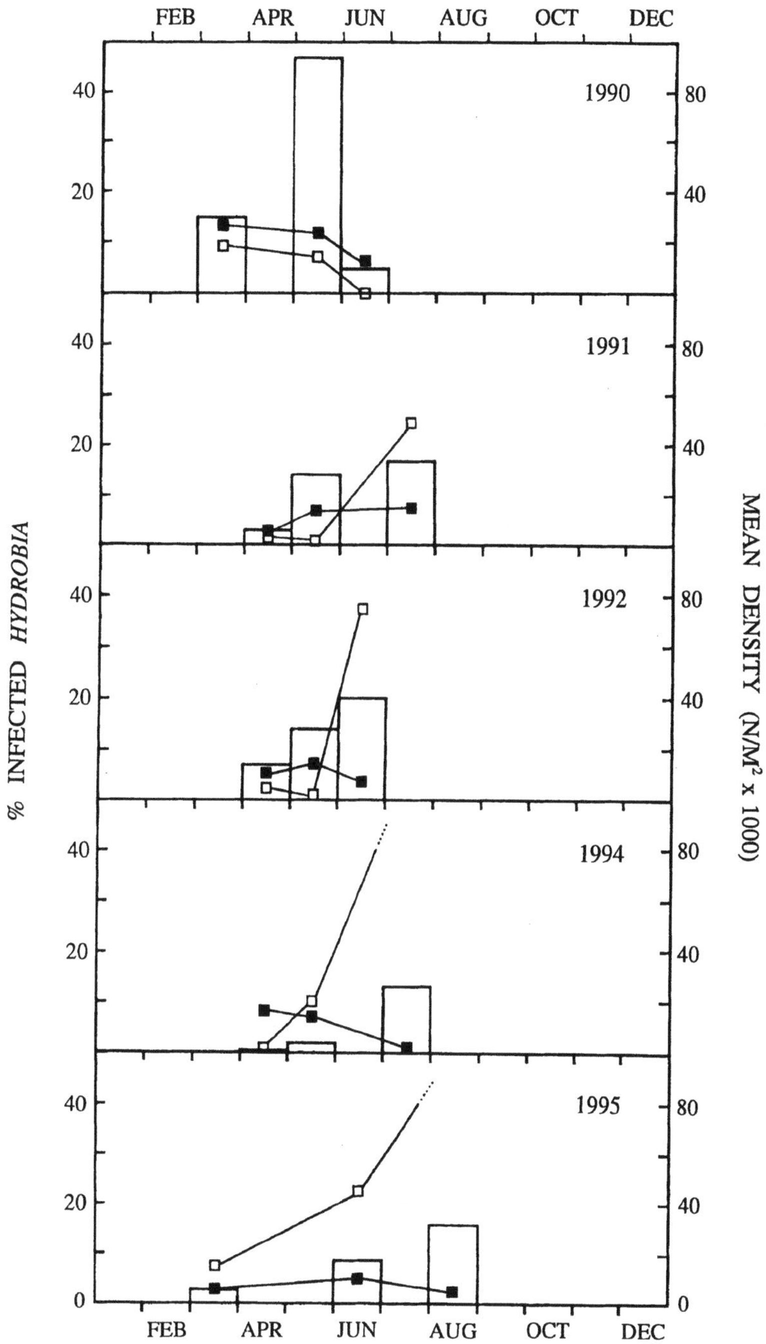

Figure 5. The mean density m^{-2} of *Hydrobia* (■) and *Corophium* (□), and the prevalence (%) of microphallid infections within the snail population (white bars) at three sampling occasions in spring and summer 1990–95. A dotted line indicates that the density is outside the range of the y-axis. During all years, the general trends of increase (decrease in 1990) in the amphipod population are statistically significant (One-way ANOVA, $F>23.6$, $p<0.001$). Also the changes in parasite prevalence from the first to the third sampling event in the years 1991–95 and between all three sampling events in 1990 are significant (Z-test, $Z>1.95$, $p<0.05$).

seek out their host actively, but rely mainly on passive transport via the ventilation current of the amphipods to obtain contact (Mouritsen & Jensen, 1997). The increase in host abundance especially during late summer is not, however, the sole explanation for the time lag. As it appears from Figure 4C, there is a significant increase from June to August in the infection intensity of the larger *Corophium* specimens, indicating that also the rate of cercariae transmission increases during this period. This is likely due to the coincident elevation of ambient temperatures (Figure 4A) causing an accelerated shedding of cercariae from the snail hosts. The final and additional explanation of the time lag is related to the development time of the larval trematodes within the snails. Although a high prevalence of trematode infected snails was observed already in early June, not all infections were mature and therefore not ready for transmission. Although the above explanations are not mutually exclusive, we expect that the rise in temperature during late summer is the most important factor contributing to the observed time lag.

During autumn (August–October) the density of both the infected first and the secondary intermediate hosts, the abundance of amphipods in general, and the infection level of *Corophium* decrease (Figure 4). The decline in amphipod density, infected or uninfected, should readily be explained by bird predation. Decreasing temperatures during autumn probably reduce the transmission of cercariae from snails to amphipods significantly, contributing to a lower infection level in the latter. A decline in the density of infected *Hydrobia* (Figure 4A) further enforces this trend. The similar trend in the parasite level of the larger amphipod specimens (Figure 4C) may also be explained by a lowered transmission rate, by predation (the larger individuals are most attractive to feeding birds (Peer et al., 1986) and usually harbour most parasites (Bick, 1994)), and by growth of smaller and less infected individuals into the larger size class. In conclusion, most of the changes of the amphipod population can be attributed to the presence of very high numbers of feeding shorebirds in the study area (not only the adult birds, but also their offspring pass through the Wadden Sea during the autumn migration period) and lower ambient temperatures.

However, the observed autumn decline in the density of infected *Hydrobia*, though the abundance of snails in general seems to be steady (Figure 4B), is a most intriguing result. Following a period of high bird presence, the density of infected snails would at least be expected to remain unaltered. The reason may be that the Danish Wadden Sea is the first main stopover during the birds' autumn migration (Smit & Piersma, 1989). Hence, the birds arrive from basically terrestrial breeding habitats and can therefore not yet be expected to harbour many marine trematodes; the parasites persist for only a few weeks in the bird host before they are lost. Accordingly, no trematode eggs are probably expelled through the droppings of the birds during autumn and consequently no new infections can be expected to accumulate. More speculations include: (1) Infected snails show changed behaviour (Mouritsen & Jensen, 1994) perhaps making them more subjected to bird predation than uninfected specimens (Dobson, 1988), (2) since it has been shown that infected snails may be less resistant to unfavourable abiotic conditions (e.g. Tallmark & Norrgren, 1976; Huxham et al., 1993; Sousa & Gleason, 1989; Jensen et al., 1996), infected specimens may experience a higher mortality rate than uninfected, and (3) infected snails may die off following intensive cercariae emission (Dobson, 1988; Jensen & Mouritsen, 1992). Additional possibilities could be imagined, but an exhaustive discussion is out of the scope of this paper.

Consequences for population dynamics

The results show that microphallid trematodes are rather successful in completing their complex life cycle. However, if the parasite level is sufficiently high, they are also able to kill their secondary intermediate host before trophic transmission is accomplished, and hence, the parasites may be expected to affect the population dynamics of the amphipods. By experimental infection of *Corophium volutator* with microphallids, we found that those individuals that died during the treatment harboured on average 22 metacercariae each, whereas the surviving (and infected) specimens contained only 12 cysts (Jensen, 1996). Considering 22 metacercariae as a kind of lethal level, the maximum of about six cysts per amphipod specimen found in the largest animals of the samples from August 1995 (Figure 4C) does not suggest any obvious parasite-induced mortality. Although some regulation in the sense of May and Anderson's theoretical framework (Anderson, 1991 and references therein) on the *Corophium* population may occur (individuals harbouring more metacercariae might have died off before sampling), no significant reduction in the abundance of *Corophium* can be ascribed to the direct effect of parasites in 1995.

Similar conclusions can be drawn for the years 1991, 1992 and 1994 which resemble quite well the situation in 1995 (Figure 5). The general picture is a gradual and significant increase in the prevalence of microphallid trematodes within the snail population during spring and summer, reaching about 20% at the most. During the same period significant increases also in the density of *Corophium* due to recruitment have been observed. However, 1990 was different. That year the parasite prevalence in the snail population increased rapidly already in spring reaching close to 50% in May, and the parasites had almost disappeared in June. The high microphallid prevalence in May corresponds to a density of infected *Hydrobia* of about 12 000 m^{-2}, a number that should be compared with 1000 m^{-2} in August 1995 (Figure 4A). During the same period the whole amphipod population vanished, and from further three sampling occasions in autumn and winter 1990 (data not shown) we know that the population of *Corophium* did not recover until spring 1991. Beside these changes in population dynamics of both host and parasites, 1990 was also unusual regarding spring temperatures. Compared with a ten year temperature-normal, the ambient temperatures in spring 1990 were several degrees higher and reached a maximum of 26 °C already in the beginning of May (see Jensen & Mouritsen, 1992 for details). We therefore suggested that the population crash in *Corophium* was a result of the high prevalence of microphallid trematodes in the *Hydrobia* population in concert with unusually high ambient temperatures causing a mass transmission of cercariae from snails to amphipods, eventually driving the population of the latter extinct. This interpretation is strongly supported by a laboratory experiment showing that microphallid trematodes are able to induce a mortality of above 50% in comparison with a control during nine days of experimentation under 1990 *in situ* conditions (Mouritsen & Jensen, 1997). As emphasized in the introduction, many reasons can be given why a species should show wide fluctuations in abundance other than parasites. However, as argued in detail elsewhere (Jensen & Mouritsen, 1992), most conceivable factors causing significant mortality in an intertidal infaunal amphipod species, including e.g. predation, strong off- or onshore winds, anoxic sediment conditions, were absent or insignificant in May and June 1990.

Besides our observation in 1990, previous studies have also reported on sudden decline or even local extinction of *Corophium* populations in which parasites have been, or could be, suspected to be the causative mortality agent (Segerstråle, 1960; Muus, 1967; Möller & Rosenberg, 1982; Reise, 1985; Ólafsson & Persson, 1986). But parasite-induced mass mortality in *Corophium*, however, does not seem to be an annual event, as it also appears from Figure 5. The underpinning reason for the mortality in 1990 should be found in the concurrence of unusually high temperatures and parasite prevalence in the snails. The high temperatures may have been responsible not only for the transmission of cercariae, but also for an accelerated egg-to-snail transmission and cercariae development within the snails, giving rise to the unexpected high spring prevalence of parasites in the snail population. It is also possible that the shorebirds feeding at our study site in spring 1990 were unusually highly infected, and hence, supplied the tidal flat with high numbers of infective trematode eggs. The reason for the latter possibility should probably be sought in the birds' wintering areas.

Conclusions

The metapopulations of microphallid trematodes within their intermediate hosts show profound seasonal patterns. They decrease during spring and autumn and increase during the summer period. The occurrence of the parasites in the secondary intermediate hosts is time delayed in relation to their occurrence in the first intermediate host population. An intimate interplay of factors is responsible for this scenario, including predation by the shorebird final hosts, the ambient temperatures and the development time of infective larval trematodes within the *Hydrobia* snails. Hence, the microphallid life cycle does not seem to be completed within a single season (spring or autumn) in which all three hosts occur together, but runs mainly in three steps separated in time: (1) transmission of trematode eggs to snails during spring, (2) transmission of cercariae to the amphipods during summer, and (3) trophic transmission of metacercaria-infected *Corophium* to the birds during autumn.

Microphallids may occasionally give rise to high mortality in the secondary intermediate host population. The prerequisites for this to happen in the Danish Wadden Sea apparently imply a high parasite prevalence in the mud snail population in addition to high ambient temperatures (at least above 20 °C).

Acknowledgments

We are grateful to Lone Thybo Mouritsen for reading an earlier version of the manuscript. The work has been supported financially by the Carlsberg Foundation. Meteorological data were kindly provided by the Danish Meteorological Institute.

References

Ambrose, W. G. J., 1991. Are infaunal predators important in structuring marine soft-bottom communities. Am. Zool. 31: 849–860.

Anderson, R. M., 1991. Populations and infectious diseases: ecology or epidemiology? J. anim. Ecol. 60: 1–50.

Barnes, R. S. K. & R. N. Hughes, 1988. An introduction to marine ecology. Black. Sci. Publ., Oxford, 351 pp.

Bick, A., 1994. *Corophium volutator* (Corophiidae: Amphipoda) as an intermediate host of larval digenea – an ecological analysis in a coastal region of the southern Baltic. Ophelia 40: 27–36.

Deblock, S., 1980. Inventaire des trematodes larvaires parasites des mollusques *Hydrobia* (Prosobranches) des côtes de France. Parasitologia 22: 1–105.

Dobson, A. P., 1988. The population biology of parasite-induced changes in host behaviour. Quat. Rev. Biol. 63: 139–165.

Dobson, A. P. & P. J. Hudson, 1986. Parasites, disease and the structure of ecological communities. TREE 1: 11–15.

Ginetsinskaya, T. A., 1988. Trematodes, their Life Cycles, Biology and Evolution. Dr Indira Kohli (translator), Amerind Publishing Co. Pvt. Ltd., New Delhi, 559 pp.

Holt, R. D., 1993. Infectious diseases of wildlife, in theory and in practice. TREE 8: 423–425.

Huxham, M., D. Raffaelli & A. Pike, 1993. The influence of *Cryptocotyle lingua* (Digenea: Platyhelminthes) infections on the survival and fecundity of *Littorina littorea* (Gastropoda: Prosobranchia); an ecological approach. J. exp. mar. Biol. Ecol. 168: 223–238.

Jensen, T., 1996. The influence of microphallid trematodes on their host organisms *Corophium volutator* and *Corophium arenarium*. Ms. thesis, University of Aarhus, Denmark.

Jensen, K. T., G. Latama & K. N. Mouritsen, 1996. The effect of larval trematodes on the survival rates of two species of mud snails (Hydrobiidae) experimentally exposed to desiccation, freezing and anoxia. Helgol. Meeresunters. 50: 327–335.

Jensen, K. T. & K. N. Mouritsen, 1992. Mass mortality in two common soft-bottom invertebrates, *Hydrobia ulvae* and *Corophium volutator* – the possible role of trematodes. Helgol. Meeresunters. 46: 329–339.

Kinne, O., 1980. Diseases of marine animals: General aspects. In Kinne, O. (ed.), Diseases of Marine Animals. Vol. 1. Wiley, New York: 13–73.

Lauckner, G., 1987. Effects of parasites on juvenile Wadden Sea invertebrates. In Tougaard, S. & S. Asbirk (eds), Proceedings of the 5th International Wadden Sea Symposium. The National Forest and Nature Agency and the Museum of Fisheries and Shipping, Esbjerg, Denmark: 103–121.

Laursen, K. & J. Frikke, 1984. The Danish Wadden Sea. Cambridge University Press, Cambridge: 214–223.

Minchella, D. J. & M. E. Scott, 1991. Parasitism: A cryptic determinant of animal community structure. TREE 6: 250–253.

Mouritsen, K. N., 1994. Day and night feeding in Dunlins *Calidris alpina*: choice of habitat, foraging technique and prey. J. Avian Biol. 25: 55–62.

Mouritsen, K. N. & K. T. Jensen, 1994. The enigma of gigantism: effect of larval trematodes on growth, fecundity, egestion and locomotion in *Hydrobia ulvae* (Pennant) (Gastropoda: Prosobranchia). J. exp. mar. Biol. Ecol. 181: 53–66.

Mouritsen, X. N. & K. T. Jensen, 1997. Parasite transmission between soft-bottom invertebrates: temperature mediated infection rates and mortality in *Corophium volutator*. Mar. Ecol. Prog. Ser. 151: 123–134.

Muus, B., 1967. The fauna of Danish estuaries and lagoons. Distribution and ecology of dominating species in the shallow reaches of the mesohaline zone. Meddr. Danm. Fisk. og Havunders., 316 pp.

Möller, P. & R. Rosenberg, 1982. Production and abundance of the amphipod *Corophium volutator* on the west coast of Sweden. Neth. J. Sea Res. 16: 127–140.

Nybakken, J. W., 1993. Marine Biology. An Ecological Approach. Harper Collins College Publishers, New York, 462 pp.

Olafsson, E. B. & L. E. Persson, 1986. The interaction between *Nereis diversicolor* O. F. Mueller and *Corophium volutator* Pallas as a structuring force in a shallow brackish sediment. J. exp. mar. Biol. Ecol. 103: 103–117.

Peer, D. L., L. E. Linkletter & P. W. Hicklin, 1986. Life history and reproductive biology of *Corophium volutator* (Crustacea: Amphipoda) and the influence of shorebird predation on population structure in Chignecto Bay, Bay of Fundy, Canada. Neth. J. Sea Res. 20: 359–373.

Posey, M. H., 1987. Influence of relative variabilities on the composition of benthic communities. Mar. Ecol. Prog. Ser. 39: 99–104.

Reise, K., 1985. Tidal Flat Ecology. An Experimental Approach to Species Interactions. Springer-Verlag, Berlin, 191 pp.

Segerstråle, S. G., 1960. Fluctuations in the abundance of benthic animals in the Baltic area. Soc. Scient. Fenn., Comment. Biol. 23: 1–19.

Smit, C. J. & T. Piersma, 1989. Numbers, midwinter distribution, and migration of wader population using the east atlantic flyway. IWRB Special Publication No. 9, Canadian Wildlife Service, Ottawa: 24–63.

Snedecor, G. & W. G. Cochran, 1989. Statistical Methods. Iowa State University Press, Ames, Iowa, 503 pp.

Sousa, W. P., 1991. Can models of soft-sediment community structure be complete without parasites? Am. Zool. 31: 821–830.

Sousa, W. P. & M. Gleason, 1989. Does parasitic infection compromise host survival under extreme environmental conditions? The case for *Cerithidea californica* (Gastropoda: Prosobranchia). Oecologia 80: 456–464.

Tallmark, B. & G. Norrgren, 1976. The influence of parasitic trematodes on the ecology of *Nassarius reticulatus* (L.) in Gullmar Fjord (Sweden). Zoon 4: 149–154.

Wilson, W. H., 1991. Competition and predation in marine soft-sediment communities. Ann. Rev. Ecol. Syst. 21: 221–241.

Worrall, D. H., 1984. Diet of the dunlin *Calidris alpina* in the Severn Estuary. Bird Study 31: 203–212.

Hydrobiologia **355**: 71–76, 1997.
A. D. Naumov, H. Hummel, A. A. Sukhotin & J. S. Ryland (eds), Interactions and Adaptation Strategies of Marine Organisms.

The association between the caprellid *Pariambus typicus* Krøyer (Crustacea, Amphipoda) and ophiuroids *

Ute Volbehr & Eike Rachor
Alfred-Wegener-Institute for Polar and Marine Research, D-27515 Bremerhaven, Germany

Key words: Pariambus typicus, Ophiura, commensalism, phoresis, epibenthos, German Bight

Abstract

Caprellid amphipods are small marine crustaceans which usually live as epibionts on a variety of substrates. Apart from mostly sessile organisms such as algae, hydroids and bryozoans, they also frequently use vagile fauna as substrates. *Pariambus typicus* (Krøyer, 1844) is a common associate of subtidal asteroids and echinoids in European seas, but has also been found free-living on the sea floor. In the German Bight, *P. typicus* has also been discovered regularly on ophiuroids (*Ophiura albida* Forbes and *Ophiura ophiura* (L.)), which had not been described before. Several aspects of the biology of both partners were investigated with major focus on their distribution and relation to different substrates, behavioural and morphological adaptations to their habitat and their modes of nutrition. Various behavioural and morphological adaptations enable *Pariambus typicus* to live on a variety of substrates, on which the amphipod settles after contact. The highly mobile ophiuroid hosts open up new habitats for the caprellid (phoresis). Extensive grooming behaviour and specialized mouthpart morphology enable *P. typicus* to use detritus as an important food source, which contributes also a great part to the ophiuroids' nutrition. The complex association between *P. typicus* and *Ophiura* is interpreted as a commensalism.

Introduction

Marine organisms frequently use other organisms as substrates. The amphipod suborder Caprellidea represents a group of about 250 extremely specialized species of which the major part live as epibionts (Mayer, 1882; Wetzel, 1932; Laubitz, 1993). In most cases their living substrates are sessile organisms like algae, sponges, colonial hydroids and bryozoans. According to the occurrence of these substrates, caprellids are often designated as part of the phytal (Mayer, 1882; McCain, 1968).

More seldom, caprellids also use vagile fauna as a substrate. They can occasionally be found on sea-turtles, majiid decapods and also scyphozoans like *Rhizostoma*, but in most cases they occur on echinoderms, usually on sea-stars and sometimes on echinoids (e.g. Mayer, 1882; Vader, 1972, 1978; Jones, 1973; Barel & Kramers, 1977; Caine, 1986). Associations between caprellids and ophiuroids, the dominating echinoderm group on soft bottoms, surprisingly have not been described before.

In European seas, the caprellid *Pariambus typicus* (Krøyer, 1844) is a common associate of subtidal echinoderms, like the sea-stars *Asterias rubens* L. or *Crossaster papposus* (L.) and also the echinoid, *Echinus esculentus* L., but has also been found free-living (Jones, 1970, 1973; Comely & Ansell, 1988). The latter generally has been described as the subspecies *P. t. armata*, while epibionts obviously belong to the subspecies *P. t. inermis* (Chevreux & Fage, 1925; Jones, 1970, 1973; Comely & Ansell, 1988).

In the sandy and muddy subtidal areas of the German Bight, southeastern North Sea, the brittle stars *Ophiura albida* Forbes and *Ophiura ophiura* (L.) represent major elements of the epibenthos and contribute a significant proportion to its total abundance and biomass (Salzwedel et al., 1985; Dahm, 1993). In this area *Pariambus typicus inermis* has been found to be a regular associate of both *Ophiura* species (Figure 1).

* This is publication no. 1287 of the AWI Bremershaven.

Figure 1. Pariambus typicus (female), photographed in a typical position on an arm of an *Ophiura*, (length of *Pariambus* = 5 mm).

This paper is intended to describe and define this association.

Material and methods

Field samples were taken between June and November 1994 at 7 stations in the inner German Bight (Figure 2) to investigate dependences of the association from environmental factors and to obtain animals for laboratory observations. Therefore, stations with different substrates and depths were chosen. Sampling was carried out by means of a 0,1 m^2 Van Veen grab with lids on its upper side through which living ophiuroids were carefully removed from the substrate surface by a forceps. Sediments taken from the grabs were analyzed for grain size and organic content.

For subsequent analyses, a proportion of the animals were fixed with 5% buffered formaldehyde and identified, measured and dissected in the laboratory. Living animals were immediately put into aquaria together with sediment from the same location and maintained at 10 °C in a dark cold-storage chamber for up to 8 months, to undertake behavioural observations (using a stereomicroscope) and morphological studies. Chemical attraction of *P. typicus* to various organisms was tested in a sediment-free aquarium divided into two parts by nylon mesh, separating the attractants from the caprellids. Substrate preference was tested by providing *P. typicus* with various substrata, such as coarse sediment particles, shell fragments, worm tubes, algae and ophiuorid fragments in petri dishes or small aquaria. Nutritional investigations comprised food preference tests with different food items, stomach content analyses of preserved animals and studies of functional mouthpart morphology of *P. typicus* including scanning electron microscopy.

Results

Abundance, substrate preference and frequency of association

The abundances of both *Ophiura* species are correlated with the sediment type. The dominating species, *O. albida*, prefers coarser, sandy sediments and greater depths than *O. ophiura*, which occurs in relatively high numbers in restricted, more shallow, near-shore waters characterized by muddy sands.

The abundance of *P. typicus* is positively correlated with the total abundance of both *Ophiura* species (Figure 3), while sediment type or depth are less meaningful. Highest densities of caprellids were found at inshore stations with mixed sediment structure and high organic content, where both ophiuroids are frequent. Here (station ST) the densities per m^2 reached more than 200 *Ophiura* spp. and 3500 *P. typicus*. At all stations examined, *P. typicus* and *Ophiura* were found together. The mean incidence of *O. albida* bearing *P. typicus* was 94,5%, but was much lower in *O. ophiura* with 65,5%.

Almost all sizes of ophiuroids were infested, the smallest having a disc diameter of 1 mm (total size about 9 mm) and bearing two *P. typicus*. The number of caprellids per ophiuroid clearly depends on the host's size (Figure 4, for *O. albida*). As a maximum, 40 *Pariambus* were found on a single host ophiuroid (disk diameter 8 mm). The highest average was 17 *P. typicus* per *Ophiura* at the mentioned near-shore station (ST).

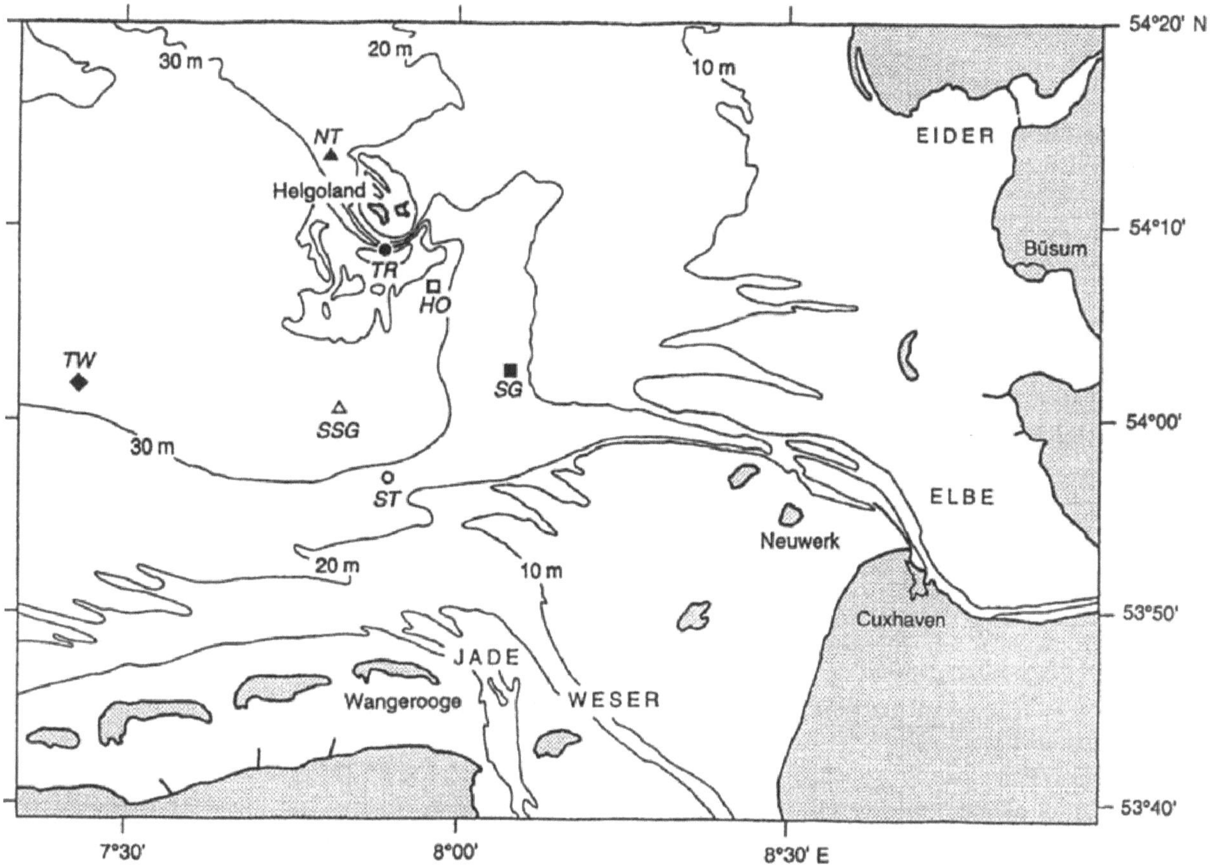

Figure 2. Sampling area of the inner German Bight (North Sea); TR is the deepest, ST and SG are the shallowest stations; SG has the finest sediments, NT the most sandy ones.

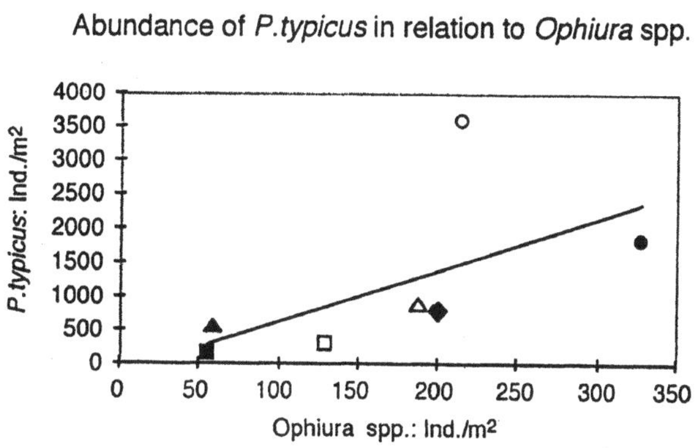

Figure 3. Abundances of *Pariambus typicus* in relation to the densities of *Ophiura* species at the different locations shown in Figure 2 (symbols of stations are the same).

Behaviour and morphology

Experiments on chemical attraction of *P. typicus* with living or dead ophiuroids, sea stars and echinoids did not bring about any clear reaction. The caprellids mostly remained passive. However, *P. typicus* showed strong reactions to tactile stimuli: every accidental contact of appendages or the body with a seizable substrate was answered by a strong clasping reflex of the gnathopods and/or pereipods. In aquaria and Petri dishes, preferences of the substrata offered did not exist. A wide range of items such as ophiuroid arms, starfishes, polychaete tubes, filamentous algae, shell fragments, stones or even glass pipettes were settled for long terms.

Aquaria observations revealed that *P. typicus* spent most of the time for grooming its appendages and body, which is usually followed by feeding on the groomed material and then rejecting the remainders. Grooming becomes more extensive when the water is more turbid. Comb-shaped structures on the mouthparts and gnathopods, as seen by SEM, may help in using settled detritus as a food source.

As shown by underwater photographs from the German Bight, individuals of both *Ophiura* species are often aggregated in a special pattern on the sea floor, where almost every single animal is in contact with its neighbouring congeners. A similar behaviour was observed in aquaria, where this more cryptic attitude (with *Ophiura* being burrowed in the sediment and holding only the tips of its arms into the water) alternates with phases of high activity, in which the brittle stars quickly move around. Thus, caprellids are accidentally picked up by their motile hosts, and, accordingly, their dispersal is in most cases passive. In this context, another observation is to be mentioned. Several times *Ophiura* individuals of both species were seen wiping with one arm over those arms which were occupied by *Pariambus*, likely to get rid of their guests.

Feeding biology of both partners

As shown by feeding experiments and stomach content analyses, *Ophiura albida* and *O. ophiura* consume a wide range of food items, such as phytoplankton, detritus, zoobenthos, carrion, and even *Pariambus* and their own congeners. More than 50 per cent of their stomach content consists of sediment, 8–10% is unidentified organic material, while all other food items are less frequent.

P. typicus also uses a variety of food sources. Microphagous feeding of suspended or sedimented material, groomed particles, and scraping off encrusting diatoms from the aquaria wall alternate with predatory and carrion feeding. Larger suspended particles are grasped by the second gnathopods, supported by the antennae, which haul particles into reach of the gnathopods. Sedimented particles including sand grains are also taken up with the gnathopods and then devoured or nibbled with the mouth parts. If ophiuroids are injured or dead, the caprellids also feed on their tissues. If a suitable food is abundant in aquaria, they even leave their ophiuroid host to feed on it.

Discussion

P. typicus has been found everywhere in soft-bottom habitats of the German Bight (e.g. Salzwedel et al., 1985). It is not accidentally living on both *Ophiura* species, but regularly, for longer terms and in high numbers. We presume that this is also the case in many other areas where both partners are found, and that their association has been unknown because of the small size of *P. typicus* (a few mm only) and the inadequate zoobenthos sampling and sieving methods used by former investigators.

The differences in the numbers of *P. typicus* per individual of either species can be explained by the higher mobility and activity of *O. ophiura* and its ability to wipe off the caprellids from its arms by pulling one arm over another.

A similar aggregation behaviour as in the *Ophiura* species investigated is known from other ophiuroids. This attitude is interpreted not as true social behaviour, but as a result of individual interaction with environmental factors such as suitable substrate or food (Reese, 1966; Warner, 1971). However, it may also be a means of information transfer between individuals. For *P. typicus*, it offers the possibility of further dispersal by moving over from one ophiuroid to another.

From the aquaria observations of ophiuroid behaviour and the remarkably variable abundances found at such stations repeatedly sampled within a short period, it can be supposed that *Ophiura* regularly migrate within their habitat. The dense aggregations and, on the other hand, high mobility of the brittle stars allow further dispersal without own effort of the tiny amphipod. With these self-motile substrates the caprellids open up the habitats of the wider benthal for themselves.

P. typicus is adapted to its life on echinoderms in many respects. The slender, cylindrical and pointed subchelate pereiopods of its var. *inermis* lack spines or setae, which is a secondary reduction frequent in starfish epibionts and which protects them from getting entangled between paxillae and tube feet (Caine, 1978). These prehensile pereiopods (Vader, 1983) can be interpreted as an adaptation to agitated water assuring a powerful hold on the highly mobile ophiuroids. They also prevent dislodgement during grooming and burrowing activities of the host. It is almost impossible to remove *P. typicus* from its host without injury of their appendages; the caprellids stay on the arm-tips of burrowed and even of dead ophiurids. Such a strong thigmotaxis is frequent in mobile epibionts, preventing them from loosing their substrate (Patton, 1965). However, *P. typicus* has to be regarded as a habitat opportunist that can successfully inhabit many other substrata and also occur free-living.

In accordance with former investigations we found that both *Ophiura* species are omnivorous, preferring the most abundant food source (Eichelbaum, 1910; Feder, 1981). In turbid waters, detritus, suspended or sedimented, becomes the main food source for the ophiuroids and also the caprellids. The diets and the

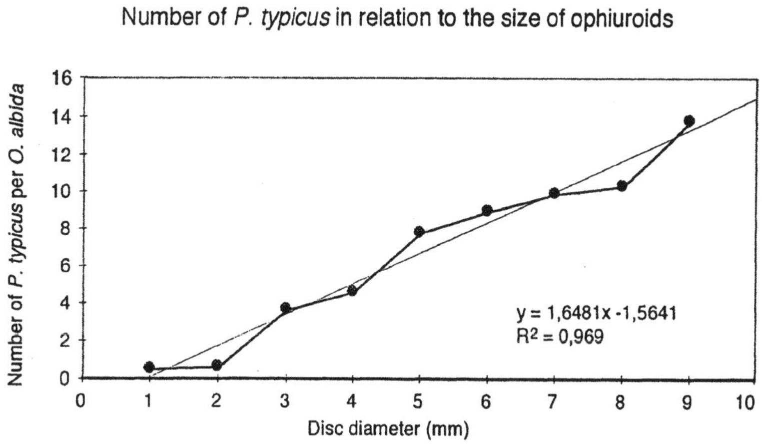

Figure 4. Correlation of mean numbers of *Pariambus typicus* (per host) and sizes (disc diameters) of *Ophiura albida.*

ways of feeding of all three species examined overlap largely. All are opportunistic omnivores consuming the most abundant food source. *P. typicus* benefits from the additional food sources, which the highly mobile ophiuroids with their similiar nutritional demands and feeding biology open up for it. This can be interpreted as a commensalism *sensu stricto* (Matthes, 1967), literally: feeding 'at one table', i.e. on the same food. However, feeding on the host's tissue, as observed in *P. typicus*, indicates that a transition from commensalism to food-parasitism is well possible (Schwerdtfeger, 1977; Wallin, 1978).

From the view of the ophiuroids, however, *P. typicus* apparently does not offer any obvious advantages to them; instead, it is noticed as a disturbance, which is shown by their attempts to wipe the caprellids off their body. Besides, the highest numbers of *P. typicus* were found on injured, weak (e.g. after autotomy of arms) or regenerating animals and may even contribute to higher mortalities of such hosts. Accordingly, generally healthy individuals of both brittle stars and the more active and mobile species *O. ophiura* are much less infested.

The described association is an epizoism, where the smaller and more sessile partner lives outside its mobile host (Odum, 1983). The high motility of the ophiuroids and their migration is used by the caprellids as a means of transport and, thus, dispersion (phoresis, s. Schwerdtfeger, 1977). Referring to nutrition, the association between *Pariambus typicus* and the *Ophiura* species is designated as a commensalism.

As in most cases of associated animals, the relationship between *Pariambus* and *Ophiura* comprises more than one type of association (Gotto, 1969). It is characterized by its high complexity which results from the various adaptations of both partners to their environment. Though they benefit from their specialized types of behaviour and morphological adaptations, *P. typicus* and the two *Ophiura* species are at the same time opportunists. This enables them to exist in high abundances almost everywhere in the subtidal soft bottom habitats of the German Bight.

References

Barel, C. D. N. & P. G. N. Kramers, 1977. A survey of the echinoderm associates of the north-east Atlantic area. Zool. Verh., Leiden 156: 1–159.

Caine, E. A., 1978. Habitat adaptions of North American caprellid Amphipoda (Crustacea). Biol. Bull. 155: 288–296.

Caine, E. A., 1986. Carapace epibionts of nesting loggerhead sea turtles. Atlantic coast of the U.S.A.. J. exp. mar. Biol. Ecol. 95: 15–26.

Chevreux, E. & L. Fage, 1925. Amphipoda. Faune de France 9. Lechevalier, Paris: 1–488.

Comely, C. A. & A. D. Ansell, 1988. Invertebrate associates of the sea urchin, *Echinus esculentus* L., from the Scottish west coast. Ophelia 28: 111–137.

Dahm, C., 1993. Growth, production and ecological significance of *Ophiura albida* and *O. ophiura* (Echinodermata: Ophiuroidea) in the German Bight. Mar. Biol. 116: 431–437.

Eichelbaum, E., 1910. Über Nahrung und Ernährungsorgane von Echinodermen. Wiss. Meeresunters. Abt. Kiel, 11: 189–275.

Feder, H. M., 1981. Aspects of the feeding biology of the brittle star *Ophiura texturata.* Ophelia 20: 215–235.

Gotto, R. V., 1969. Marine animals. Partnerships and Other Associations. The English Universities Press, London, 96 pp.

Jones, M. B., 1970. The distribution of *Pariambus typicus* var. *inermis* Mayer (Amphipoda, Caprellidae) on the Common Starfish *Asterias rubens* L. Crustaceana 19: 89–93.

Jones, M. B., 1973. Geographical and ecological distribution of *Pariambus typicus* (Krøyer) (Amphipoda, Caprellidae). Crustaceana 25: 204–210.

Laubitz, D. R., 1993. Caprellidea (Crustacea: Amphipoda) towards a new synthesis. J. nat. Hist. 27: 965–976.

Matthes, D., 1967. Die Terminologie interspezifischer (heterotypischer) Beziehungen. Zool. Anz. 179: 313–319.

Mayer, P., 1882. Die Caprelliden des Golfes von Neapel und der angrenzenden Meeres-Abschnitte. Fauna u. Flora des Golfes von Neapel 6: 1–201.

McCain, J. C., 1968. The Caprellidae (Crustacea: Amphipoda) of the Western North Atlantic. Bull. U.S. Natn. Mus. 278: 1–147.

Odum, E. P., 1983. Grundlagen der Ökologie. 2 vols, 2nd edn. G. Thieme, Stuttgart, 836 pp.

Patton, W. K., 1965. Commensal Crustacea. Marine Biol. Ass. of India, Proc. Symposium Ernakuleam 1965, Part III, Symposium 2: 1228–1243.

Reese, E. S., 1966. The complex behaviour of Echinoderms. In Boolootian, R. A. (ed.), Physiology of Echinodermata. Interscience, New York: 157–218.

Salzwedel, H., E. Rachor & D. Gerdes, 1985. Benthic macrofauna communities in the German Bight. Veröff. Inst. Meeresforsch. Bremerh. 20: 199–26.

Schwerdtfeger, F., 1977. Autökologie. Parey, Hamburg, Berlin, 458 pp.

Vader, W., 1972. Associations between gammarid and caprellid amphipods and medusae. Sarsia 50: 51–56.

Vader, W., 1978. Associations between amphipods and echinoderms. Astarte 11: 123–134.

Vader, W., 1983. Prehensile pereopods in Gammaridean Amphipoda. Sarsia 68: 139–148.

Wallin, D., 1978. Togetherness underwater. Sea-Front 24: 47–54.

Warner, G. F., 1971. On the ecology of a dense bed of the brittle-star *Ophiothrix fragilis*. J. mar. biol. Ass. U.K. 51: 267–282.

Wetzel, A., 1932. Studien Über die Biologie der Caprelliden. I. Bewegung, Nahrungserwerb, Aufenthaltsort. Z. wiss. Zool. 141: 347–398.

Hydrobiologia **356**: 77–89, 1997.
A. D. Naumov, H. Hummel, A. A. Sukhotin & J. S. Ryland (eds), Interactions and Adaptation Strategies of Marine Organisms.

Chemically-mediated interactions in benthic organisms: the chemical ecology of *Crambe crambe* (Porifera, Poecilosclerida)

Mikel A. Becerro[1], María J. Uriz[1] & Xavier Turon[2*]
[1]*Centre for Advanced Studies (CSIC). Camí de Sta. Bàrbara s/n. 17300 Blanes (Girona), Spain*
[2]*Dept. of Animal Biology (Invertebrates). Faculty of Biology. University of Barcelona. 645, Diagonal Ave., 08028 Barcelona, Spain (*author for correspondence)*

Key words: Chemical bioactivity, defence, toxicity quantification, toxicity variation, sponges

Abstract

We studied the chemically-mediated interactions of the encrusting sponge *Crambe crambe*, one of the most toxic and widespread species in rocky sublittoral habitats in the Northwestern Mediterranean. Guanidine alkaloids accounted for *C. crambe*'s toxicity, which seems to have multiple functions in nature, as evidence has been found for antifouling, antipredation, and space competition roles.

We investigated the factors underlying the chemical defence strategy of this species by assessing variation in the production of toxic substances as a function of different biological and environmental variables. The working hypothesis was that the production of these metabolites should be optimized according to the biological features (morphogenesis, reproduction, growth, life history) and ecological conditions (biotic pressures and abiotic factors) of the particular specimens.

One cell type, the spherulous cell, which was concentrated near the sponge's surface, accumulated the toxic substances. Within-specimen analyses showed that toxicity was higher in the ectosome than in the choanosome of the sponges. There was a seasonal pattern of change in the toxicity of the species. Life-history stage also proved significant in the production of toxic substances: larvae were non-toxic, and feeding-deterrence experiments showed that larvae and newly metamorphosed individuals were not protected from predation, while two-week-old recruits already showed strong feeding deterrence. Overall, toxicity increased from small to medium-sized adult sponges, and decreased again in larger individuals. Variation in toxicity was also found at an ecological level: the values at a highly competitive site dominated by slow-growing animal species were higher than those at an adjacent, well-lit site with algal dominance. The relative investment in structural material (collagen, fibres, spicules...) was also higher in the shaded habitat, thus a positive relationship was found between investment in chemical and physical defences. In the two habitats compared, allocation to defence correlated negatively with reproduction and growth, and positively with survival.

The results showed that *C. crambe* can adjust, at organismal and population levels, the production of bioactive substances to different environmental and physiological situations. Space competition emerged as a key factor explaining the variation found in the production of bioactive substances.

Introduction

Marine organisms have been a rich source of novel chemical compounds (see Faulkner, 1996, and previous reviews by the same author), which have boosted the development of marine natural products chemistry for about 3 decades. Pharmacologists soon joined this field because of the applied interest of many compounds. Only over the last fifteen years, however, has the incorporation of marine biologists and ecologists resulted in the development of the interdisciplinary field of marine chemical ecology (Faulkner, 1993; Pawlik, 1993; Hay, 1996), which focuses on the ecological functions of marine organisms' secondary metabolites and on the selective forces modeling their evolution (for reviews on different aspects, see Paul, 1992; Coll, 1992; Hay & Steinberg, 1992; Pawlik, 1992, 1993; Clare, 1996).

In seaweeds, models explaining evolution and selection for different types of chemical defence originate largely from models developed for terrestrial plants. Plant apparency models, resource availability models and spatial variation-in-herbivory models have been analysed as potential descriptors of algal chemical defence evolution (e.g. Hay & Steinberg, 1992; Yates & Peckol, 1992). The concepts of quantitative vs. qualitative, or those of constitutive, inducible, and activated defences have been applied to marine plants (Van Alstyne, 1988; Paul & Van Alstyne, 1992; Hay & Steinberg, 1992; Steinberg, 1994). Underlying all these models and concepts is the notion of a cost associated with the production of bioactive metabolites and a need for optimization of this cost (the optimal defence theory, Fagerström et al., 1987). Clear demonstration and quantification of costs, however, remains as elusive for marine organisms as it has been for terrestrial ones (Hay & Steinberg, 1992; Adler & Karban, 1994).

Possibly because of the roots of marine chemical ecology in terrestrial plant studies, most developments in this field encompass the idea of adaptation to herbivory (or, more broadly, to predator-prey) interactions as the main factor explaining the evolution of chemical defences. Thus, predation-oriented studies are common in the literature on seaweeds (Steinberg & Van Altena, 1992; Hay et al., 1994; Pennings et al., 1996, to cite some examples) and also on benthic animals (*e.g.* Paul et al., 1990; Pennings et al., 1994; Van Alstyne et al., 1994; Pawlik et al., 1995). Other factors have received less attention. Competition for space or antifouling mechanisms may also be chemically-mediated (Coll, 1992) and, therefore, may be important factors affecting both production and evolution of chemical defence, particularly in sessile forms. It is generally acknowledged that many herbivore-susceptible marine species contain potentially defensive compounds, and that there are a large number of metabolites which appear not to have a role in antipredation (Hay, 1984; Steinberg & Van Altena, 1992; Pawlik, 1993). Notwithstanding, studies considering functions other than antipredation as the role of chemical defences or as the underlying selective theme are scarce and usually focused on benthic invertebrates (e.g. Sullivan et al., 1983; Coll et al., 1987; Porter & Targett, 1988; Davis et al., 1991; Sammarco & Coll, 1988; Maida et al., 1993, 1995; Wahl et al., 1994; Teo & Ryland, 1995). As there are diverse selective pressures, chemical defences may have evolved as a multiple-purpose response. Such a possibility is relevant for evolutionary scenarios as it hinders co-evolution of particular pairs of interacting species (Schmitt et al., 1995).

We are not able at present to make generalizations about the applicability of the proposed models of chemical defence evolution. Likely, there is no generalization that can be made, different factors being important for different types of organisms. We need more data on the ecological roles of bioactive metabolites and, crucial to this issue, on variation of bioactivity at several levels (intra- and inter-specimen, spatial, temporal, etc.) and its relationship with other biological parameters of the organisms. Without estimates of this variation, we can do little to increase our understanding of processes behind the patterns observed (Cronin et al., 1995). The present contribution is intended to provide such a framework of variation of bioactivity and its correlates in a model organism.

The aim of this work is to study the ecological functions and variability of the chemical bioactivity of the encrusting sponge *Crambe crambe* (Schmidt). This variability is addressed from several points of view and at several scales, and combined with variability in other biological parameters. This allowed us to correlate differences in chemical bioactivity with other traits of the life-history. It is the first instance, to our knowledge, in which the bioactivity of a species is studied in a comprehensive manner from the cell level to the population level. The particular questions we wanted to answer were: what function or functions does toxicity fulfil in the biology of this species? are there any patterns of toxicity variation at the intra- and inter-individual levels? what information on the underlying selective pressures can be gleaned from the patterns observed? how does investment in chemical defence correlate with investment in other biological functions? As we will review studies performed during the last few years, we will draw partially on data published by the authors since 1994, but we will present them within a common framework and add new evidence to provide a general picture of the chemical ecology of this species.

Material and methods

The poecilosclerid sponge *Crambe crambe* (Schmidt) is a red encrusting form, which can attain surface areas of 0.5 m^2 in the study zone (Northwestern

Mediterranean). This species was selected because it is one of the most abundant sponges in Mediterranean littoral communities (Uriz et al., 1992b), where it is found in a wide range of habitats. At the same time, *C. crambe* featured strong bioactivities in previous, pharmacologically-oriented, screenings (Jares-Erijman et al., 1991; Berlinck et al., 1992; Uriz et al., 1992a). *C. crambe* also possesses an array of potentially active metabolites, grouped into two types of guanidine alkaloid: crambines and crambescidins (Berlinck et al., 1990, 1992; Jares-Erijman et al., 1991). Besides, this species is a thinly encrusting form and therefore highly surface-dependent, which implies strong space competition with neighbours. All these features make *C. crambe* a suitable species for the purposes of assessing ecological roles and the variation of its chemical bioactivity.

All specimens studied were collected by SCUBA-diving at Blanes (NE Spain, Western Mediterranean). For chemical analyses, sponge pieces were blotted on paper towels and freeze-dried. Unless stated otherwise, secondary metabolites were obtained by three successive extractions (for 5, 15, and 30 min, respectively) with dichloromethane (DCM). Preliminary studies showed that this method extracts all compounds of interest (Becerro, 1994).

In some of the experiments we quantified the bioactive properties of the sponges. We chose to use a toxicity test instead of chemically analysing the active compounds, and this point requires some justification. First, we were unable to quantify (by chromatographic methods or by magnetic nuclear resonance) all the potentially active compounds (up to four crambines, plus homologues, and four crambescidins have been described). Second, the relative activity of these compounds is poorly known, and the potential synergistic effects between them are unknown, so trying to quantify one or several of them may not provide the information sought. We looked for a quick and precise tool for measuring bioactivity (i.e. the end-product of all these compounds and their interactions), and so we resorted to a standard toxicity test, the Microtox bioassay (Ribo & Kayser, 1987). This method is based on measurements of bioluminiscence of the deep sea bacterium *Photobacterium phosphoreum*. Although of no ecological meaning in themselves, the results of this test correlate well with those of other, more ecologically relevant analyses, and Microtox performed best in terms of repeatability and accuracy (Becerro, 1994; Becerro et al., 1995b). Pastorok & Becker (1990) also found this method to be the most sensitive in a comparison of marine species used in bioassays. Therefore, toxicity analyses were carried out using the Microtox device and crude DCM extracts, homogeneously resuspended in water (through sonication), and at an initial concentration of 250 ppm relative to initial sponge dry weight. We present the results in Toxicity Units (TU), which equal $100/EC_{50}$ (EC_{50} = concentration which resulted in 50% decrease in light production). For details see Becerro et al. (1995b).

The methods employed in data collection and analysis not reported elsewhere are explained in full in the following sections, while we will only outline the methods used in previously published experiments; readers may refer to the references given for more details on the techniques and the statistical tests used.

Study of ecological roles

Antifouling

For the study of an antifouling effect, we analysed the ability of *C. crambe* to prevent the formation of microbial film on its surface. We quantified the amount of epibiotic bacteria in five living sponges by swabbing surfaces with sterile cotton, seeding culture media with the swabs and counting the bacteria that developed in the cultures. We also tested the antimicrobial effect of crude acetone extracts on seven bacterial strains, isolated from the field in the vicinity of *C. crambe* specimens, by the paper disk diffusion method. For details see Becerro et al. (1994).

A direct test of the effect of *C. crambe* extracts on naturally developing microbial films was performed by adding crude acetone extracts of five sponge individuals (20 cm^2 pieces were cut from each) to five marine agar (2216, Difco, 5% w/v) plates 20 cm^2 in area (5 ml of acetone with the extract was added when the agar temperature fell below 50 °C). Five further plates served as controls, and only solvent was added to them. The plates were left for three days in individual 30 l flow-through cages submersed in large aquaria supplied with running seawater. After this period, the plates were rinsed with filtered seawater and gently swabbed with sterile cottons. These swabs were used to seed agar plates which were incubated in an oven at 21 °C for 21 days; the area occupied by bacteria colonies was then measured.

We also analysed the inhibition of settlement of larvae of *Bugula neritina* (Linné), a common bryozoan in the study zone. This was studied by placing from 12 to 34 larvae in each of three Petri dishes containing 50 ppm of three subfractions (aqueous, DCM, and butanolic) of a DCM:methanol (1:1) extract of *C. crambe*. Three more Petri dishes served as controls. The number of larvae that settled relative to the controls was measured after 1 h 30 min and 5 h 30 min. The number of ancestrula relative to the controls was measured after 9 h 30 min and 20 h.

Competition for space

For the study of space competition mechanisms, we used both an observational and an experimental approach (Turon et al., 1996b). We studied the small-scale interspecific associations of *C. crambe* with its neighbours in a sublittoral community, and tested the significance of the associations found and the strength of the associations at increasing distances from the contact borders. The experimental approach evaluated regeneration rates of one of the main space colonizers in the community studied, the encrusting sponge *Scopalina lophyropoda* Schmidt. We scraped to rock replicate circular holes (about 6 cm^2) in specimens of this sponge, and the holes were then rubbed with either *C. crambe* fragments, *S. lophyropoda* fragments (rubbing controls), or not rubbed at all (absolute controls). We then surveyed the regeneration of the sponge (i.e. sealing off of the holes) in the different treatments.

We also performed a test of photosynthesis inhibition using as an assay organism the alga *Ulva rigida* Agardh and the same three subfractions of crude DCM/methanol (1:1) extracts of *C. crambe* as in the *B. neritina* experiment. Oxygen production by algal pieces in the presence of extracts (at 50 ppm) was measured in replicate ($n = 3$) samples after 90 min in an incubator and compared to controls.

We assessed whether the active substances of *C. crambe* were found on the sponge surfaces: we gently swabbed an area of 20 cm^2 on each of five large sponge specimens with ca 1 g of glass wool fibre (Sigma) for 1 min, while another five wool pieces (controls), were taken underwater and taken out of their containers at the same time, but not used for swabbing. Control and treated wool was extracted twice with DCM (10 min in 13 ml of DCM each time), the two extracts were pooled, the solvent was allowed to evaporate, and the extracts were tested with Microtox.

Antipredation

We also investigated the predator-deterrent properties of larvae, juvenile and adult sponges. We selected the benthic fish *Parablennius incognitus* (Bath) for predation tests on larvae and rhagons (functional sponge recruits), and the sea urchin *Paracentrotus lividus* (Lamarck) for tests of consumption of artificial food with chemicals and materials from adult sponge specimens. Details of these tests are given in Uriz et al. (1996b).

Study of variation in chemical bioactivity

Within-individual variation

Within-individual location of toxicity was also investigated to provide clues to the function of the active substances. This was done (Uriz et al., 1996a) by separately analysing (Microtox) in five specimens the two layers that comprise the sponge structure: the basal choanosome and the distal ectosome. Further work addressed the identification of the cell type(s) responsible for the toxic properties of the sponges. Cells were separated by depositing cell suspensions (obtained through stirring of small sponge pieces in calcium- and magnesium-free artificial sea water) in a gradient of four densities made with decreasing concentrations of Ficoll (Merck). The cells accumulated in one of the three density interfaces, and in this way we obtained three fractions, each enriched in a different cell type (quantified with a haematocytometer), whose toxicity was analysed by the Microtox procedure.

Intra-individual variation of toxicity was also studied during the seasonal cycle (Turon et al., 1996a). To this end, we selected five large specimens and took small samples from the center and the periphery of the sponges over 15 months (January 1993–March 1994). Samples were extracted and analysed with the Microtox.

Between-individual variation

Inter-individual variation was studied as a function of size and habitat (Becerro et al., 1995a). To this end, we chose a sampling site with two parallel vertical walls between 6 and 12 m in depth, which were only 3 m from each other. They were identical in all respects (including trophic and physical conditions) except in the orientation: one wall faced North, the other faced South. As a result, the former received much less irradiance (relatively shaded wall) than the latter (well-illuminated wall). Communities in these walls showed remarkable differences in species composition: a space-limited community (which will be hereafter called the sciaphilous assemblage) mainly dominated by encrusting animal species was found on the shaded wall; whereas the well-lit habitat was dominated by erect algae interspersed with patches of bare substrate (photophilic assemblage). On the latter wall, *C. crambe* was the only conspicuous macroinvertebrate (at the landscape level). Our design included these two sites and three size classes of sponges: small (<1000 mm^2 in area), medium (1000 to 10 000 mm^2) and large (>10 000 mm^2) specimens. Ten specimens for each category of size and habitat were selected at random, samples were taken and extracted, and their toxicity was quantified.

Biological parameters

We characterised the pattern of resource allocation following the same design of habitat and size by studying the following parameters in individuals chosen at random on the two walls: thickness, biomass cm^{-2}, organic matter cm^{-2}, spicule content cm^{-2}, porosity, relative amount of spongin fibres, collagen, cells and matrix, and investment in reproduction (number of larvae incubated cm^{-2}). Several techniques were used to analyse these parameters, and they are explained in detail in Uriz et al. (1995).

We also studied the growth rates and mortality on both walls. To this end, in November 1994, we selected small specimens (average area less than 100 mm^2) on each wall. Every month we drew their outlines underwater on acetate sheets. The outlines were then digitized and their surface areas were calculated. Since a high mortality was found on the well-illuminated wall from the beginning of the study, new individuals from this wall were included in the monitoring during the first 4 months of study. Final numbers of sponges monitored were 24 on the shaded wall, 51 on the well-illuminated wall. The survey lasted until January 1997.

Results

Results reported here for the first time are explained in full, while we summarize the results already reported in previous papers by the authors. The reader may refer to them for full details and ANOVA and statistical tables, which will not be presented here.

Study of ecological roles

Antifouling

When the surfaces of individuals of *Crambe crambe* were swabbed with a sterile cotton, we found (Figure 1) that the estimated number of bacteria was about 60 ± 40 cm^{-2} (all results are mean ± SE), which was of the same order as the number found by swabbing sterile Petri dishes used as controls (30 ± 25 bacteria cm^{-2}), and much lower than the figures obtained by swabbing glass slides immersed in water for three weeks (mean 18 940 ± 4303 bacteria cm^{-2}). The surfaces of *C. crambe*, therefore, were almost axenic. To test for a possible antimicrobial effect of the secondary metabolites of the sponge, cultures of seven bacterial strains (four Gram + and three Gram –) isolated from the field were assayed with paper disks (6 mm in diameter) soaked in 25 mm of crude *C. crambe* extract. A significant inhibition effect was found on the seven strains. The diameter of the inhibition zone that developed after 24 h varied from 7.7 ± 0.25 mm to 24.2 ± 0.45 mm (including the disk) depending on the bacterial strain, in contrast to the non-inhibition found in control disks (Figure 1). In the experiment of inhibition of natural bacterial films by extracts of *C. crambe*, bacteria had occupied 5.37 ± 2.51 cm^2 of the culture plates seeded with cottons swabbed in control plates, while only 0.37 ± 0.19 cm^2 of those seeded with swabs from the treatment plates, and this difference was significant (p = 0.028, Mann-Whitney U test).

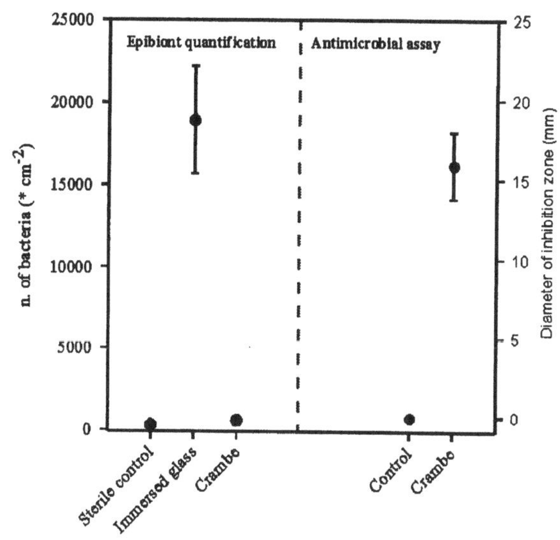

Figure 1. Summary of results from the microepibiont quantification experience and the bacterial inhibition assay. Bars are standard errors.

All subfractions studied showed an inhibitory effect on larval settlement of *Bugula neritina* larvae (Figure 2), although we present only the results for the DCM subfraction (which is the most comparable with the DCM extracts used in the other parts of this study).

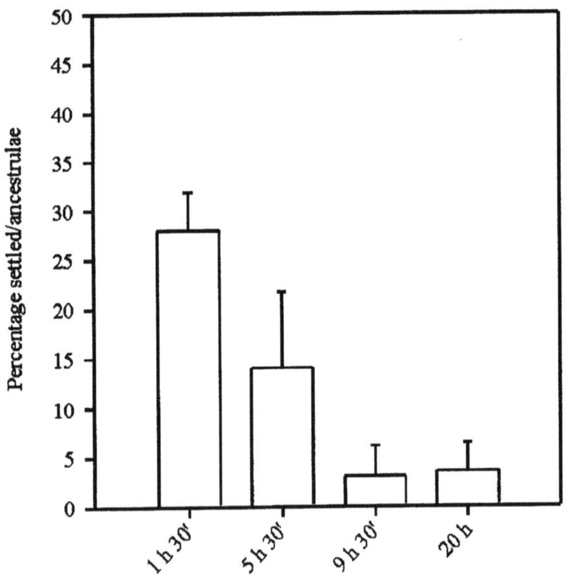

Figure 2. Percentage (relative to controls) of settlers (first and second observation times) or ancestrulae (third and fourth observation times) of *Bugula neritina* in the presence of extract of *C. crambe*. Bars are standard errors.

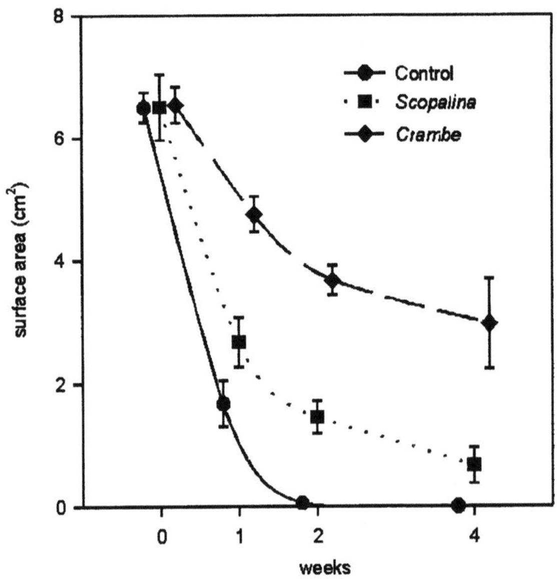

Figure 3. Time course of the area of holes scraped in colonies of *Scopalina lophyropoda*. Treatment holes were rubbed at the end of one and two weeks with either *C. crambe* or *S. lophyropoda*. Bars are standard errors.

There was a significant inhibitory effect at all observation times.

Competition for space

Small-scale association measurements of *C. crambe* were made in a community in which this species was abundant. This community was located on the shaded wall of the ecological variation study. The results showed that, out of ten main states identified (bare rock, nine different species and a miscellaneous group including all species with low abundances), *C. crambe* had negative associations with the five zoobenthic species recorded (3 sponges, 1 ascidian, and 1 bryozoan) and a crustose alga. Monte Carlo analyses revealed that these associations were significant in the case of the interactions with the three sponge species present. In contrast, *C. crambe* was positively and significantly associated with bare rock. Interestingly, when the associations were studied at increasing distances from the contact borders, their intensity fell drastically over the first few centimeters. On the other hand, when holes scraped in *S. lophyropoda* were rubbed with *C. crambe*, the regeneration rates were significantly lower than those of holes rubbed with *S. lophyropoda* (Figure 3). A rubbing effect was also apparent, as at week four non-rubbed holes were sealed off, while the rubbed ones were not. On the other hand, none of three subfractions (aqueous, DCM, and butanolic) of a DCM:MeOH extract of *C. crambe* exerted, at 50 ppm of concentration, a significant effect on *U. rigida* oxygen production rates, so no interference with photosynthesis was detected.

As for the results of the swabbing of *C. crambe* surfaces with glass wool fibre, very small amounts of substances (from 300 to 600 μg, extract dry weight) were recovered from the swabs, and a toxicity value could not be calculated with the Microtox device, as the EC_{50} was greater than the highest concentrations tested in all cases. We were, however, able to compare bioluminescence readings between controls and treatments to identify the presence of a toxic substance, even if in very small amounts. We compared the luminescence decrease (after 5 min of incubation) at the highest concentration possible (60 ppm with respect to extract weight) of the replicate for which we had least material. The variable analysed was Gamma Units, which measured the ratio of light expected from a non-toxic sample to that observed, minus 1. The Gamma Units were significantly higher (*t*-test, $p = 0.0112$) in the swabs from *C. crambe* (0.370 ± 0.051) than in the control swabs (0.129 ± 0.053). Another evidence of the presence of bioactive substances on sponge surfaces came from the histological observations of spherulous cells accumulating and being released through the surfaces of the sponges (see below).

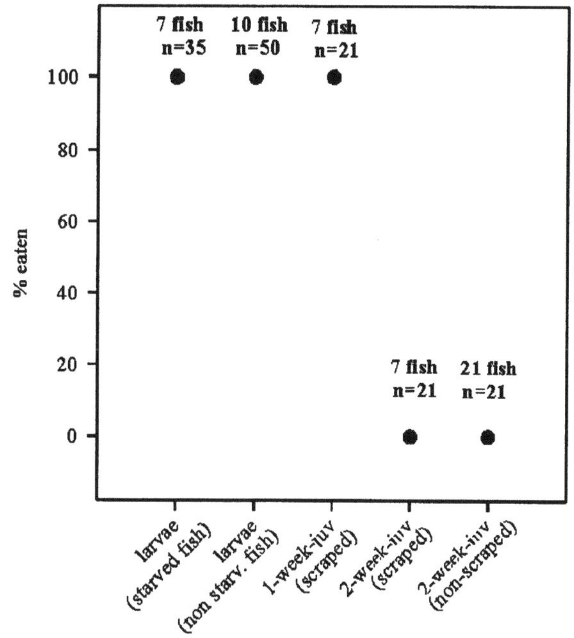

Figure 4. Summary of results from the feeding experiment with *Parablennius incognitus*. The number of fish used and the total number of larvae (n) offered to them is indicated. For the non-scraped 2-week-old juveniles, 7 fish were added to each of 3 Petri dishes with 7 sponge juveniles.

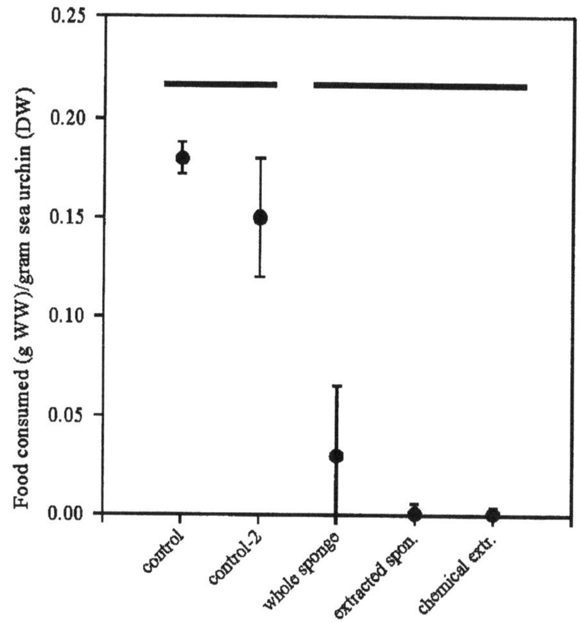

Figure 5. Artificial food consumption by *Paracentrotus lividus*. Horizontal lines join treatments non-significantly different (Tukey test). Bars are standard errors.

Antipredation

The results of fish predation on larvae and juvenile sponges were clearcut (Figure 4): all larvae offered were immediately eaten by *P. incognitus*, irrespective of whether the fish were starved or not. One-week-old juveniles (scraped from the substrate) were also readily eaten in all trials, whereas no two-week-old juvenile was eaten in any case, whether scraped from the substratum or still attached to it. The consumption of larvae is consistent with the finding that the DCM extract of 300 larvae showed hardly any toxicity in the Microtox test (0.24 TU). Furthermore, the experiment with the sea urchin *P. lividus* also showed a distinct effect of all treatments with respect to controls (Figure 5): untreated sponge material, extract from sponge, and the sponge material remaining after extraction (i.e. with all physical structures but without toxic metabolites) all significantly deterred the sea urchins from feeding on one of their preferred algae. We used sponge fragments of the same size as the food plates to keep realistic concentrations of chemicals. Control plates made with fresh (control-1) or extracted algae (control-2) were eaten at similar rates (Figure 5), so no effect due to extraction *per se* could be substantiated.

Study of variation in chemical bioactivity

Within-individal variation

The toxicity was significantly higher in the distal part of the sponge (ectosome, 12.54 ± 1.4 TU) than in the basal part (choanosome, 2.58 ± 0.92 TU) (Figure 6). There are many spherulous cells in the ectosome, which were frequently observed in histological sections clustered beneath the exopinacoderm, releasing their vacuole contents, or being shed themselves to the outside.

The Ficoll procedure gave three cell fractions. Fraction 1 (interface between 2% and 5% Ficoll) contained $90 \pm 0.9\%$ spherulous cells, the 10% of other cell types consisted of choanocytes and non-identified sponge cells or debris. Fraction 2 (interface 5–8% Ficoll) was enriched in choanocytes ($70 \pm 0.95\%$), and also had spherulous cells ($12 \pm 0.74\%$), archeocytes ($6.2 \pm 0.74\%$) and unidentified cells. Fraction 3 (interface 8–11% Ficoll) mainly contained archeocytes ($75 \pm 0.66\%$), a few spherulous cells ($7 \pm 0.74\%$) and cell aggregates ($18 \pm 0.41\%$). Fraction 1 was the most toxic (mean 9.08 TU), while Fraction 2 was mildly toxic (mean 0.48 TU) and Fraction 3 did not show any toxicity (Figure 6). There was, therefore, a good correlation between presence of spherulous cells and toxicity, both at the cellular level and between sponge

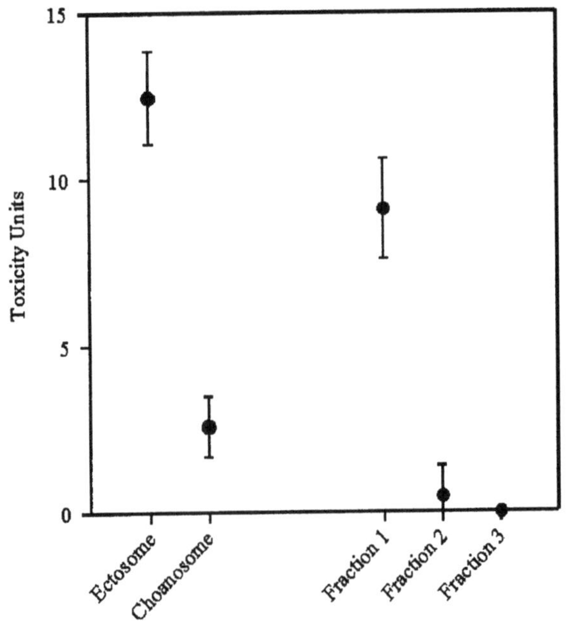

Figure 6. Toxicity readings from the ectosome and choanosome of sponges, as well as from the three cellular fractions obtained by a Ficoll gradient. Bars are standard errors.

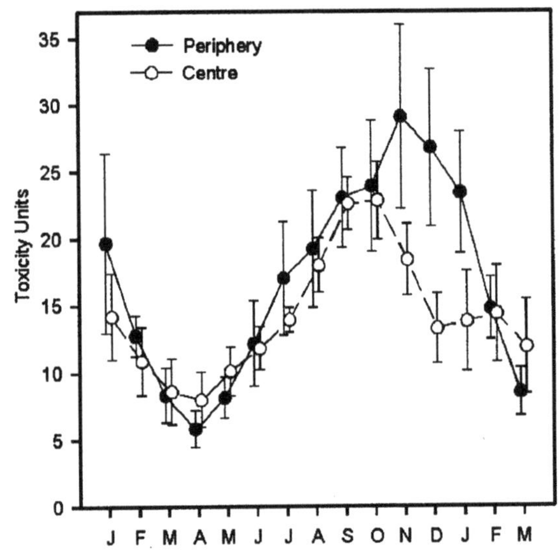

Figure 7. Time course of the toxicity values in the centre and at the periphery of the colonies monitored. Bars are standard errors.

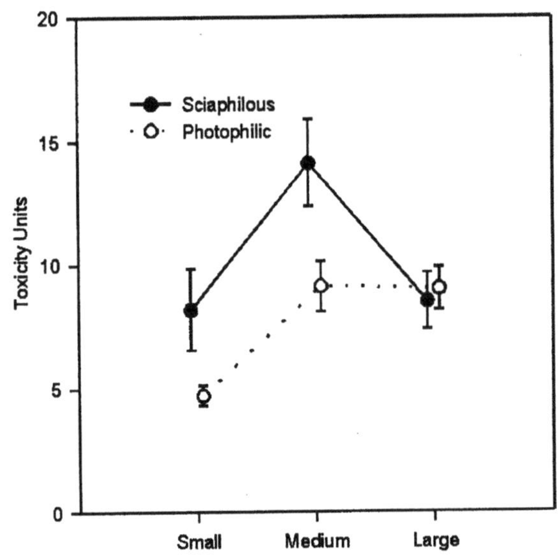

Figure 8. Toxicity values as a function of habitat (sciaphilous or photophilic assemblages) and size class. Bars are standard errors.

layers. The spherulous cells were identified as those that contained the toxic metabolites in this species.

The seasonal variation in toxicity is presented in Figure 7. Two aspects are noteworthy: firstly, there was a clear seasonal pattern, with a minimum in April and maxima at the end of summer-autumn (centre of sponges) and autumn-winter (periphery of sponges). Secondly, toxicity was significantly higher at the periphery than in the centre during the period of high toxicity.

Between-individual variation

When toxicity was analysed as a function of habitat (shaded versus well-lit community) and size class (Figure 8), both factors proved significant (two-way ANOVA), while the interaction was not. Overall, toxicity was higher in the shaded (sciaphilous) community, and in both habitats toxicity was higher in medium-sized specimens (although in the photophilic assemblage, medium- and large-sized sponges showed similar values).

Biological parameters

Table 1 summarizes the results of the analyses performed for the biological and morphological parameters studied following the habitat-size design. Many patterns found proved significant in two- and three-factor (with specimen as a nested factor) analyses of variance. Multiple comparisons were made by Ryan's *Q* method (Day & Quinn, 1989), and the results are also reported in Table 1. Sponges were thicker in the photophilic assemblage. In general, variables associated with structural materials (collagen, fibres, spicules) were higher on the shaded wall, at least (spicules) for large sponges. In contrast, more matrix material was found in photophilic specimens, and also more organic matter (in large sponges) and more cells (in medium-sized sponges). There was a higher production of larvae

Table 1. Summary of the significant effects found in ANOVA analyses of the morphological and biological parameters studied (**=significant at 0.05 probability level; ns=non significant; –=main effect not tested because of significant interactions). Whenever an effect was found significant, *post hoc* tests were made by Ryan's *Q* method. Only significant comparisons are reported. If the interaction size-habitat was not significant, levels of a significant main factor were compared pooling across levels of the other factor. If the interaction was significant, levels of each factor were compared within levels of the other factor (ph, photophilic specimens; sc, sciaphilous specimens).

Variable	Effect Habitat	Size	Habitat-Size	Comparisons
Thickness	**	ns	ns	photophilic>sciaphilic
Porosity	ns	ns	ns	
Biomass	ns	**	ns	large>small & medium
Silica content	–	–	**	ph small>sc small
				sc large>ph large
				sc large>sc medium>sc small
Organic matter	–	–	**	ph large>sc large
				ph large>ph small & medium
Amount of cells	–	–	**	ph medium>sc medium
				sc small>ph small
				sc medium>sc small
Amount of collagen	**	ns	ns	sciaphilous>photophilic
Amount of fibres	**	**	ns	sciaphilous>photophiic
				small>medium
Matrix	**	ns	ns	photophilic>sciaphilous
Larvae	**	**	ns	photophilic>sciaphilous
				large>medium
				large>small

in sponges from the well-illuminated wall, and there was a trend towards increased investment in reproduction with size in both habitats.

As for the growth rates, sponges grew more in the photophilous assemblage. Figure 9 shows the cumulative growth rates (final area minus initial area divided by initial area) at the end of the first and the second years of monitoring. The values in the photophilic habitat were higher, especially during the second year, resulting in a final mean cumulative growth rate of about 2 (i.e. initial areas had been, on average, trebled), while in the shaded wall the mean cumulative growth rate was ca 1 (i.e. initial areas had been, on average, doubled). However, high variances resulting from high inter-individual variability prevented these final cumulative growth rates from being statistically different in the two habitats (Mann-Whitney U test). When growth rates were considered on a monthly basis and compared between habitats, higher growth rates were found in the photophilic habitat in 21 out of 26 months surveyed, and the differences were significant (Mann-Whitney U-test, with Bonferroni correction for the number of comparisons) in May, June, and July 1996. Mortality, on the other hand, was higher in the well-illuminated habitat, in which only 31% of sponges survived by the end of the study, against 62% survival on the shaded wall.

Discussion

The surfaces of *Crambe crambe* were almost axenic, and the antimicrobial properties of its toxic metabolites could explain why development of the microbial film (and hence subsequent steps in the fouling sequence) was prevented. Moreover, antilarval effects of *C. crambe* extracts were also demonstrated. The pattern of small-scale associations of this sponge was consistent with the presence of a short-range inhibition mechanism which may serve for space-competition. Moreover, the rubbing experiment showed that some substance from *C. crambe*, which remained in the substrate for some time, prevented growth of one of the main space competitors of *C. crambe*.

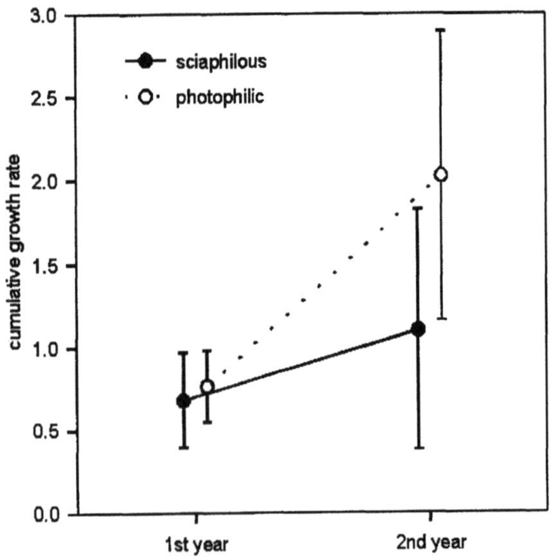

Figure 9. Cumulative growth rates of sponges at the end of the first (Jan 96) and second (Jan 97) years of the growth monitoring in the two habitats considered. Bars are standard errors.

The commonest grazer in the zone (*P. lividus*) was prevented from feeding by all sponge treatments. Feeding deterrence experiments should consider all the potential defences, not only the effects of single defences in isolation (Van Alstyne et al., 1994). We separated the effect of chemicals (extracts) from that of structural materials (extracted sponge), and found that the sponge material deterred feeding by the sea urchin, but that either the chemicals (at an ecologically relevant dose) or the physical structures, considered separately were enough to account for this deterrence. The metabolites of *C. crambe*, however, did not affect photosynthesis rates of the alga assayed. This suggests that these compounds do not constitute an antialgal competition mechanism.

Our results, therefore, showed evidence of a range of ecological functions for the bioactive metabolites of *C. crambe*. Van de Vyver et al. (1990) and Becerro (1994) also found a broad spectrum of bioactivities in sponges and other benthic invertebrates, suggesting that their toxic substances may be multiple-purpose defensive weapons. This may be the rule in benthic, encrusting (i.e. surface dependent) organisms which are subject to high fouling and space competition pressures, as well as to predation.

A key aspect of the ecological roles of bioactive chemicals is that they must be released to the medium if they are to work in antifouling or space competition. Conversely, release is not necessary if their function is solely to prevent predation. The continuous release of a highly diffusible (i.e. polar) compound to the sea would be wasteful, so active compounds that are released are most likely non-polar molecules that may remain on the surfaces and diffuse very slowly. This is consistent with the structure of crambines and crambescidins, which have a mixture of lipophilic (due to the non-polar chain) and hydrophilic (due to the polar head) properties. Although the results of our swabbing experiment were not clearcut, since we recovered very small amounts of material and obtained low activities, our *C. crambe* swabs were significantly more active than the controls, indicating the presence of bioactive substances. A second line of evidence came from the histological studies, in which the spherulous cells, those that accumulate the toxic substances, were shown to be concentrated near the exopinacoderm and to be released through the sponge surface (Uriz et al., 1996a). The available evidence, therefore, points to the release of toxic substances, possibly at small doses, to the medium.

Insight into the processes modeling the defence strategy in this sponge may come from the study of variation at several scales. There are differences between cell types, as the spherulous cells were responsible for the accumulation (and probably production) of the active substances. The same pattern has been found in other sponge species (Bretting et al., 1983; Thompson et al., 1983, but see Uriz et al., 1996c, for contrasting results in another species). The high abundance of spherulous cells in the ectosome of the sponge explains the high toxicity of the distal layer.

We also detected ontogenetic variation in chemical defences. The larvae were not toxic (Microtox), and neither the larvae nor the one-week-old recruits had predation-deterrent properties. Deterrency developed somewhere between one and two weeks after settlement. This may correlate with the time of differentiation of the first spherulous cells in the juveniles. This result is consistent with the hypothesis that rapidly developing juvenile tissues cannot produce bioactive metabolites (growth-differentiation balance hypothesis, Cronin & Hay, 1996), although many exceptions have been reported (e.g. in ascidians: Lindquist et al., 1992). Larvae of another sponge species, *Dysidea avara* (Schmidt), were found to be defended against predation (Uriz et al., 1996b). Seasonality imposes strong periodicity on the biological parameters of marine organisms in temperate seas. Investment in toxin production was also found to fluctuate seasonally in *C. crambe*. Differences between centre and periphery of the sponges were also substantiated in this study. These differences were evident during the season of high toxicity. The time course of toxicity in the centre may be modulated by internal parameters (e.g. investment in reproduction); while at the periphery external interactions (e.g. space competition) may be the dominant factor (Turon et al., 1996a). In this sense, it is indicative that the season of high peripheral bioactivity coincides with the season in which many invertebrates reactivate growth after the aestivation period. Fouling pressure varies seasonally, but its maxima are in spring in the study zone, which cannot explain our autumn peak. Mature larvae are present within *C. crambe* at the end of spring and summer, so gametogenesis and associated changes occur in spring. It seems, therefore, that the timing of allocation to reproduction and to toxin production is reversed, thus suggesting a trade-off between reproduction and chemical defence production. A combination of internal and external pressures probably determined the cycles found.

Habitat-related variation in toxicity was found, with sponges from a space-saturated (sciaphilous) community dominated by slow-growing animal species featuring more toxicity than sponges from an adjacent habitat (photophilic community) dominated by algae and with patches of bare space being continuously produced. It seems likely that space competition pressures explain the differences. This does not imply, however, that algae do not compete for space. More specifically, the key factor may be the different turnover and growth rates of the interacting species. A slow growing form such as *C. crambe* would hardly outcompete fast-growing algal species which appear and disappear seasonally, so there would be no selective advantage in investing in costly defences (physical and chemical) that would not prevent mortality in this habitat anyway. On the other hand, allocation to defensive chemicals would be more advantageous in a community dominated by perennial, surface-dependent organisms.

C. crambe was very plastic in most biological parameters. Many of the features studied varied significantly among size classes and habitats, indicating an ability to adjust its relative energy allocation in response to physiological and environmental changes. We did not find any trade-off between allocation to chemical and physical defences, as reported in other studies (between species: Sammarco et al., 1987; Coll, 1992; at the intra-individual level: Harvell & Fenical, 1989; Pennings et al., 1996). Our results on an intraspecific scale agree with those of Chanas & Pawlik (1995) in an interspecific comparison among Caribbean sponges. In our study, sponges that invested more in chemical defences (sciaphilous sponges) also allocated more resources to tough, structural materials (collagen, spongin fibres, spicules) able to serve as physical defences. It may also be that the combined effect of both types of defence is greater than their separate sums (as found by Hay et al., 1994).

The question arises as to the allocation of the energy expenditure spared by specimens (photophilic) that invest less in defence; obvious responses are growth and reproduction (Paul & Van Alstyne, 1988; Chanas & Pawlik, 1995). Reproductive output, as measured by the number of larvae incubated, is indeed higher on the photophilic habitat (Table 1). Growth rates are also higher in this habitat, although large variances prevented this effect from being statistically significant in some of the analyses. Preferential allocation to growth in small specimens, and to reproduction in larger ones, may also explain why toxicity, overall, is higher in medium-sized individuals (Figure 8). Mortality, on the other hand, was higher in the well-illuminated wall, where sponges invest less in physical and chemical defences.

In summary, the results are consistent with the idea that chemical defence is costly in *C. crambe*, and that it is optimized with respect to within-individual, ontogenetic, seasonal, and ecological constraints. Clonal construction is an energetically favourable situation for the production of costly defences (Adler & Harvell, 1990; Harvell, 1990), which may explain the lack of a dichotomy between allocation to physical and chemical defences in our between-habitat comparison. We found, however, a trade-off between allocation to defences and to reproduction and growth. What can be said about the factors determining the variation found? Predation can apparently be avoided with just physical defences, and no indication of an effect of variable fouling pressures was apparent (peaks of toxicity did not correspond to recruitment peaks of foulers, and we did not expect much difference, at least at the microfouling level, between our two adjacent walls). Differences in food availability are also unlikely between the walls studied, so resource limitations cannot explain the between-habitat pattern observed. The modulation of defence levels seems, therefore, best explained by differences in space competition pressures.

In conclusion, the optimal defence theory seems to apply in the case of *C. crambe*, and a model of variation-in-space competition pressures is consistent with most patterns observed. This may prove to be a general feature of indeterminate-growing organisms, especially those that are more dependent on free space (e.g. those of encrusting morphology), and which feature a strategy of slow growth and perennial life span.

Acknowledgments

Dr P. Steinberg proposed the glass wool swabbing method. J. Galera, J. Lozano, I. Tarjuelo and J. M. Tur helped with the field and laboratory work. G. Benito and the staff of the 'Junta de Sanejament' of the Catalan Government provided Microtox and laboratory facilities. This study was funded by projects PB94-0015 and MAR95-1764 from the Spanish Government and by project MAST-CT91-004 from the European Union.

References

Adler, F. R. & C. D. Harvell, 1990. Inducible defenses, phenotypic variability and biotic environments. Trends Ecol. Evol. 5: 407–410.

Adler, F. R. & R. Karban, 1994. Defended fortresses or moving targets? another model of inducible defenses inspired by military metaphors. Am. Nat. 144: 813–832.

Becerro, M. A., 1994. Chemically mediated bioactivity of the encrusting sponge *Crambe crambe* and its ecological implications. Ph.D. Thesis. University of Barcelona, 250 pp.

Becerro, M. A., N. I. López, X. Turon & M. J. Uriz, 1994. Antimicrobial activity and surface bacterial film in marine sponges. J. exp. mar. Biol. Ecol. 179: 195–205.

Becerro, M. A., X. Turon & M. J. Uriz, 1995a. Natural variation of toxicity in encrusting sponge *Crambe crambe* (Schmidt) in relation to size and environment. J. chem. Ecol. 21: 1931–1946.

Becerro, M. A., M. J. Uriz & X. Turon, 1995b. Measuring toxicity in marine environments: critical appraisal of three commonly used methods. Experientia 51: 414–418.

Berlinck, R. G. S., J. C. Braeckman, I. Bruno, D. Daloze, S. Pem, R. Riccio, S. Spampinato & E. Speroni, 1990. Two new guanidine alkaloids from the Mediterranean sponge *Crambe crambe*. Tetrahedron Lett. 31: 6531–6534.

Berlinck, R. G. S., J. C. Braeckman, D. Daloze, I. Bruno, R. Riccio, D. Rogeau & P. Amade. 1992. Crambines C1 and C2: two further ichthyotoxic guanidine alkaloids from the sponge *Crambe crambe*. J. nat. Prod. 55: 528–532.

Bretting, H., G. Jacobs, C. Donadey & J. Vacelet, 1983. Immunohistochemical studies on the occurrence and the function of the D-galactose-specific lectins in the tissue of the sponge *Axinella polypoides* Schmidt. Cell Tissue Res. 229: 551–572.

Chanas, B. & J. R. Pawlik, 1995. Defenses of Caribbean sponges against predatory reef fish. II. Spicules, tissue toughness, and nutritional quality. Mar. Ecol. Prog. Ser. 127: 195–211.

Clare, A. S., 1996. Marine natural product antifoulants: status and potential. Biofouling 9: 211–229.

Coll, J. C., 1992. The chemistry and chemical ecology of octocorals (Coelenterata, Anthozoa, Octocorallia). Chem. Rev. 92: 613–631.

Coll, J. C., I. R. Price, G. M. König & B. F. Bowden, 1987. Algal overgrowth of alcyonacean soft corals. Mar. Biol. 96: 129–135.

Cronin, G. & M. E. Hay, 1996. Within-plant variation in seaweed palatability and chemical defenses: optimal defense theory versus the growth-differentiation balance hypothesis. Oecologia 105: 361–368.

Cronin, G., N. Lindquist, M. E. Hay & W. Fenical, 1995. Effects of storage and extraction procedures on yields of lipophilic metabolites from the brown seaweeds *Dyctiota ciliolata* and *D. menstrualis*. Mar. Ecol. Prog. Ser. 119: 265–273.

Davis, A. R., A. J. Butler & I. Van Altena, 1991. Settlement behaviour of ascidian larvae: preliminary evidence for inhibition by sponge allelochemicals. Mar. Ecol. Prog. Ser. 72: 117–123.

Day, R. W. & G. P. Quinn, 1989. Comparisons of treatments after an analysis of variance in ecology. Ecol. Monogr. 59: 433–463.

Fagerström, T., S. Larsson & O. Tenow, 1987. On optimal defence in plants. Funct. Ecol. 1: 73–81.

Faulkner, D. J., 1993. Marine natural products chemistry: introduction. Chem. Rev. 93: 1671–1672.

Faulkner, D. J., 1996. Marine natural products. Nat. Prod. Rep. 13: 75–125.

Harvell, D., 1990. The ecology and evolution of inducible defenses. Quart. Rev. Biol. 65: 323–340.

Harvell, D. & W. Fenical, 1989. Chemical and structural defenses of Caribbean gorgonians (*Pseudopterogorgia* spp.): intracolony localization of defense. Limnol. Oceanogr. 34: 382–389.

Hay, M. E., 1984. Predictable spatial scapes from herbivory: how do these affect the evolution of herbivore resistance in tropical marine communities? Oecologia: 64: 396–407.

Hay, M. E., 1996. Marine chemical ecology: what's known and what's next? J. exp. mar. Biol. Ecol. 200: 103–134.

Hay, M. E., Q. E. Kappel & W. Fenical, 1994. Synergisms in plant defenses against herbivores: interactions of chemistry, calcification, and plant quality. Ecology 75: 1714–1726.

Hay, M. E. & P. D. Steinberg, 1992. The chemical ecology of plant-herbivore interactions in marine versus terrestrial communities. In Rosenthal, G. A. & M. R. Berenbaum (eds), Herbivores: their Interactions with Secondary Plant Metabolites. V. II. Evolutionary and Ecological Processes. Academic Press, San Diego: 371–413.

Jares-Erijman, E. A., R. Sakai & K. L. Rinehart, 1991. Crambescidins: new antiviral and cytotoxic compounds from the sponge *Crambe crambe*. J. org. Chem. 56: 5712–5715.

Lindquist, N., M. E. Hay & W. Fenical, 1992. Defense of ascidians and their conspicuous larvae: adult vs. larval chemical defenses. Ecol. Monogr. 62: 547–568.

Maida, M., A. R. Carroll & J. C. Coll, 1993. Variability of terpene content in the soft coral *Sinularia flexibilis* (Coelenterata: Octocorallia), and its ecological implications. J. chem. Ecol. 19: 2285–2296.

Maida, M., P. W. Sammarco & J. C. Coll, 1995. Effects of soft corals on scleractiian coral recruitment. I: directional allelopathy and inhibition of settlement. Mar. Ecol. Prog. Ser. 121: 191–202.

Pastorok, R. A. & D. K. Becker, 1990. Comparative sensitivity of sediment toxicity bioassays at three superfund sites in Puget Sound. In Landis, W. G. & W. H. van der Schalie (eds), Aquatic Toxicology and Risk Assessment. Vol. 13. American Society for Testing and Materials, Philadelphia: 123–139.

Paul, V., 1992. Ecological Roles of Marine Natural Products. Comstock Pub. Assoc., Ithaca, 245 pp.

Paul, V., N. Lindquist & W. Fenical, 1990. Chemical defenses of the tropical ascidian *Atapozoa* sp. and its nudibranch predators *Nembrotha*. J. exp. mar. Biol. Ecol. 119: 15–29.

Paul, V. & K. L. Van Alstyne, 1988. Chemical defense and chemical variation in some tropical Pacific species of *Halimeda* (Halimedaceae; Chlorophyta). Coral Reefs 6: 263–269.

Paul, V. & K. L. Van Alstyne, 1992. Activation of chemical defenses in the tropical green algae *Halimeda* spp. J. exp. mar. Biol. Ecol. 160: 191–203.

Pawlik, J. R., 1992. Chemical ecology of the settlement of benthic marine invertebrates. Oceanogr. Mar. Biol. annu. Rev. 30: 273–335.

Pawlik, J. R., 1993. Marine invertebrate chemical defenses. Chem. Rev. 93: 1911–1922.

Pawlik, J. R., B. Chanas, R. J. Toonen & W. Fenical, 1995. Defenses of Caribbean sponges against predatory reef fish. I. Chemical deterrency. Mar. Ecol. Prog. Ser. 127: 183–194.

Pennings, S. C., M. P. Puglisi, T. J. Pitlik, A. C. Himaya & V. J. Paul, 1996. Effects of secondary metabolites and $CaCO_3$ on feeding by surgeonfishes and parrotfishes: within-plant comparisons. Mar. Ecol. Prog. Ser. 134: 49–58.

Pennings, S. C., S. R. Sonia, V. Paul & J. E. Duffy, 1994. Effects of sponge secondary metabolites in different diets on feeding by three groups of consumers. J. exp. mar. Biol. Ecol. 180: 137–149.

Porter, J. M. & W. M. Targett, 1988. Allelochemical interactions between sponges and corals. Biol. Bull. 175: 230–239.

Ribo, J. M. & K. L. E. Kayser, 1987. *Photobacterium phosphoreum* toxicity bioassay. I. Test methods and procedures. Toxicol. Assess. 2: 305–323.

Sammarco, P. W. & J. C. Coll, 1988. The chemical ecology of Alcyonarian corals. Coelenterata: Octocorallia. In Scheuer, P. J. (ed.), Bioorganic Marine Chemistry. Vol. II, Springer-Verlag, Berlin: 87–116.

Sammarco, P. W., S. La Barre & J. C. Coll, 1987. Defensive strategies of soft corals (Coelenterata: Octocorallia) of the Great Barrier Reef. III. The relationship between ichthyotoxicity and morphology. Oecologia 74: 93–101.

Schmitt, T. M., M. E. Hay & N. Lindquist, 1995. Constraints on chemically mediated coevolution: multiple functions for seaweed secondary metabolites. Ecology 76: 107–123.

Steinberg, P. D., 1994. Lack of short-term induction of phlorotannins in the Australasian brown algae *Ecklonia radiata* and *Sargassum vestitum*. Mar. Ecol. Prog. Ser. 112: 129–133.

Steinberg, P. D. & K. L. Van Altena, 1992. Tolerance of marine invertebrate herbivores to brown algal phlorotannins in temperate Australasia. Ecol. Monogr. 62: 189–222.

Sullivan, B., D. J. Faulkner & L. Webb, 1983. Siphonodictine, a metabolite of the burrowing sponge *Siphonodictyon* sp. that inhibits coral growth. Science 221: 1175–1176.

Teo, S.L.-M. & J. S. Ryland, 1995. Potential antifouling mechanisms using toxic chemicals in some British ascidians. J. exp. mar. Biol. Ecol. 188: 49–62.

Thompson, J., K. D. Barrow & D. J. Faulkner, 1983. Localization of two brominated metabolites, aerothionin and homoaerothionin, in spherulous cells of the marine sponge *Aplysina fistularis* (= *Verongia thiona*). Acta Zool. 64: 199–210.

Turon, X., M. A. Becerro & M. J. Uriz, 1996a. Seasonal patterns of toxicity in benthic invertebrates: the encrusting sponge *Crambe crambe* (Poecilosclerida). Oikos 75: 33–40.

Turon, X., M. A. Becerro, M. J. Uriz & J. Llopis, 1996b. Small-scale association measures in epibenthic communities as a clue for allelochemical interactions. Oecologia 108: 351–360.

Uriz, M. J., M. A. Becerro, J. M. Tur & X. Turon, 1996a. Location of toxicity within the Mediterranean sponge *Crambe crambe* (Demospongiae: Poecilosclerida). Mar. Biol. 124: 583–590.

Uriz, M. J., D. Martín & D. Rosell, 1992a. Relationship of biological and taxonomic charactersitics to chemically mediated bioactivity in Mediterranean littoral sponges. Mar. Biol. 113: 287–297.

Uriz, M. J., D. Rosell & D. Martín, 1992b. The sponge population of the Cabrera archipelago (Balearic Islands): characteristics, distribution and abundance of the most representative species. P.S.Z.N.I. Mar. Ecol. 113: 101–117.

Uriz, M. J., X. Turon, M. A. Becerro & J. Galera, 1996b. Feeding deterrence in sponges. The role of toxicity, physical defenses, energetic contents, and life-history stage. J. exp. mar. Biol. Ecol. 205: 187–204.

Uriz, M. J., X. Turon, M. A. Becerro, J. Galera & J. Lozano, 1995. Patterns of resource allocation to somatic, defensive, and reproductive functions in the Mediterranean encrusting sponge *Crambe crambe* (Demospongiae, Poecilosclerida). Mar. Ecol. Prog. Ser. 124: 159–170.

Uriz, M. J., X. Turon, J. Galera & J. M. Tur, 1996c. New light on the cell location of avarol within the sponge *Dysidea avara* (Dendroceratida). Cell Tissue Res. 285: 519–527.

Van Alstyne, K. L., 1988. Herbivore grazing increases polyphenolic defenses in the intertidal brown alga *Fucus distichus*. Ecology 69: 655–663.

Van Alstyne, K. L., C. R. Wylie & V. J. Paul, 1994. Antipredator defenses in tropical Pacific soft corals (Coelenterata: Alcyonacea). II. The relative importance of chemical and structural defenses in three species of *Sinularia*. J. exp. mar. Biol. Ecol. 178: 17–34.

Van de Vyver, G., J. Huysecom, J. C. Braeckman & D. Daloze, 1990. Screening and bioassays for toxic substances in sponges from western Mediterranean Sea and North Brittany. Vie Milieu 40: 285–292.

Wahl, M., P. R. Jensen & W. Fenical, 1994. Chemical control of bacterial epibiosis on ascidians. Mar. Ecol. Prog. Ser. 110: 45–57.

Yates, J. L. & P. Peckol, 1993. Effects of nutrient availability and herbivory on polyphenolics in the seaweed *Fucus vesiculosus*. Ecology 74: 1757–1766.

Hydrobiologia **355:** 91–100, 1997.

A. D. Naumov, H. Hummel, A. A. Sukhotin & J. S. Ryland (eds), Interactions and Adaptation Strategies of Marine Organisms.

Fauna associated with detached kelp in different types of subtidal habitats of the White Sea

A. B. Tzetlin[1], V. O. Mokievsky[3], A. N. Melnikov[1], M. V. Saphonov[2], T. G. Simdyanov[1] & I. E. Ivanov[1]
[1] *Department of Invertebrate Zoology, Moscow State University, 119899, Vorobievy Gory, Moscow, Russia*
[2] *White Sea Biological Station, Moscow State University*
[3] *Shirshow Institute of Oceanology, Moscow*

Key words: kelp decomposition, White Sea, benthic communities, sea urchins, cyanobacteria

Abstract

The fauna, associated with Laminaria and other large brown macroalgae was studied by using SCUBA and dredging in two different types of underwater habitats of the White Sea.

In shallow water fjords and bays, with a depth of no more than 30–40 m, detached kelp (mainly *Laminaria saccharina, L. digitata* and *Alaria esculenta*) formed large accumulations. One of these benthic accumulations, which has existed more than 20 years, was studied. It covers about 2000 m^2, and is about 2 m thick. The upper layer of the accumulation of fronds is characterized by high turbulence and is well aerated. The lower layer is characterized by anoxic conditions. Mats of sulphur bacteria were not observed, although fronds in the middle layer were covered by layers of cyanobacteria. About 50 species of macroinvertebrates were found, mainly species that are normally associated with living kelp, such as the detritivorous species *Ophiura robusta* and *Gammarus oceanicus*, and few species that are specific inhabitants of organic-rich biotopes in the White Sea such as *Capitella capitata, Ophryotrocha irinae* and *Nebalia bipes*. It was remarkable that in the shallow water basins of the White Sea, the process of decomposition of brown algae in the sublittoral takes place without sea urchins, and no other macrofaunal form plays an ecological role in the mechanical breakdown of the plant substratum, even not in the large accumulations of detached kelp.

Along the open rocky shoreline, communities associated with dead detached kelp were situated at a depth of 60–90 m, 40–50 m below the belt of living kelp. In this deep zone, no macroinvertebrates typical of the kelp community in the photic zone were found. During the passage from the shoreline to the deeper benthic community, where sea urchins were dominant, all plant debris became fragmented. These deeper benthic communities appeared to be the zone for decomposition of the detached kelp.

Introduction

In waters and on rocky coasts of temperate areas large brown algae (Phaeophyta) are the dominant sublittoral macrophytes. The fronds of these algae resemble moving belts of tissue (Mann, 1972). Erosion from the senescent distal region of the algal frond is compensated for by meristematic activity at the frond base. In a single year, these fronds may replace their length up to five times (Mann, 1972), contributing large quantities of dissolved organic matter and particulate detritus to the sediments and water column. In addition to these processes, winter storms can detach large numbers of intact kelp plants (Bedford & Moore, 1984). Such plants can either be washed ashore, fuelling intertidal detritus food chains, or sink to the sea bed below the zone of brown algae.

Fauna associated with sea bed accumulations of decomposing brown algae is well studied and normally resembles the epifauna on living attached kelp with the addition of a few specialized detritivorous species. Sea urchins (Diademata), such as *Psammechinus* spp. or *Strongylocentrotus* spp. are dominant (Bedford & Moore, 1984; Breen & Mann, 1976; Ivanjushina et al., 1991)

In the sublittoral zone of the White Sea, brown algae (Phaeophyta) form well developed belts. The average wet biomass of kelp is estimated at 10–15 kg m^{-2} (Vozhinskaya, 1986). Macrofaunal invertebrates associated with kelp in the sublittoral zone of the White Sea are described by Derugin (1928) and Golikov et al. (1985a). Unlike the communities associated with kelp in the boreal zone of the Pacific and Atlantic, similar communities in the White Sea are devoid of sea urchins. Although two species of Diademata are described for the White Sea, specimens of sea urchins, such as *Strongylocentrotus pallidus* which is common in the White Sea, do not live above 40–50 m because of the low salinity in the upper part of the sublittoral zone (Golikov et al., 1985a, b; Bazhin, 1995). As a result, the macrofaunal communities of shallow sea bed accumulations of detached kelp here show remarkable differences from similar ones in the North Atlantic or even with those in the neighbouring Barents Sea.

Therefore, in this study we present a first description of these communities at different types of (shallow versus deeper open) coasts of the White Sea, indicating the different interactions between habitat, macrofaunal composition, and kelp decomposition.

Material and methods

The study is based on more than twenty years of SCUBA diving observations in the vicinity of the White Sea Biological Station of Moscow State University along the Karelian Coast of the White Sea. During this period we studied the sublittoral zone along the coast of Kandalaksha Bay (Figure 1). During the two last summer seasons (1995–1996) a set of 42 qualitative samples were taken by means of SCUBA diving. Besides these samples, about one hundred hours of SCUBA underwater observations were made by the authors during these two years. Samples of living and detached kelp were taken by divers. Fronds were packed into plastic bags with a volume of 10 litres, and analysed for faunal content in the laboratory directly after transportation to the surface. In summer 1995, three quantitative samples in the accumulations of detached kelp situated at depths of 12–15 m were taken with a special sampling device, consisting of a nylon cylinder (mesh size 0.5 mm, 3 m long) with a heavy metal ring of 50 cm diameter near the opening. Divers pushed the cylinder down into the 1.5 to 2 m thick swell of kelp and cut fronds along the perimeter of the ring. After transportation to the surface, wet fronds from these samples and the total of all extracted animals were weighed. Animals were then fixed in formalin for species identification.

In the deeper areas (from 40 to 130 m) 43 benthic samples were taken by dredging with a Sigsby trawl of 1,25 m width (mesh size 1.0 mm). This zone is too deep for divers and because of the hard substrate, corers or grabs could not be used successfully for quantitative sampling.

The direction of tidal currents was measured by a Vallyport current meter of the propeller type with distant result indication. Redox values were measured by a Hanna Instrument Redox potential value tester.

Results

Open shoreline

Large brown algae *Laminaria saccharina*, *L. digitata* and *Alaria esculenta* inhabit the upper 12 m of rocky shores. A list of fauna species associated with the living Laminaria is given in Table 1. Detached thalli moved down the slope and could be found to a depth of 60–90 m (Figure 2). Here accumulations of detached kelp occurred and the associated fauna was well developed. Thirty three species of macroinvertebrates were found (Table 1, Figure 3); *Strongylocentrotus pallidus* dominated. In the deeper community it was notable that species typical for the epifaunal community of kelp in the photic zone were absent (Figure 3). Sea urchins were not found below a depth of 90 meters.

Shallow water fiords and basins

The shoreline of Kandalaksha Bay and Onega Bay is characterized by shallow water fiords and bays, with depths of 30–40 m and a salinity of 24–28‰ (Figure 2B). Specimens of sea urchins (*S. pallidus*) can be found only occasionally. In these basins, large accumulations of kelp, produced by tidal currents and local irregularities of the bottom, are typical at a depth of about 10–30 meters. One of these benthic accumulations is situated near the White Sea Biological Station of the Moscow State University (Figure 1) in the strait between Veliky Island and Cape Kindo (Figure 4).

Table 1. Macrofauna associated with kelp in the White Sea. Frequencies of species in the samples: + = rare (less then 5% of samples), ++ = frequent (5 to 50% of samples), +++ dominant (> 50%).

	Living kelp (depth 0–12 m)	Detached kelp Open shoreline (depth 60–90 m)	Detached kelp Shallow fiord/basin (depth 12–15 m)
PORIFERA			
Amphoriscus kukentali	+		
Haliclona gracilis	+		
Halichondria sitiens	+		
Polymastia mammillaris	+		
Suberites domuncula	++		+
Sycon sp.	+		
CNIDARIA/Coelenterata			
Dynamaena pumila	+++		+
Lucernaria quadridentata	+++		+
Obelia longissima	+++		+
TURBELARIA			
Notoplana atomata	+++		+++
NEMERTINI			
Lineus sp.			+
PRIAPULOIDEA			
Priapulus caudatus			+
ANNELIDA			
Ampharete sp.		+	
Anobothrus gracilis	+		
Autolytus prismaticus		++	
Arenicola marina	+		
Aricidea nolani	+		
Brada granulata	++		
Bylgides sp.		++	
Capitella capitata			+++
Chaetosone setosa	+	+	
Cirratulus sp.	++		
Eteone longa	+		+
Euchone analis		+	
Eulalia viridis			+
Flabelligera affinis	+++		+
Glycera capitata	+		
Harmothoe imbricata	+++		+++
Laena abranchiata		+	
Lepidonotus squamatus			+
Myriochele oculata		+	
Neoamphitrite figulus	+++		
Nereimyra punctata	+++		+++
Nereis pelagica			+
Nicomache minor	+		

Table 1. Continued

	Living kelp (depth 0–12 m)	Detached kelp Open shoreline (depth 60–90 m)	Detached kelp Shallow fiord/basin (depth 12–15 m)
Nicomahce lumbricalis		+	
Ophelia limacina	+		
Pectinaria hyperborea		+	
Pectinaria koreni	+		+++
Pholoe minuta		++	
Phyllodoce maculata	+++		
Phyllodoce groenlandica			+++
Phyllodoce mucosa	+		
Polycirrus medusa	+	+	
Polydora quadrilobata	++		
Potamilla reniformis	+++		+
Pterocirrus finmarchica			+
Pterosyllis finmarchica	+++		
Spio cf. *filicornis*	+		
Spirorbis spirillum	+		
Syllis sp.		+	
Travisia forbesii	+		
Tubificoides benedeni	+++		
MOLLUSCA			
Admete couthouji	+		
Ariadnaria borealis	+		
Astarte crenata		+	
Ciliatocardium ciliatum	+		
Coryphella rufibranchialis			++
Coryphella verrucosa	+++		
Cryptonatica clausa	+++		
Cylichna alba	++		
Dendronotus arborescens	+		
Dendronotus robustus			+
Elliptica elliptica		+	
Epheria vincta	+++		+++
Hyatella arctica		+	
Leda pernula		+	
Lepeta caeca		+	
Littorina littorea	+++		
Macoma balthica			++
Margarites groenlandicus			+
Margarites helicinus	+		+
Musculus discors	++		+++
Mytilus edulis	+++		++
Oenopota harpularioides	+		
Portlandia arctica		+	
Testudinalia tesselata	+++		+
Tonicella marmorea		+	
Velutina velutina	+		

Table 1. Continued

	Living kelp (depth 0–12 m)	Detached kelp Open shoreline (depth 60–90 m)	Detached kelp Shallow fiord/basin (depth 12–15 m)
CRUSTACEA			
Acanthostepheia malmgteni		+++	
Amphithoe rubricata	+		+
Amphithopsis longicaudata	+		
Anonyx nugax	+		
Apherusa bispinosa			+
Apherusa tridentata			+
Caprella septentrionalis			++
Corophium bonelli	+++		+++
Diastylis glabra	+		
Eualus gaimardi	+	+	+
Gammarellus homari			+
Gammarus oceanicus			+++
Haligradia fulvocinatus	+++		
Hyas araneus	+		+++
Ischyrocerus anguipes	+		
Lamprops fuscata	+		
Lysianassidae gen. sp.	+		
Monoculoides borealis	++		
Munna fabricii	+++		+
Mysis oculata	+		
Nebalia bipes			+++
Oediceridae g. spp.		+	
Orchomenella minuta	+		
Paroediceros sp.		+	
Paroediceros lynceus	++		
Pleustes panoplus			+++
Pleustidae g.sp.		+	
Praunus inermis			+
Rhachotropis aculeata		+++	
Sabinea sp.		+	
Sclerocrangon boreas		+++	
Spirontocaris turgida	+		
Unciola planipes			+
PANTOPODA			
Nymphon mixtum	+		
Phoxichilidium femoratum	+++		+++
Pseudopolene spinipes			+
ECHINODERMATA			
Asterias rubens	+++		+++
Ophiacanta bidentata		+	
Ophiopholis aculeata			+
Ophiura robusta	+	+++	+++
Stegophiura nodosa	+		+
Strongylocentrotus pallidus		+++	
Urasterias lincki		+++	
BRACHIOPODA			
Rhynchonella psittacea		+	

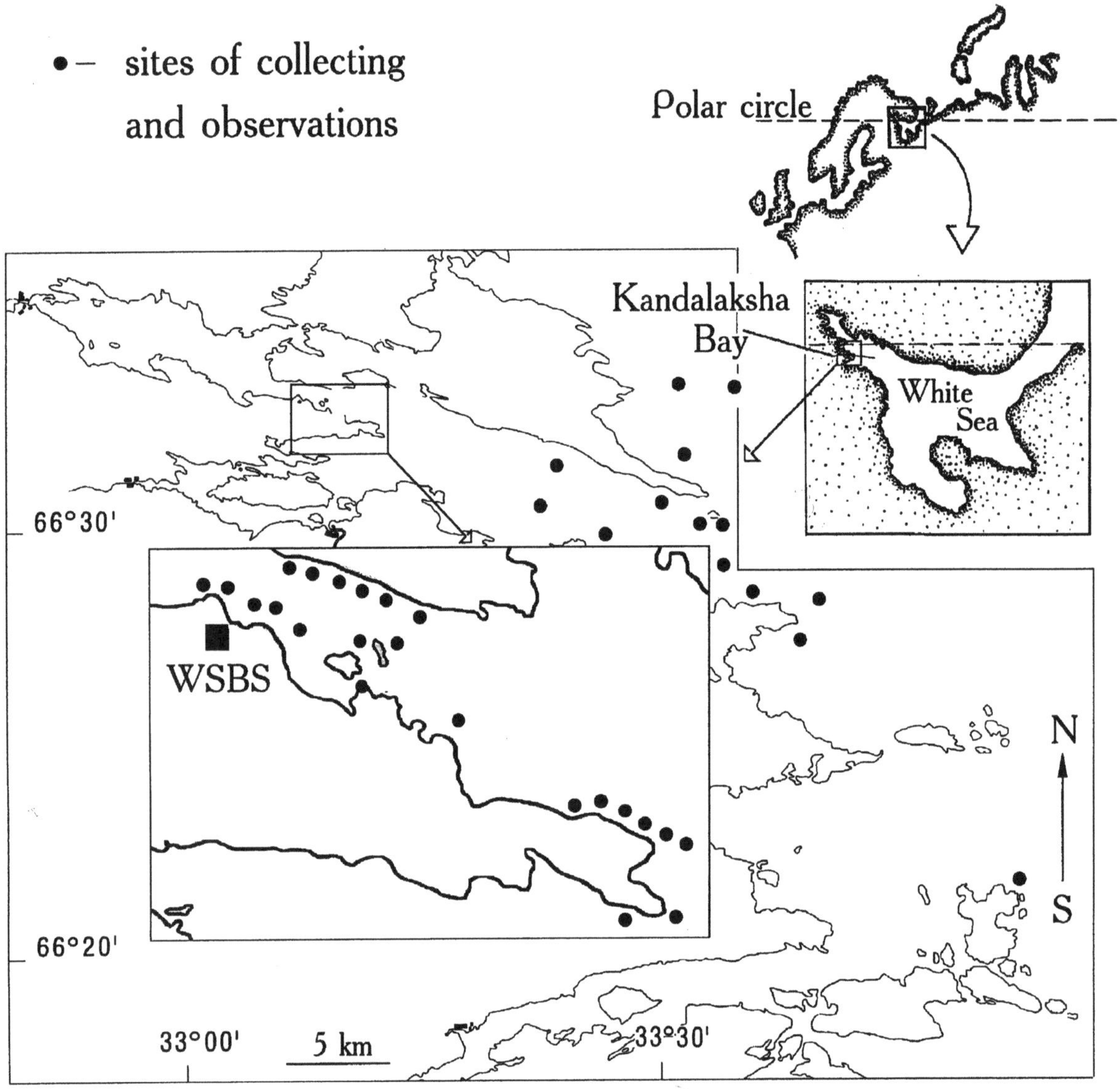

Figure 1. Sampling stations in the Kandalaksha Bay. WSBS: The White Sea Biological Station of Moscow State University.

According to our SCUBA underwater observations this accumulation has existed for at least 20 years. The accumulation (about 100 m in length, 25 m in width, and 2 m thick) is situated on the slope of the strait at a depth of 12 to 15 m, just below the zone of living kelp (Figures 4, 5). The average wet biomass of kelp in the accumulation is 90 kg m^{-2}; the total biomass of the invertebrates in the accumulations is 401,5 g m^{-2}.

Measurements of currents in the area of the accumulation showed the presence of local circulations of tidal currents and constant water upflow from greater depths (Figure 5). The accumulation appears to be formed by these currents.

The accumulation of detached kelp consisted of three layers without distinctive borders (Figure 6). The upper layer, formed by fresh and in most cases living fronds, was 1 to 1.5 m thick and well aerated (Eh = +45). The middle layer (about 20 cm thick) was anoxic (Eh = −131 to −345) and consisted of small pieces of fronds covered by layers of cyanobacteria. Seventeen species of cyanobacteria were found in this layer (Table 2). The lower layer was anoxic (Eh = −345 to −360) with a strong smell of hydrogensulfide. This layer was black with semi-liquid organic matter and remains of kelp stalks. Its thickness was difficult to define, but in all cases it was more than 0.7 m. Coverage by cyanobacteria was not found in the lower layer.

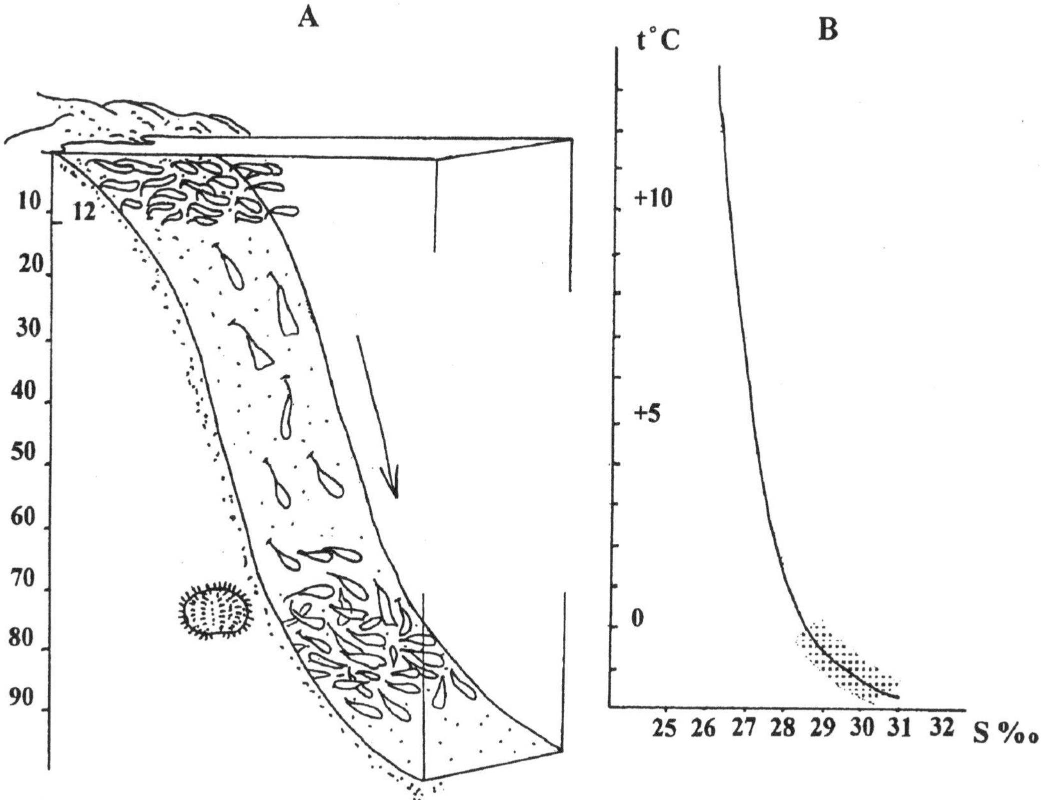

Figure 2. Scheme of the open shoreline in the White Sea. A. Localization of communities, associated with (detached) kelp on the slope. B. Average position of near bottom TS curves during the summer (after Pantulin, 1990). The dotted area is the zone inhabited by *Strongylocentrotus pallidus*.

Mineral particles were practically absent in this layer. Mats of sulfur bacteria were found neither in the middle nor in the lower layers of the accumulation.

Form and borders of the weed accumulation were different every day because its border lines could move several meters, and fronds of *Laminaria* and other large brown algae in the upper aerated layer were constantly mixed. The intensity of these processes is so high, that several attempts to estimate the rate of new algae sedimentation failed. For this estimation, part of the accumulation (about 20 m^2) was covered by nylon net, but moving algae carried the net away together with anchors inside it.

In the accumulation 49 macroinvertebrate species were found (Table 1) Several of them were typical inhabitants of living and attached kelp communities (Table 1, Figure 3), probably transported to the accumulation together with the substrata. Detritivorous species, such as *Ophiura robusta* and *Gammarus oceanicus*, were abundant. The surface of fronds, covered by cyanobacteria in the middle layer, was mainly inhabited by mobile fauna, such as the amphipods *Gammarus oceanicus* and *Pleustes panoplus*. Crustaceans in this zone were also covered by layers of cyanobacteria. Crustaceans, covered by layers of cyanobacteria, could be found not only in the midlayer, but also in the upper one. At the border of the upper and lower layer numerous crabs (*Hyas araneus)* were present.

Gastropods, polychaetes and *Ophiura robusta* were practically absent in the middle layer of the accumulation. Species, typical for organic-rich substrata, such as *Capitella capitata, Ophryotrocha irinae* and *Nebalia bipes*, inhabited the periphery of the accumulation and did not enter the mass of kelp. Information on the distribution of meiofauna through the accumulation was limited, but the middle layer was inhabited by a single species of nematodes *Monchystea* cf. *disjuncta*. Neither macrofaunal nor meiofaunal animals were found in the lower layer.

Discussion

In the shallow basins of the White Sea, the decomposition of brown algae in the sublittoral proceeded without

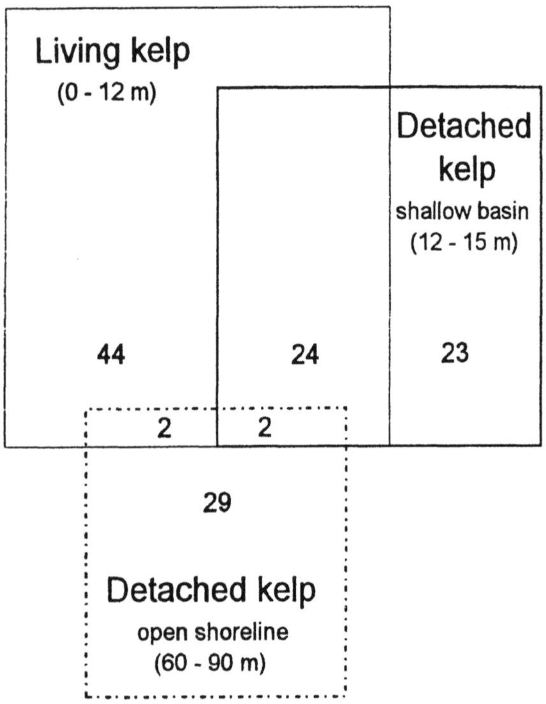

Figure 3. Scheme on the number of species in each of the different habitats. The surface of the squares is proportional to the number of species occurring in one or more of the habitats.

Table 2. Cyanobacteria found on the surface of fronds in the middle layer of the accumulation of detached kelp at a depth of 12 to 15 m in the shallow water strait near the White Sea Biological Sation of the Moscow State University.

Calothrix sp.
Lingbya nordgaardi Wille
L. holdenii Forti
L. infisa Fremy
L. perelegans Lemmermann
Oscillatoria pulchra Lindstrom
O. subuliformis Kutzing
Phoemidium crossbyanum Tilden
Ph. ectocarpi Gomont
Ph. endoliticum Ercegegovic
Ph. hoemoides Setchell et Gardner
Ph. laysanense Lemmermann
Ph. minimum Lindstrom
Ph. mycoudeum Fremy
Ph. papyraceum (Agardh) Gomont
Ph. subsalsum Gomont
Symphoca hydneides Kutzing

sea urchins and no other macrofaunal form occupied their ecological role of mechanical breakdown of the plant substratum, not even in the large accumulations of detached kelp. Mats of sulfur bacteria, typical for accumulations of detached brown macroalgae in waters with low salinity and absence of sea urchins (Norkko & Bonsdorf, 1996), were absent in our study area. Yet, in the middle anoxic layer of the accumulations, the fronds of detached kelp were covered with well developed layers of cyanobacteria, as has not been described before. Amphipods, covered with layers of cyanobacteria were found in the middle layer, in the upper layer and at the periphery of the accumulation. Not determined is whether these crustaceans migrated from one layer to another, or whether this is a result of general disturbance of the fronds.

Few species appear to be specific inhabitants of the shallow subtidal weed accumulations. The polychaete *Ophryotrocha irinae* (Dorvilleidae) normally inhabits bacterial mats under stones in the low tidal zone (Tzetlin, 1980). The first biotope where this species is now found in the subtidal zone is in the accumulation of kelp. The leptostracan *Nebalia bipes* is a rather rare species in the bottom communities of the White Sea (Golikov et al., 1985a, b) and seems also to be associated here with subtidal weed beds. The nematode *Mochistera* cf.*disjuncta* belongs to the small group that inhabits organic-rich substrata in tidal and subtidal areas (Riemann, 1995). They are also found in gelatinous phytodetritus aggregations, which follow the phytoplanktonic spring bloom at depths of 4000 m in the mid-oceanic region of the north-eastern Atlantic (Riemann, 1995).

Along the open rocky shoreline, communities, associated with dead detached kelp, are situated at a depth of 60–90 m, being 40–50 meters below the belt of living kelp. During the passage through the zone with sea urchins all plant debris becomes fragmentated. In the deep basin of the White sea only detritus particles were found, and no pieces of algae (Nevessky et al., 1977). Previously, Golikov et al. (1985a, b) found high densities of *S. pallidus* along the south-east part of the Karelian Coast and in the open part of Onega Bay at the same depth of 60–90 m. So, along the open shoreline, the zone of the benthic community, situated at a depth of 60–90 m where sea urchins were dominant, appears to be the zone of decomposition of detached kelp. It is difficult to estimate how many species, which were found in the samples from this zone, are really associated with the detached kelp communities. The collected data are based only on samples taken by trawl. Most of

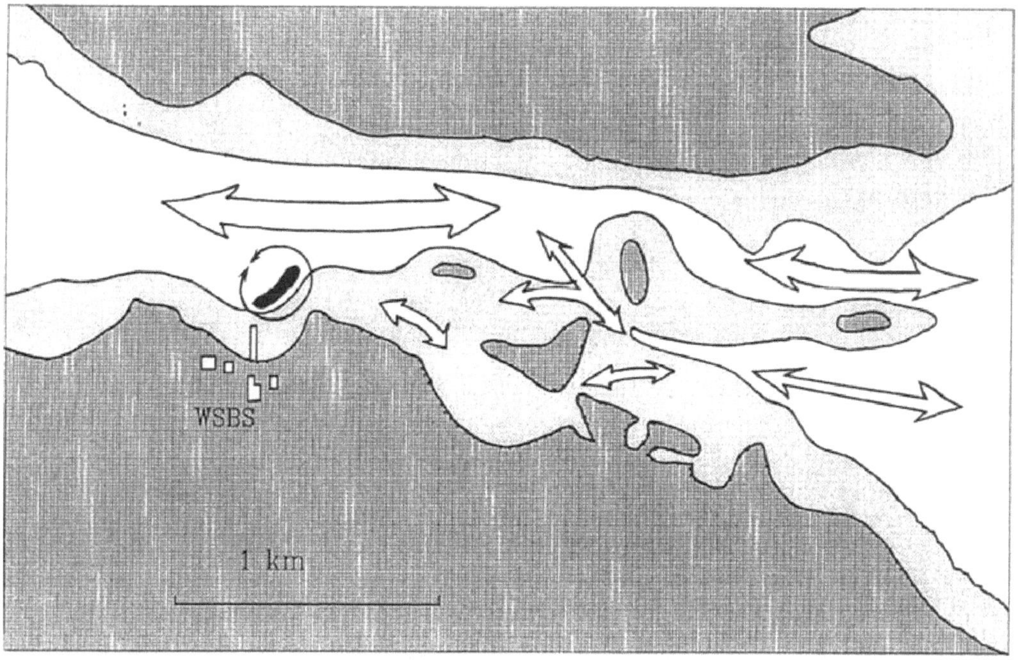

Figure 4. Position of the studied accumulation of detached kelp near the White Sea Biological Station of Moscow State University. The dark dotted area is the coast of Cape Kindo, Island Veliky, and some little islands; the light dotted area is the zone occupied by living kelp, the black area is the position of the accumulation of detached kelp. Thick arrows indicate the main streams of tidal currents in the strait between Cape Kindo and Island Veliky, small arrows indicate local circulation.

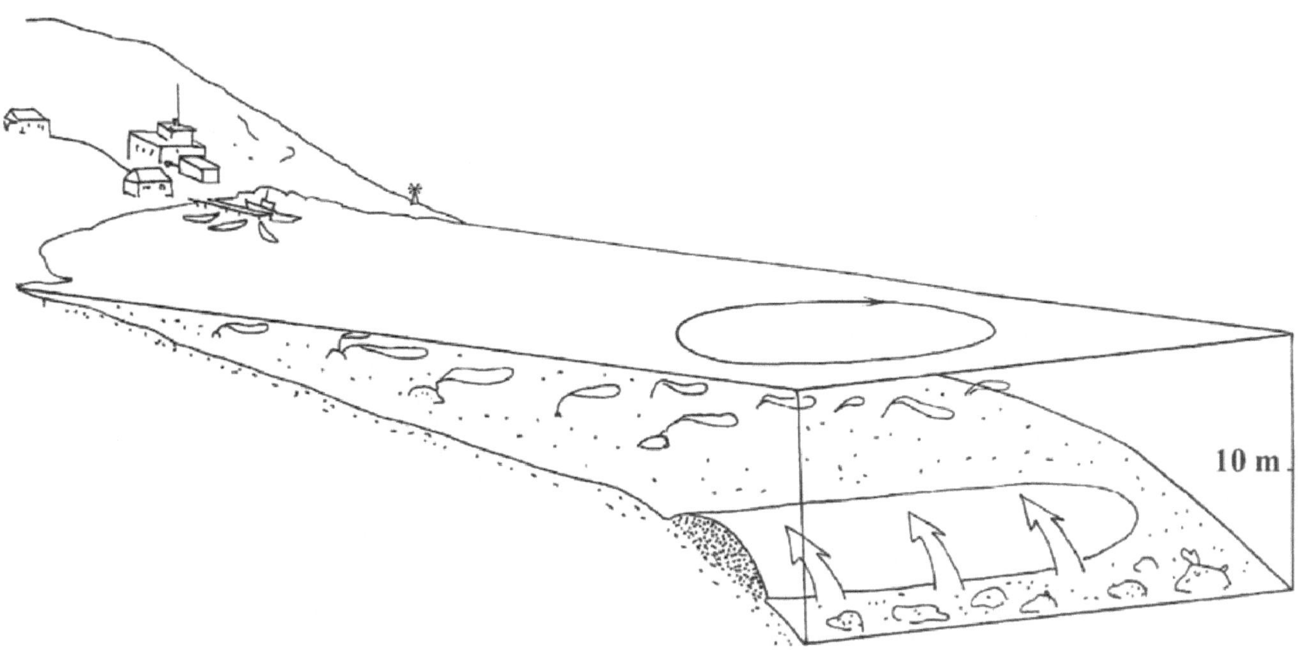

Figure 5. General appearance of the benthic accumulation of detached kelp. Only half of the accumulation is shown. Thick arrows show the water flow coming up from large depths; small arrows indicate the position of the local circulation of tidal currents.

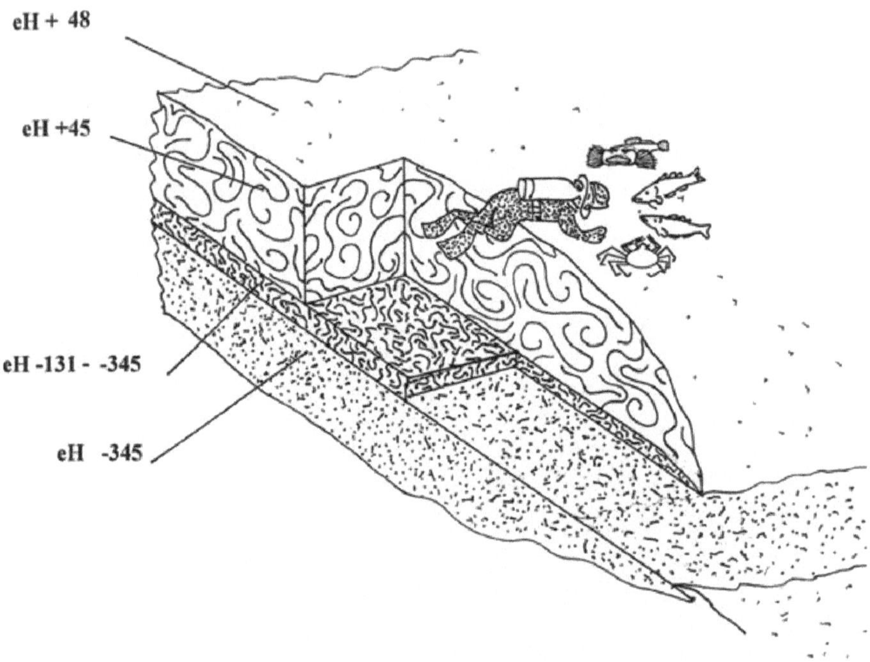

Figure 6. Scheme for the structure of the benthic accumulation of detached kelp (see text for further explanation).

the species found are inhabitants not only of detached kelp but also of different biotopes at the slope, and non of the species, except some eurybiotic species, inhabit the belt of living kelp.

Acknowledgments

The authors are very grateful to Dr Anatoly Pantulin for help with measurements of tidal currents, Greg Kolbasov and Dmitry Zhadan for participation in sampling, and Phyllip Sapognikov for identifying cyanobacteria. Special thanks to Dr Herman Hummel and three anonymous reviewers for fruitful discussion, comments on and corrections of the English version of the manuscript. The study was supported by the State Program in Biodiversity of the Russian Federation (grant N 2-1-93) and the Russian Fund for Basic Researsches (grant N 95-04012737).

References

Bazhin, A. G., 1995. Taxonomy, ecology and distribution of sea urchins of the genus *Strongylocentrotus* in Russian Seas. Ph.D. Thesis, Petropavlovsk, Kamchatsky, 125 pp. (in Russian).

Bedford, A. P. & P. G. Moore, 1984. Macrofaunal involvement in the sublittoral decay of kelp debris: the detritivore community and species interactions. Estuar. coast. Shelf Sci. 18: 97–111.

Breen, P. A. & K. H. Mann, 1976. Destructive grazing of kelp by sea urshins in Eastern Canada. J. Fish. Res. Bd Can. 33: 1278–1283.

Derugin, K. M., 1928. Fauna of the White Sea and condition of its existence. Isseldovania Morey 7–8, 511 pp. (in Russian).

Golikov, A. N., O. A. Skarlato, V. V. Galtsova & T. V. Menshutkina, 1985a. Ecosystems of the Chupa Bay of the White Sea and their seasonal dynamics. Isseldovania Fauny Morey 31: 5–83 (in Russian).

Golikov, A. N., I. A. Babkov, A. A. Golikov & O. K. Novikov, 1985b. Ecosystems of Onega Bay of the White Sea. Isseldovania Fauny Morey 33: 20–87 (in Russian).

Ivanjushina, E. A, A. V. Rzawsky, O. N. Selivanova & V. V. Oshurkov, 1991. Structure and distribution of benthic communities of the shallow water zones of the Commander Islands. In Sokolov, V. E. (ed.), Natural Resources of the Commandor Islands. Moscow State University, Moscow: 155–170 (in Russian).

Mann, K. H., 1972. Macrophyte production and detritus food chains in coastal waters. Mem. Ist. ital. Idrobiol. 29: 353–383.

Nevessky, E. N., V. S. Medvedev & V. V. Kalinenko, 1977. The White Sea. Sedimentogenesis and history during Holocene. Nauka, Moscow, 236 pp. (in Russian).

Norkko, A. & E. Bonsdorff, 1996. Rapid zoobenthic community responses to accumulations of drifting algae. Mar. Ecol. Prog. Ser. 131: 143–157.

Pantulin, A. N., 1990. Formation and changeability of water structure in the White Sea. In Matekin, P. V. (ed.), Biological Resources of the White Sea. Moscow University Press, Moscow: 7: 9–16.

Riemann, F., 1995. The deep-sea nematode *Thalassomonhystera bathislandica* sp. nov. and microhabitats of nematodes in flocculent surface sediments. J. mar. biol. Ass. U.K. 75: 715–724.

Tzetlin, A. B., 1980. Two new species of the fam. Dorvilleidae from the White and the Barents Seas. Zoologichesky Journal 59: 1817–1822 (in Russian).

Vozhinskaya, V. B., 1986. Bottom Macrophytes of the White Sea. Nauka, Moscow, 191 pp.

Hydrobiologia **355**: 101–113, 1997.

A. D. Naumov, H. Hummel, A. A. Sukhotin & J. S. Ryland (eds), Interactions and Adaptation Strategies of Marine Organisms.

Soft-bottom macro invertebrate fauna of North Norwegian coastal waters with particular reference to sill-basins. Part one: Bottom topography and species diversity

Lars-Henrik Larsen
Akvaplan-niva, N-9005 Tromsø, Norway

Key words: Northern Norway, sill-fjords, soft bottom benthos, diversity

Abstract

The soft bottom macrofauna of three fjords in Northern Norway is presented and compared. One fjord is open, with gradually decreasing water depth from mouth to head, and two fjords possess shallow sills, splitting the fjords into two and three basins respectively. The most southerly samples were collected from Rombaken in the innermost part of the Vestfjord, at approximately 68° 23′ N; 17°32 ′ E, close to the city of Narvik. The northernmost samples were collected from Akkarfjord at 70° 46′ N; 23°25 ′ E, near North Cape, Europe's northernmost tip.

The ten numerically most abundant taxa at each station are listed, together with a description of bottom topography, sediment composition and the extent of anthropogenic impact on the fjord. Data are presented on species diversity related to water depth, sedimentary TOC content and presence/absence of shallow sill(s). The dominant faunal groups present are the Polychaeta and Mollusca. In sill basins echinoderms are absent or poorly presented. Species composition generally appears to be most influenced by bottom topography and sediment composition. Waste discharge appears to have only local effects on the faunal assemblages investigated. The TOC content of the bottom sediments tends to increase from exposed, outer parts to the landward parts near the head of the fjord basins. Faunal diversity decreased from open areas (mouth of fjords) towards the sill basins.

Introduction

The North Norwegian coastline is characterised by large and deep fjords, often penetrating hundreds of km inland. Rivers drain into the inner parts of most fjords and, particularly in summer, a temperature and salinity gradient is observed in the upper water masses from open coastal areas to the innermost parts of the fjords. Many North Norwegian fjords have one or more shallow sills, which reduce horizontal water exchange along the longitudinal axis of the fjord. Thus, variations in bottom topography and water depth at the sill (sill depth) between fjords lead to very different water exchange conditions from one fjord to another. It has been shown that sill depth is an important feature regulating the hydrographic conditions of fjords in Western and Central Norway (Aure & Stigebrandt, 1989).

Total or partial renewal of the water in North Norwegian sill basins normally takes place during autumnal/winter inflows of denser coastal water, when passage over the sills is facilitated by the weakening of the stratification as a result of the cooling of the surface waters and the reduced fresh water inflow from the rivers (Eilertsen et al., 1981). However, in some fjords the basin water (the water mass below the sill depth on the landward side of the sill) may not be renewed every year. Over such periods, basin water can potentially become hypoxic or even anoxic due to oxygen-consuming degradation of organic material, but it has been suggested that sub-pycnocline hypoxia is unlikely to occur naturally in Northern Norway (Oug, 1988), and that total renewal of basin water and vertical mixing normally takes place all the way down to the bottom every year. The validity of this suggestion is discussed further in the article.

As decomposition of organic material proceeds in the basin, the oxygen content of the bottom water is reduced during the stagnant period. Minimum oxygen levels are often recorded in August or September. A puzzling observation, made in several fjords, is the presence of a high level of organic material in the bottom sediment, and at the same time a rich benthic fauna and high level of oxygen in the bottom water (present work and own unpublished observations). The present article will introduce the benthic macrofauna of two different fjord environments, with and without sill(s), illustrated by data from three fjords. Further, the diversity of the soft bottom fauna at 54 stations from 14 fjords is related to topographic features in order to describe the general effects of bottom topography on the faunal diversity.

Materials and methods

Survey area

The coastal waters of northern Norway are influenced by the Norwegian coastal current, which flows in a north-easterly direction along the Norwegian west coast. The salinity of the coastal waters varies between 33 and 34.5‰ and the surface water temperature varies from approx. 3 °C in winter (March) to 10 to 12 °C in July/August. The continuous inflow of water from lower latitudes is responsible for all-year ice free conditions in the outer coastal waters of Northern Norway.

The location of the investigated fjords is shown in Figure 1. Faunal data are presented from Akkarfjord with no sills, Rombaken with one shallow sill and Kvænangen with two shallow sills. The longitudinal section of bottom topography in each fjord is obtained from the official sea charts.

Akkarfjord is a relatively small (4 km long) fjord oriented along a northwest-southeast axis with the mouth facing southeast with a sill-free connection towards Sørøysund and the Barents Sea (Figure 2). The bottom topography is relatively flat, with only a few m difference between the innermost parts and the mouth. At the time of the survey (June 1990), two fish farms producing Atlantic salmon were located at the southern bank of the fjord, south of station 2. The fish farms had been in operation for eight and five years respectively. Sludge from a fish processing plant (1000 ton yr^{-1}) had been dumped in the fjord west of station 1 for nearly 50 years.

Rombaken with Rombaksbotn is the inner part of the more than 200 km long Vestfjord/Ofotfjord complex (Figure 3). This very deep fjord (345 m) is located north of the city of Narvik and is approximately 25 km long. The eastern part, Rombaksbotn, is separated from Rombaken by an 18 m deep sill. The basin depth in Rombaksbotn is 113 m. Rombaken receives approx. 12 000 person equivalents (pe.) of municipal waste from the city of Narvik, and approx. 3000 pe. of seepage from a municipal rubbish dump. Both discharges are located on the southern bank of the fjord, at approx. 25 m depth. In December 1990, four stations (G1–G4 in Figure 3) were sampled at intermediate depths (18–35 m) along the southern bank of the fjord. In July 1992, four stations were sampled along the central longitudinal axis of the fjord (stations 4, 5, 6 and 13 in Figure 3).

Kvænangen is a more than 80 km long fjord with several side fjords and branches (Figure 4). Kvænangen is oriented along a northwest-southeast axis. Two shallow sills (7 m and 3 m) separate the fjord into three distinct basins and, furthermore, a relatively deep sill (160 m) separates the fjord from the open sea. Kvænangen has a maximum depth of 200 m, and the two basins have maximal depths of 108 m and 56 m, respectively. In September 1990, four stations were sampled in the deepest parts of the three fjord basins in Kvænangen (Figure 4). The area around Kvænangen is very sparsely populated, and only insignificant amounts of sewage are discharged to the fjord. The catchment area of Inner Kvænangen is 800 km^2 and the precipitation from 40% of this area is conveyed to the fjord through a hydroelectric plant. Inner Kvænangen is normally ice-covered from December to mid-May. The release of fresh water through the power plant during winter increases the duration of the ice cover, and increases the thickness of the ice.

Sediment samples

Samples of surface sediment (0–1 cm) for analysis of total organic carbon (TOC) and sediment grain size were collected using a 0.1 m^2 lead weighted Van Veen grab with hinged, lockable rubber covered inspection flaps of 0.5 mm mesh size. From one grab sample, the top cm of the sediment was sampled. Grain size was determined as fraction of coarse (i.e. >0.063 mm) and fine (i.e. <0.063 mm) particles by wet sieving and weighing each fraction. TOC was determined by two different methods. The samples from Rombaken (1992) were analysed by a method where a dry sample was burnt in oxygen saturated helium gas at approx. 1800 °C. The amount of organic carbon in the initial sample was calculated from the amounts of N_2 and

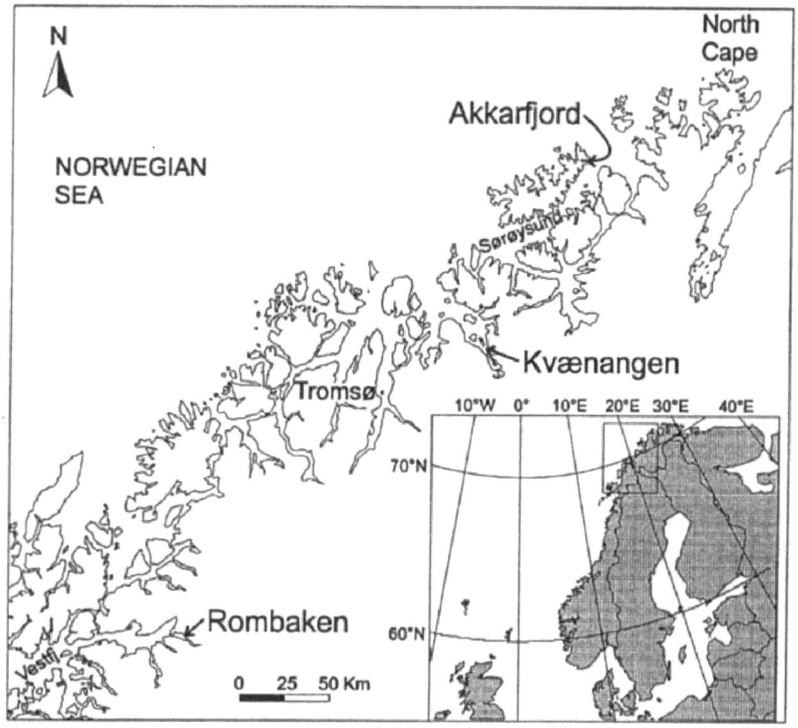

Figure 1. Location of Akkarfjord, Kvænangen and Rombaken, Northern Norway. In addition, material from Beisfjord (1992), Herjangen (1992), Gratangen (1989), Selfjord (1990), Kaldfjord (1990), Ramfjord (1994), Sørfjord (1994), Tverfjord (1989), Stjernsund (1993), Altafjord (1994) Rypefjord (1995) and Forsøl (1995), all located within the indicated area, is included in the delta depth – diversity regression (Figure 5).

CO_2 produced. The sediment samples from Rombaken (1990), Kvænangen and Akkarfjord were analysed by a Leco IR carbon analysator, which measures the amount of CO_2 produced after burning the HCl washed sample at 480 °C.

Fauna samples

The fauna samples were collected using a 0.1 m^2 lead weighted Van Veen grab with hinged, lockable rubber covered inspection flaps of 0.5 mm mesh size. At most stations four replicates were taken (Table 2). All fauna samples were sieved through a 1 mm round hole screen. Material larger than one mm was then preserved in 4% formaldehyde solution, stained with rose bengal. Upon return, the animals were sorted out from the remaining detritus, identified and counted.

The ten numerically most abundant (top-ten) benthic macroinvertebrate taxa are presented from each station. The density of animals per m^2 was calculated for each station. Only benthic, individually living, macrofauna are included, leaving out groups like Foraminifera, Bryozoa and colonial cnidarians, together with copepods, euphausiids and fish. The macrofauna has been identified to the lowest possible taxon, preferably species. For each station, the total number of individuals and taxa were recorded, and the Shannon-Wiener diversity index H' (Shannon & Weaver, 1949) and ES_{100} (Hurlbert, 1971) are presented.

The faunal diversity differences are illustrated by plotting the Shannon-Wiener index vs. delta depth (the difference between the sampling depth at the station and the deepest passage from the station towards the open ocean (Buhl-Jensen, 1986)), this being an index of vertical openness between a station sampled and the open sea.

A clustering analysis is used to illustrate the differences in faunal composition among the stations.

Results and discussion

Investigated fjords

Akkarfjord

The sediment at the three sampling stations consisted of coarse sand, with low TOC content (Table 1). The benthic macrofauna communities were rich in species and

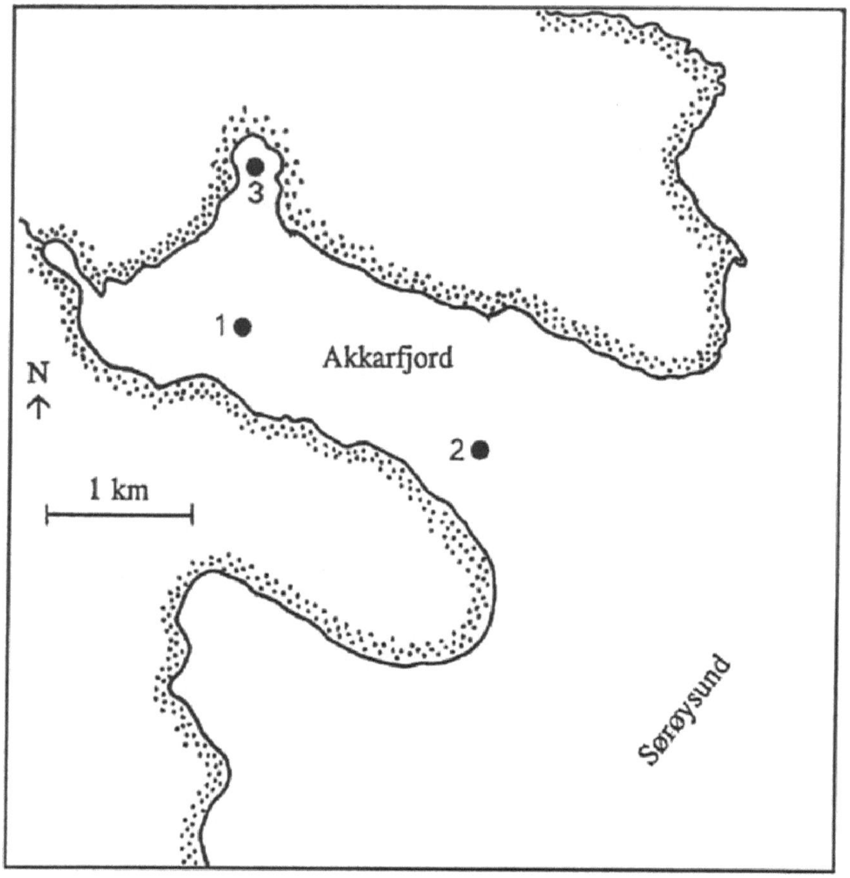

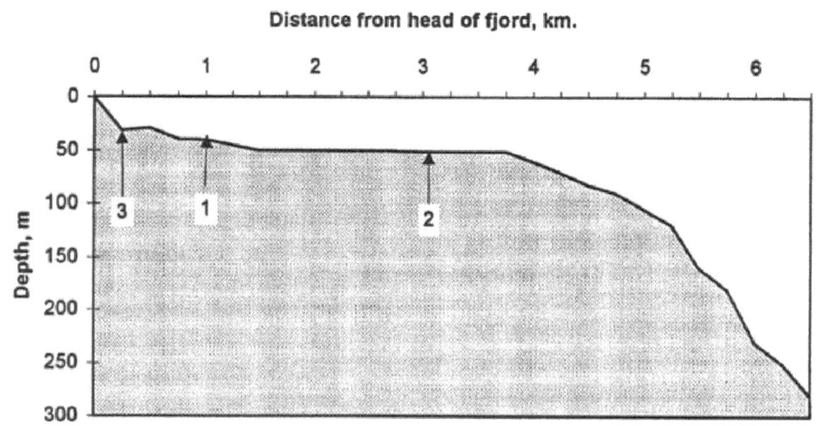

Figure 2. Location and bottom topography along the central axis of Akkarfjord with the three sampling stations indicated.

individuals. The diversity was high at all three stations (Table 2) and the values were well above the lower limit for normal diversity (H'>3.1) according to Norwegian environmental quality criteria (Rygg & Thélin, 1993). Twenty-one different taxa were recorded among the top-ten at the stations in Akkarfjord (Table 3). Eleven taxa belonged to the annelids (all polychaetes), together with 3 crustacean taxa, 3 molluscs, 2 echinoderms, 1 sipunculid and 1 cnidarian.

The most abundant species at station 1 was the burrowing tanaid crustacean *Apseudes spinosus* M. Sars (Table 3). However, this species was not among the ten

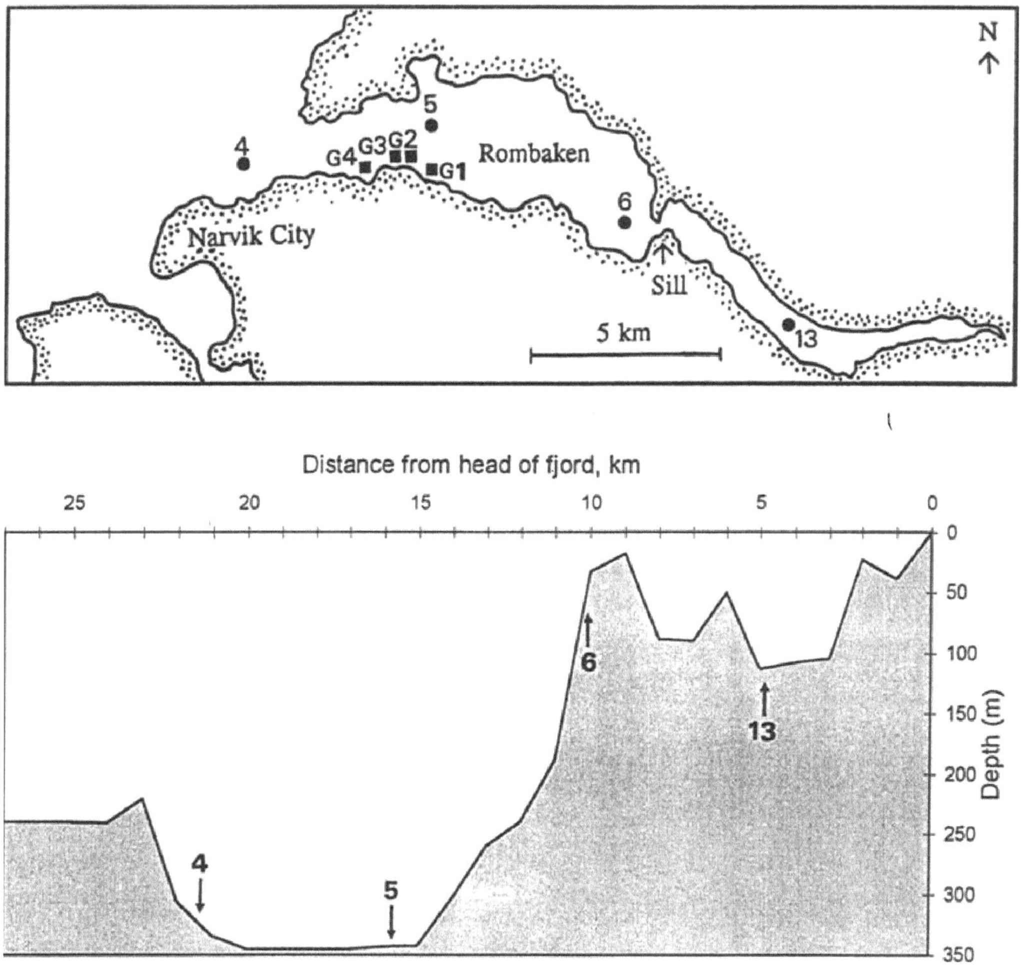

Figure 3. Location and bottom topography along the central axis of Rombaken with the eight sampling stations and the sill indicated. Stations sampled in December 1990 are marked with black squares (G1–G4), while stations sampled in July 1992 are marked with black circles.

most abundant taxa at station 2, and was not recorded at station 3 at all. The most numerically abundant species at station 2 was the polychaete *Onuphis conchylega* M. Sars, which also was recorded among the ten most abundant taxa at station 1, but was not found at station 3 at all. Another polychaete, the sub surface deposit feeding *Scoloplos armiger* O. F. Müller was recorded in highest numbers at station 3, while still another polychaete, the deposit feeding *Chaetozone setosa* Malmgren was the only species recorded among the ten dominant taxa at all the three stations in Akkarfjord (Table 3).

Rombaken and Rombaksbotn

In the 1990 survey, the station G1 was located less than 20 meters from the discharge point for the seepage from the municipal rubbish dump, but the sediment composition and sediment characteristics differed only slightly from the other three stations of that survey (Table 1), located with increasing distance from the outlet. Moreover, compared to the stations of the 1992 survey, the TOC content at G1, G2 and G3 is at the same level (Table 1).

The bottom sediment at the deeper stations (4 and 5 of the 1992 survey) in Rombaken was finer, and had a higher TOC content compared to the intermediate depth station 6 of the same survey, while station 13 had the highest TOC content and the finest sediment of the 1992 stations.

The stations sampled in 1990 had high faunal diversity, even the station G1 located closest to seepage outlet. This station had, however, the lowest diversity of the four stations sampled in 1990 (Table 2), but was still well above the lower limit for normal diversity

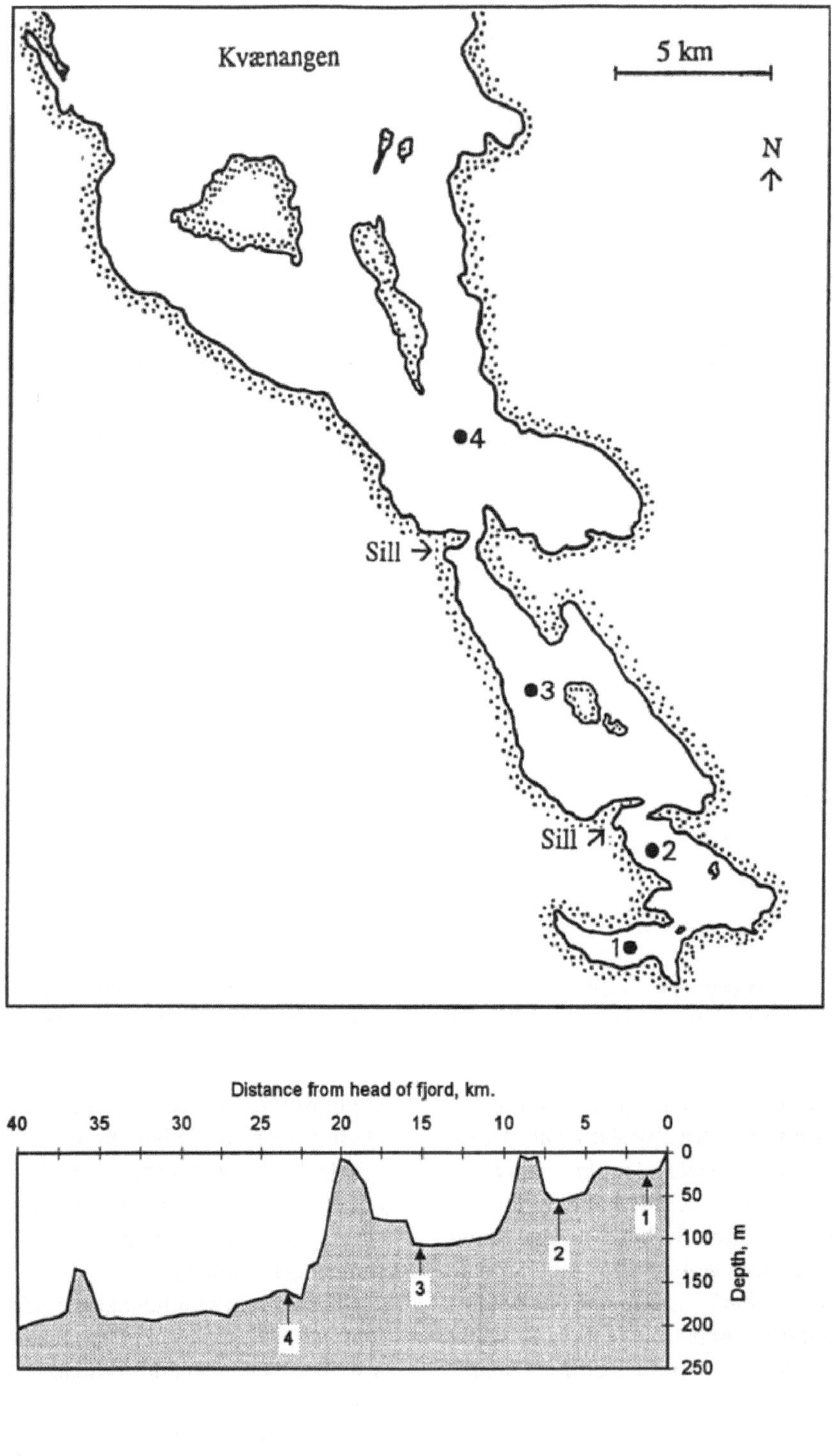

Figure 4. Location and bottom topography along the central axis of Kvænangen with the four sampling stations and the sills indicated.

(H'>3.1) according to Rygg & Thélin (*op cit.*). Of the two deep water stations in the 1992 survey, which had lower diversity compared to the intermediate depth stations, and the station in Rombaksbotn (13) only the latter fell below the quality criteria defined in Rygg & Thélin (*op cit.*). Station 13 had the lowest H' and ES_{100}, and at the same time the highest number of individuals per m^2 of all stations in Rombaken (Table 2).

Table 1. Description of the bottom sediments at the three stations in Akkarfjord, June 1990, the eight stations in Rombaken (December 1990 and July 1992) and the four stations in Kvænangen September 1990.

Station	Water depth (m)	Total Organic Carbon (%)	Pelite (%) *	Sand (%) **	Sediment description
Akkarfjord 1	40	0.8	11	74	Coarse sand with some pebbles and frag-ments of bivalve shells and chalk incrusted redalgae *Lithotamnion* sp. No smell
Akkarfjord 2	56	0.6	5	94	as above
Akkarfjord 3	32	0.5	7	91	as above
Rombaken G1	35	0.8	9	87	Grey sand with fragments of shells covered by a 2-3 mm brown surface layer. No smell
Rombaken G2	18	0.5	n.d.	n.d.	Muddy sand without smell
Rombaken G3	28	0.9	18	82	Muddy sand with shell fragments, no smell
Rombaken G4	25	1.2	9	52	Muddy sand with a lot of shell fragments. No smell
Rombaken 4	311	1.1	63	34	One cm of light brown surface layer upon blue-grey clay. No smell
Rombaken 5	350	1.1	77	23	One cm of light brown surface layer upon blue-grey clay. No smell
Rombaken 6	37	0.3	28	64	Grey sand/stones and fragments of bivalve shells. No smell
Rombaken 13	113	1.7	82	18	Dark brown silt without smell.
Kvænangen 1	24	1.2	82	18	Dark grey silt. Black below 3 cm depth, weak smell of H_2S from deeper parts. Some shell fragments and polychaete tubes on the surface.
Kvænangen 2	56	1.6	78	22	Dark fine silt, deeper than 2 cm black mud. Weak smell of H_2S
Kvænangen 3	108	1.3	52	47	Brown-grey clay-silt. No dark coloration and no smell
Kvænangen 4	103	0.7	38	62	Fine greyish sand with some shell fragments. No smell.

$* = \Sigma$ particles $< 63\ \mu$m.
$** = \Sigma$ particles $> 63\ \mu$m and < 2 mm.
n.d. = not determined.

Station 6 of the 1992 survey had a species diversity comparable to the stations of the 1990 survey (Table 2), which can be explained by comparable water depth and sediment composition (Table 1). At the four stations sampled in Rombaken in 1990, twenty-three different taxa were recorded among the top ten at one or more stations (Table 4), while the number was twenty-six for the stations sampled in 1992. Station 6 of the 1992 survey had 5 'top-ten' list taxa in common with the stations of the 1990 survey, which is not surprising when comparing location, depth and sedimentary conditions (Table 1). The two deep stations of the 1992 survey only had the deposit feeding polychaete *Prionospio cirrifera* Wirén in common with the stations from 1990 and station 6 from 1992. The fauna at station 13 was strongly dominated by the tube dwelling oweniid polychaete *Myriochele oculata* Zachs. This species alone made up 66% of the total number of individuals recorded at this station (Table 2). The fauna at station 13 only had three 'top-ten list' species in common with the deep-water stations on the sea-ward side of the sill (Table 4). The polychaetes are the numerically dominant group in the intermediate depth coarse sediment stations (G1–G4 in 1990 and 6 in 1992), where they made up 15 of 23 taxa in the 1990 survey and 7 of 10 taxa at station 6 in the 1992 survey. In the deeper, fine sediment relatively TOC rich stations on the sea-ward side of the sill (4 and 5, 1992) the polychaetes made up 4 of 12 taxa at the top-ten list, while bivalve molluscs dominated with 6 different bivalve species represented at the top-ten list (Table 4).

Kvænangen

The sediment gradually got coarser when moving from the innermost station (1) to the outermost (4), despite the increasing water depth (Table 1). The three innermost stations in Kvænangen all have H' values around the lower limit for the classification 'good environ-

Table 2. Total number of benthic macrofauna-taxa and individuals recorded per station, together with diversity indices (H' and ES_{100}) at the three stations in Akkarfjord, June 1990, the eight stations in Rombaken December 1990 and July 1992, and the four stations from Kvænangen, September 1990.

Station	Sampled area	Number of taxa	Number of individuals	Number of individuals per 1 m^2*	ES_{100}	H'
Akkarfjord 1	0.4	148	1129	2823	45.6	5.70
Akkarfjord 2	0.4	133	740	1850	51.5	6.04
Akkarfjord 3	0.1	58	200	2000	40.0	5.01
Rombaken G1	0.4	88	967	2418	33.2	4.83
Rombaken G2	0.3	96	670	2231	43.2	5.49
Rombaken G3	0.4	111	1285	3213	42.9	5.56
Rombaken G4	0.4	125	1440	3600	43.1	5.52
Rombaken 4	0.4	83	1315	3288	27.1	4.29
Rombaken 5	0.4	53	1163	2908	23.4	3.98
Rombaken 6	0.4	107	938	2345	41.6	5.40
Rombaken 13	0.4	47	1788	4470	14.0	2.22
Kvænangen 1	0.4	46	606	1515	20.3	3.15
Kvænangen 2	0.4	41	801	2003	16.2	3.04
Kvænangen 3	0.4	85	2258	5645	19.7	3.08
Kvænangen 4	0.4	92	1015	2538	33.5	4.57

* Calculated values

mental quality' according to Rygg & Thélin (*op cit.*) (Table 2). The faunal assemblages were rich in individuals, particularly at station 3, where the polychaete *M. oculata* made up 44% of the more than 5500 individuals per m^2 (Table 2).

Twenty-seven different taxa are recorded among the 'top-ten' at one or more of the stations in Kvænangen. Fourteen taxa were annelids (all polychaetes), while 6 mollusc taxa, 4 crustacean, 2 echinoderm and 1 cnidarian taxa were found (Table 5). The polychaete dominance in the Kvænangen material was most pronounced at station 2 and 3, while relatively few polychaetes (3 taxa) were recorded at station 1 and 4. *M. oculata* was the most abundant species, and together with a species of the bivalve family Thyasiridae, it was the only taxon occurring at all four stations (Table 5).

Importance of human impact

The human impact on the North Norwegian fjords arises mostly from discharges of organic effluents from sewage, aquaculture, dumping of fish processing waste and from fishing and manipulations of the hydrographic regime through operation of hydroelectric power stations. The latter increase the formation and duration of ice-cover on the inner parts of the fjords. The discharges generally affect the organic content of the sediments directly by adding organic material or nutrients. The fishing activities do not exert any direct physical impact to the bottom communities, as the use of trawls is prohibited, and the major impact arises form the manipulation of the biomass of benthic feeding fish like cod (*Gadus morhua* L.), haddock (*Melanogrammus aeglefinus* L.), and different flatfish species. The data in the present article have been collected in the course of environmental monitoring studies, assessing the human impacts on the fjord environments, which is reflected in the location of the stations.

Rombaken and Akkarfjord are the fjords which at the time of the surveys received the largest direct anthropogenic discharges, but still the faunal diversities were high. Even station G1 in Rombaken, located twenty meters from a discharge of 3000 pe seepage had high species diversity, and normal TOC content in the sediment. Station 4 of the 1992 survey is located some 200 m from the discharge of 12 000 pe of sewage, and still maintains a faunal diversity higher than at station 5, located outside the expected range of any discharges. Station 13 in the sill basin Rombaksbotn is located more than 10 km from any major discharges, but has still the lowest species diversity (Table 2). Thus, comparing the stations in Rombaken, the largest influence

Table 3. The ten numerically most abundant taxa at each station in Akkarfjord (June 1990) with number of specimens per m^2.

Taxon	Akkarfjord 1	2	3
Polychaeta:			
Chaetozone setosa	172	80	150
Nereimyra punctata	134	70	
Harmothoe imbricata	100		50
Typosyllis armillaris	70		
Owenia fusiformis		85	
Scoloplos armiger		73	220
Euclymene praetermissa		48	130
Pholoe minuta			170
Spio filicornis			170
Goniada maculata			60
Onuphis conchylega	92	170	
Mollusca:			
Ichnochiton albus	70		
Macoma calcarea			90
Thyasira spp			60
Echinodermata:			
Ophiura robusta	154	48	
Ophiopholis aculeata	160	43	
Crustacea:			
Urothoe elegans		93	
Apseudes spinosus	398		
Ostracoda indet.	68		
Sipunculida:			
Phascolion strombi		68	
Cnidaria:			
Cnidaria indet.			110
Top ten% of total	50.2	42.1	60.5

on the species composition and the diversity of faunal assemblages seems to be exerted by the fjord bottom topography, e.g. presence or absence of a shallow sill, compared to the recorded anthropogenic discharges.

Generally, organic input will act as a fertiliser, stimulating an increase in the populations of benthic macrofauna, but comparing the fauna of station 13 in Rombaksbotn with station 3 in Kvænangen (both behind one sill, water depth 113 and 108 m resp.) receiving almost no anthropogenic discharges, reveals a nearly identical faunal density and diversity and the highest faunal densities among all stations (Table 1 and 2). This indicates that the stimulating effect from organic discharges has less influence on faunal density than the presence of a shallow sill.

Kvænangen receives only insignificant amounts of sewage, and the discharge from a minor aquaculture plant to the inner part of the fjord is the only anthropogenic input of organic material. The two stations in Inner Kvænangen (1 and 2) had low numbers of taxa, but approximately the same number as station 13 in Rombaksbotn (Table 2). At station 3 in Kvænangen, nearly twice as many taxa were recorded compared to the station in Rombaksbotn (Table 2). The polychaete *M. oculata* dominated the fauna both at station 3 in Kvænangen and station 13 in Rombaksbotn. The sedimentary conditions in Rombaken were comparable to station 2 in Kvænangen, while station 3 in Kvænangen had a somewhat coarser sediment.

Characteristics of the fauna in fjord basins

On basis of the faunal assemblages at the investigated stations a cluster analysis was made (Figure 5). The results are similar when all species are included (Figure 5a), or when only the top ten species are included (Figure 5b). The largest similarity in species composition is seen among stations in the same fjords, and in each fjord the sill basin stations form individual clusters. Only station 6 in Rombaken falls outside this pattern, with a fauna more comparable to the fauna recorded in Akkarfjord (Figure 5).

Thus, the data presented from the three North Norwegian fjords show marked differences in fauna diversity. Inter fjord comparison reveals that only four taxa are common for the top-ten lists from Akkarfjord and Kvænangen, while 11 species are common for Akkarfjord and Rombaken (Table 4). Kvænangen and Rombaken are both large fjords with a variety of habitats, and should expectedly share a relatively high number of taxa at the top-ten lists. However, only 7 taxa are shared by these two large fjords (Table 4 and 5). The dominant groups of animals in sill basins are polychaetes and bivalve molluscs, while the echinoderms, which are strongly represented in outer parts of the fjords, are very poorly represented in the sill basins. No echinoderms are among the dominant taxa at the station in Rombaksbotn (13), and a single specimen of the mud sea star *Ctenodiscus crispatus* (Retzius) was the only echinoderm in the entire material. In Kvænangen, only station 4 on the sea ward side of the sills had echinoderms represented at the top-ten list. The oweniid polychaete *M. oculata* has a nearly alternating dominance compared to the echinoderms, and is recorded in great numbers in the sill basins in Kvænangen and Rombaksbotn, but is less represented

Table 4. The ten numerically most abundant taxa at each station in Rombaken (December 1990 and July 1992), with number of specimens per m^2.

	Rombaken 1990				Rombaken 1992			
Taxon	G1	G2	G3	G4	4	5	6	13
Polychaeta:								
Chaetozone setosa	223		148	148				308
Onuphis conchylega		130					125	
Owenia fusiformis			73				183	
Scoloplos armiger	305	57	183	123				40
Pholoe minuta		107	150	140				
Spio filicornis	80							
Goniada maculata	90	53						
Tauberia gracilis	378							
Myriochele oculata	128	117	350				245	2963
Prionospio cirrifera	85	263	158		100	423	55	55
Capitella capitata	70							
Nephtys ciliata			145	153				
Petaloproctus tenuis			113	118				
Cossura longocirrata				103				
Heteromastus filiformis					773	668		153
Asychis biceps					70	80		
Streblosoma intestinale						63		
Spio decoratus							265	
Myriochele danielsseni								95
Melinna cristata							88	
Maldane glebifex								270
Sabellidae indet.		60	85				130	
Mollusca:								
Macoma calcarea				465				
Astarte montagui	83							
Parvicardium minimum		173						
Thyasira obsoleta					398	120		
Thyasira minuta					328	365		138
Thyasira equalis					218	285		150
Thyasira eumyaria						70		
Abra nitida					193	110		
Nucula tumidula					123			
Cardium minimum							60	
Echinodermata:								
Ophiura robusta		140	165	378				
Labidoplax buskii					65			
Echinus sp.				118				
Crustacea:								
Hemilamprops rosea							48	
Leuchon nasica								48
Apseudes spinosus				108				
Diastylis sp.							80	
Ostracoda indet.		63						
Nematoda:								
Nematoda indet.	153							
Sipunculida:								
Onchnesoma steenstrupi					275	168		
Top ten% of total	66.0	52.1	48.9	51.5	77.3	80.9	54.5	94.4

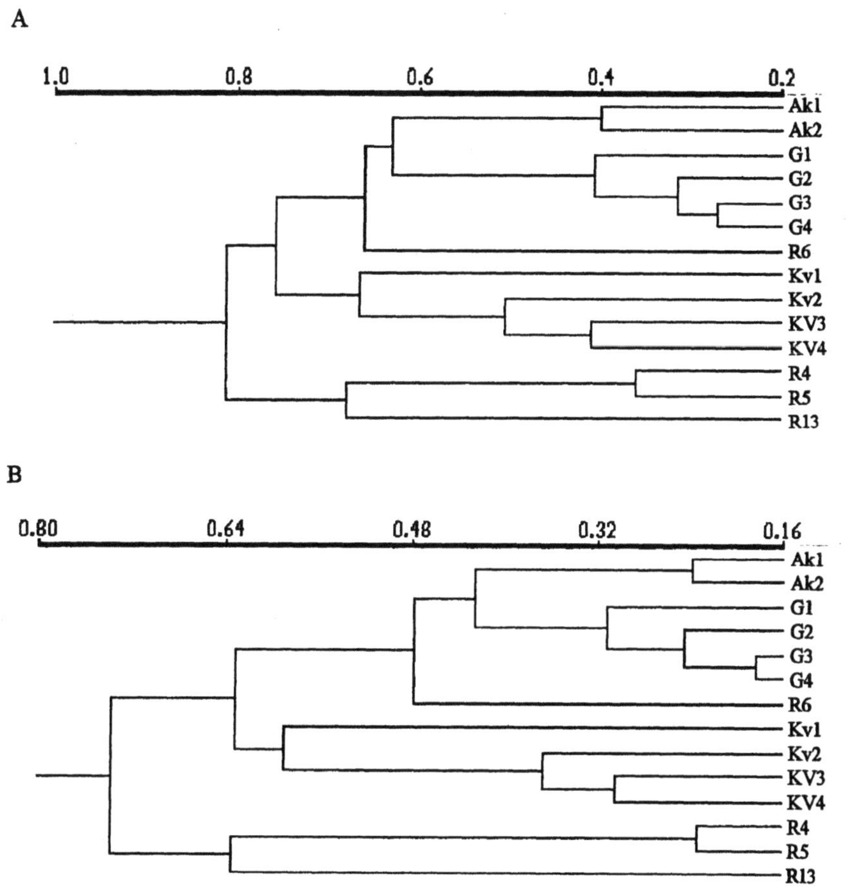

Figure 5. A: Dendograms of the classification of the faunal assemblages at the 14 stations in Akkarfjord (Ak1, 2), Kvænangen (Kv1–4) and Rombaken (G1–G4 and R4, 5, 6, 13), Northern Norway, based on Bray-Curtis similarity measures and group average sorting. A: All species included, matrix correlation $r=0.913$, B: only top 10 species included, matrix correlation $r=0.883$.The stations Kv1–3 and R13 are located in sill basins. Station 3 in Akkarfjord, consisting of only one grab sample, is excluded. The horizontal axis indicates dissimilarity (value 0 = total similarity, value 1 = no species in common, total dissimilarity).

in outer fjord areas. In Akkarfjord, *M. oculata* was not recorded at all, while this species was dominant at three of four stations in Kvænangen (Table 5).

The bottom topography of the fjords exerts a major influence on water exchange and sedimentation conditions, which in turn is permissive for the species composition and macrofauna diversity. In the two sill basins presented, H′ index attained a value of around or less than three, while sill free areas are characterised by higher diversities (H′ normally above 4). The values of the ES_{100} shows the same variation as the H′, generally with values below 20 for stations in sill basins, and values above 20 for stations on the seaward side of the sills (Table 2). The general validity of the influence exerted by the presence of a sill is illustrated in Figure 6. This shows the diversity at 52 stations sampled in 14 different fjords (Figure 1) of Northern Norway as function of 'topographic isolation' (delta depth sensu Buhl-Jensen (1986)) of the stations. The topographic isolation is calculated as the difference between the depth of a sampling station and the depth of the deepest passage from the station to the open sea. Sampling stations in fjord basins on the landward side of shallow sills will appear to the right along the x-axis (Figure 6). The correlation is significant ($n=52$, $r=-0.57$, $p<0.001$). The diversity H' attains a value of between 4 and 6 for open area stations. These stations are all located near the left vertical axis in Figure 6. Stations at intermediate depths in sill basins should seem to take values for H' of between 3 and 4, while stations in deeper parts of sill basins have H' values around 3 and lower. The tendency towards reduced diversity in sill basins compared to open sea sites has been described for fjords of western Norway by Buhl-Jensen (1986), working on amphipods and by Buhl-Mortensen & Høisæter (1993) working on molluscs. A much stronger correlation is

Table 5. The ten numerically most abundant taxa at each station in Kvænangen (September 1990), with number of specimens per m^2.

Taxon	Kvænangen 1	2	3	4
Polychaeta:				
Chaetozone setosa			173	
Onuphis conchylega				324
Owenia fusiformis	43			
Scoloplos armiger				87
Myriochele oculata	98	946	3326	1006
Prionospio cirrifera				107
Nephtys ciliata		24		
Maldane sarsi	693	266	1449	
Pectinaria koreni		204		
Lumbrineris fragilis		150	230	
Laphania boecki		31	230	
Sabellides borealis			124	
Myriochele heeri			133	
Praxillella sp.		27		
Mollusca:				
Astarte montagui	45			
Musculus niger	38			
Yoldiella fraterna				58
Yoldiella sp.				98
Thyasira spp.	33	477	88	128
Caudofoveata indet.		80	65	
Echinodermata:				
Labidoplax buski				118
Ophiura sp.				65
Crustacea:				
Eudorella emarginata	33			
Diastylis rathkei	255			
Bathymedon obtusifrons	43			
Protomedeia fasciata	30			
Cnidaria:				
Anthozoa indet.	0	20	215	65
Top ten% of total	86.6	90.6	81.9	66.0

recorded between delta depth and diversity in fjords of Western Norway compared to the present material (Buhl-Jensen 1986). This might indicate a generally better water exchange in North Norwegian fjords, compared to western Norway, allowing a more diverse fauna in deeper sill basins.

The fauna recorded in the sill basins of the present survey includes several long lived species like e.g. *Maldane sarsi*, a dominant in the Kvænangen material, but also the *Sabellides borealis* and *Lagis (Pectinaria) koreni* Malmgren have a life span covering more than one year. The presence of these organisms indicates that hypoxia is unlikely to have occurred in the investigated sill basins during the latest years prior to the surveys. During sampling of the sediment and fauna at all fifteen stations, a smell of H_2S from the sediment was only recorded once. This observation, together with the records of large, deep burrowing macrofauna, also indicates that hypoxia or anoxia is unlikely to have occurred.

However, the low number of echinoderms, which generally are sensitive to hypoxia, recorded in the sill basins indicates that sub-pycnocline hypoxia might occur naturally in the fjords. On the other hand, the distribution pattern of the echinoderms might be influenced to a larger extent by the sedimentary conditions than by the oxygen contents of the water column, and can thus not be used as evidence for occurrence of hypoxia. Besides facilitating the summer stratification, the outflow of fresh water causes a general reduction in salinity in the upper layers of the water column, and the absence of echinoderms in the sill basins might be due to the lower salinity of the upper waters, being lethal to the echinoderm larvae. However, the present material does not allow a thorough investigation of this suggestion.

Buhl-Mortensen (1994) concludes that the most important feature, leading to a reduced diversity in molluscs and amphipods in sill basins is the finer grain size of the sediment, arising due to the sill's reduction of current speed, which in turn leads to a less heterogeneous sediment with less different habitats. Hampering the water exchange, a sill is also responsible for the keeping of stagnant basin water in sill basins.

Besides the generally cooler summer and thus less heating of the surface water in Northern Norwegian fjords compared to the fjords of Western Norway, the tidal amplitude is larger in Northern Norway. Average tidal amplitude in Narvik is 182 cm, compared to 90 cm in Bergen, Western Norway. The tidal water exchange is thus a factor contributing to a larger exchange of water in North Norwegian fjords, and an expected shorter duration of a stagnant situation during summer.

In conclusion, the investigations of the benthic macrofauna of three North Norwegian fjords indicate, that the bottom topography exerts a far larger influence on the species composition and species diversity in sill basins, compared to anthropogenic discharges of sewage. The results also indicate that hypoxia or anoxia is unlikely to have occurred in the investigated sill basins the latest two–three years before the surveys.

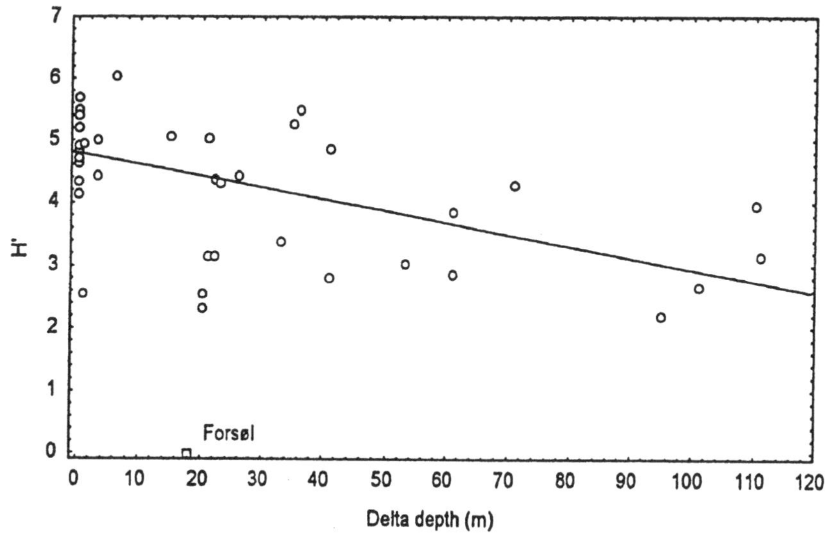

Figure 6. Benthic macrofaunal diversity (H') versus delta depth (index of topographic isolation) for 52 stations in coastal areas of Northern Norway (regression line: $y=4.8-0.019x$; $r=-0.57, p<0.001$).

Acknowledgments

The material has been collected by Akvaplan-niva in connection with a number of different commissions, each with a specific purposes and extent. All material has been reported to the commissioners in Norwegian. Field work, identification and treatment of the material has been carried out in co-operation with colleagues at Akvaplan-niva. The compilation and presentation of the results has received financial support from the Norwegian Ministry of the Environment through a grant to Akvaplan-niva. The author is grateful to the following colleagues for carrying out field work, identification, and statistical analyses of the material: Sabine Cochrane, Salve Dahle, Rosie Evans, Reinhold Fieler, Bjørn Gulliksen, Børge Holte, Hans-Petter Mannvik, Rune Palerud, Wim Vader, Roger Velvin, Anders Wàren and Ursula Witte. The manuscript has benefited from constructive comments and suggestions from Eivind Oug, Norwegian Institute for Water Research, and from the anonymous reviewers of the manuscript.

References

Aure, J. & A. Stigebrandt, 1989. On the influence of topographic factors upon the oxygen consumption rate in sill basins of fjords. Est. Coast Shelf. Sci. 28: 59–69.

Buhl-Jensen, L., 1986. The benthic amphipod fauna of the west Norwegian continental shelf compared with the fauna of five adjacent fjords. Sarsia 71: 193–208.

Buhl-Mortensen, L., 1994. Are fjord-basins habitat islands? Dr Sci. thesis, University of Bergen, Norway: 28 pp.

Buhl-Mortensen, L. & T. Høisæter, 1993. Mollusc fauna along an offshore-fjord gradient. Mar. Ecol. Prog. Ser. 97: 209–224.

Eilertsen, H. C., S. Falk-Petersen, C. C. E. Hopkins & K. Tande, 1980. Ecological investigations on the plankton community of Balsfjorden, Northern Norway. Program for the project, Study Area, Topography and Physical Environment. Sarsia 66: 25–34.

Hurlbert, S. N., 1971. The non-concept of species diversity. A critique and alternative parameters. Ecology 52: 577–586.

Oug, E., 1988. Sill-fjords, suitable or not for aquaculture? Presentation at World-Aqua symposium, Tromsø, 5 pp.

Rygg, B. & I. Thélin, 1993. Klassifisering av Miljøkvalitet i fjorder og kystfarvann. (Classification of environmental quality in fjords and coastal waters). Norwegian State Pollution Control Agency, Guideline 93:05, 16 pp (in Norwegian with English summary).

Shannon, C. E. & W. Weaver, 1949. The Mathematical Theory of Communication. Univ. Illinois Press, Urbana, 117 pp.

Hydrobiologia **355:** 115–126, 1997.
A. D. Naumov, H. Hummel, A. A. Sukhotin & J. S. Ryland (eds), Interactions and Adaptation Strategies of Marine Organisms.

Mechanisms of salinity adaptations in marine molluscs

V. J. Berger[1] & A. D. Kharazova[2]
[1] *Zoological Institute of the Russian Academy of Sciences, St. Petersburg, Russia*
[2]*St. Petersburg State University, St. Petersburg, Russia*

Key words: adaptation, salinity, molluscs, resistance, tolerance

Abstract

A review on salinity adaptation of marine molluscs based on mainly Russian scientific literature is presented. The existence of two relatively independent systems of adaptation to extreme (resistance level) and moderate (tolerance level) changes of environmental salinity was shown. The resistance of molluscs is based mainly on an impeded water-salt exchange with the external medium due to mantle cavity hermetization. The tolerance of molluscs is determined by cellular mechanisms of adaptation. Reversible changes of protein and RNA synthesis, alteration of the pattern of multiple molecular forms of different enzymes, and the regulation of ionic content and cell volume were shown to be of importance for the above mentioned mechanisms. The efficiency of resistance and tolerance adaptations to salinity changes may vary in different species and in different colour phenotypes of the same species (intrapopulational polymorphism). Parasites (trematodes) may suppress the resistance of the mollusc-host to extreme salinity changes without effecting the host's capacity for adaptive changes in salinity tolerance.

Introduction

Salinity is one of the most important environmental factors. Being relatively constant in open seas, it varies considerably in intertidal zones, estuaries and other biotopes. The ability to exist at varying salinity, euryhalinity, depends on different adaptations. The most effective one is osmoregulation based on active ion transport mechanisms, well investigated in fishes, crustaceans and polychaetes. Many osmoconformers, lacking the ability to regulate the osmotic pressure of the internal medium, nevertheless demonstrate considerable euryhalinity.

Marine molluscs may live in a wide range of salinities from 4–5 up to 75–80‰ (Fretter & Graham, 1962; Khlebovich, 1962, 1974; Hedgpeth, 1967; Golikov & Kusakin, 1978; Berger, 1986; Berger et al., 1995). They possess numerous salinity adaptations investigated to a much lesser extent than those of osmoregulators. This report is an attempt to review some data on adaptations of marine molluscs presented mainly in Russian scientific literature.

Water salt balance and resistance to extreme salinity changes

In addition to burrowing into bottom sediments, actively choosing the environment and escaping from unfavourable conditions, marine molluscs may close their shells at abnormal salinity. Such reactions take place due to the activity of peripheral detectors located on head tentacles, mantle ridges and siphon surfaces (Freeman & Rigler, 1957; Vasilieva et al., 1960; Davenport, 1979, 1981). There are two kinds of these detectors: osmoreceptors, sensitive to osmotic pressure, and special receptors, sensitive to changes in sodium ion concentration (Natochin, 1966, 1976). Marine molluscs may exist in wide salinity ranges without hermetization of their mantle cavity. Exposure to extremely diluted or concentrated sea water initiates an isolating reflex, firstly in the more sensitive individuals and then in all experimental animals. The sensitivity of this isolation reaction may be shifted by previous acclimation to high or low salinity (Figure 1).

The blood of marine invertebrates is nearly isoosmotic to the environment in normal conditions (Krogh,

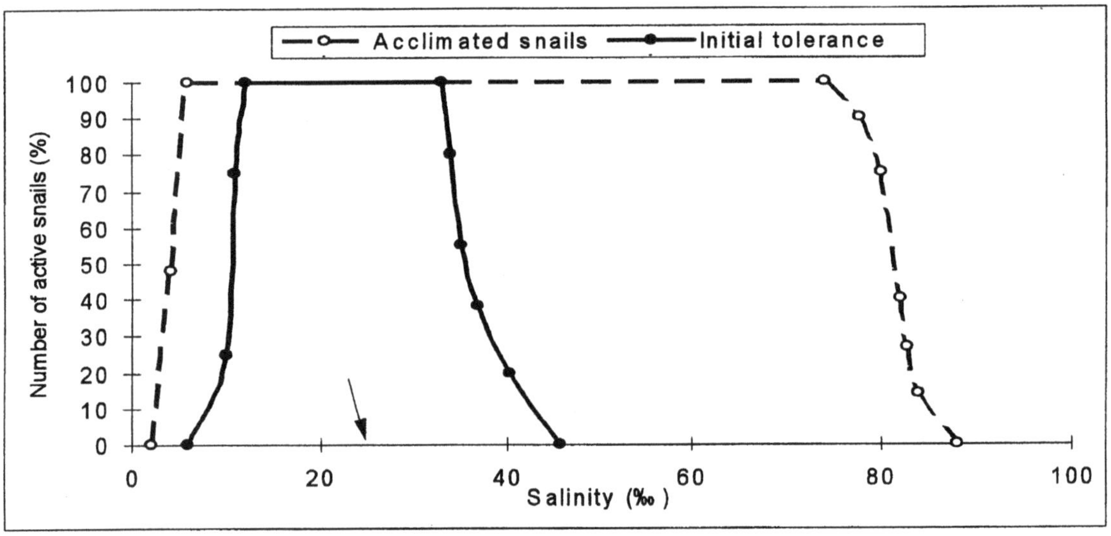

Figure 1. Changes of salinity tolerance after stepwise acclimation to different salinities in *Hydrobia ulvae* from original populations (initial tolerance) and populations acclimated to low and high salinity (from Khlebovich & Kondratenkov, 1973). The arrow indicates the initial salinity.

1939; Beliaev, 1951, 1957; Ginetsinskiy, 1963; Potts & Parry, 1964; Todd, 1964; Gilles, 1972, 1979; Berger, 1986). A basically different situation is typical for marine molluscs hermetizing the mantle cavity at extreme changes of sea water salinity. Osmotic balance with the environment has been observed only at salinities at which all the animals were active. This range of salinities differs in molluscs inhabiting different biotopes. For example, in *Littorina littorea* from the Baltic Sea, osmotic equilibrium with the environment was observed only at salinity values over 5‰ (Figure 2). In the same species from the White Sea osmotic equilibrium was registered at salinities higher than 12 to 14‰. At lower salinity the osmotic concentration of extravisceral fluids was higher than that of the environment (Figure 2). Such a situation is due to a low level of salt loss and impeded water penetration in closed molluscs. Comparative data on salt loss rate in molluscs and other invertebrates support this idea. Shelled molluscs show a minimal rate of salt loss of about 0.2 to 0.4 mg g^{-1} h^{-1}, while all other investigated animals (coelenterates, echinodermates, chitons, pteropods and nudibranchs) lose salt rapidly (15 to 40 mg g^{-1} h^{-1}) in distilled water (Berger, 1986). Thus, due to a highly effective hermetization, shelled molluscs can survive in fresh water 100–500 times longer than animals lacking such abilities.

But the low rate of salt-loss is not the only mechanism determining the resistance to fresh water in marine shelled molluscs. It is also connected with the capacity to withstand prolonged asphyxia and accumulation of acidic products of anaerobic metabolism typical for 'closed' molluscs (Aliakrinskaya, 1972). Therefore, it is the result of a suite of nonspecific resistance adaptations. Adaptation of molluscs to increased salinity is provided by similar mechanisms. During the first period 'closed' molluscs maintain the osmotic pressure at a level lower than that of the external medium by blocking the water-salt exchange with the environment (Figure 3).

Osmoregulation of homoiosmotic animals, based on active transport of ions in specialised organs, allows the level of the internal osmotic concentration to be maintained for an unlimited period of time. On the other hand, the maintenance of the hetero-osmotic state by molluscs is possible only for restricted periods, though rather long, up to 1 to 2 months, in some species. Bearing in mind these pecularities of water-salt exchange in osmoconforming marine molluscs, it appears reasonable to name the capacity for temporary maintenance of the hetero-osmotic concentration in the internal medium as 'facultative pseudo-osmoregulation' (Berger, 1986). It implies that molluscs keep up the osmotic disbalance with the environment not by active ion transport, as typical osmoregulators do, but only by means of temporary inhibition of the water-salt exchange as a result of mantle cavity hermetization. The relevant term 'short term osmoregulators' was introduced by Davenport (1979).

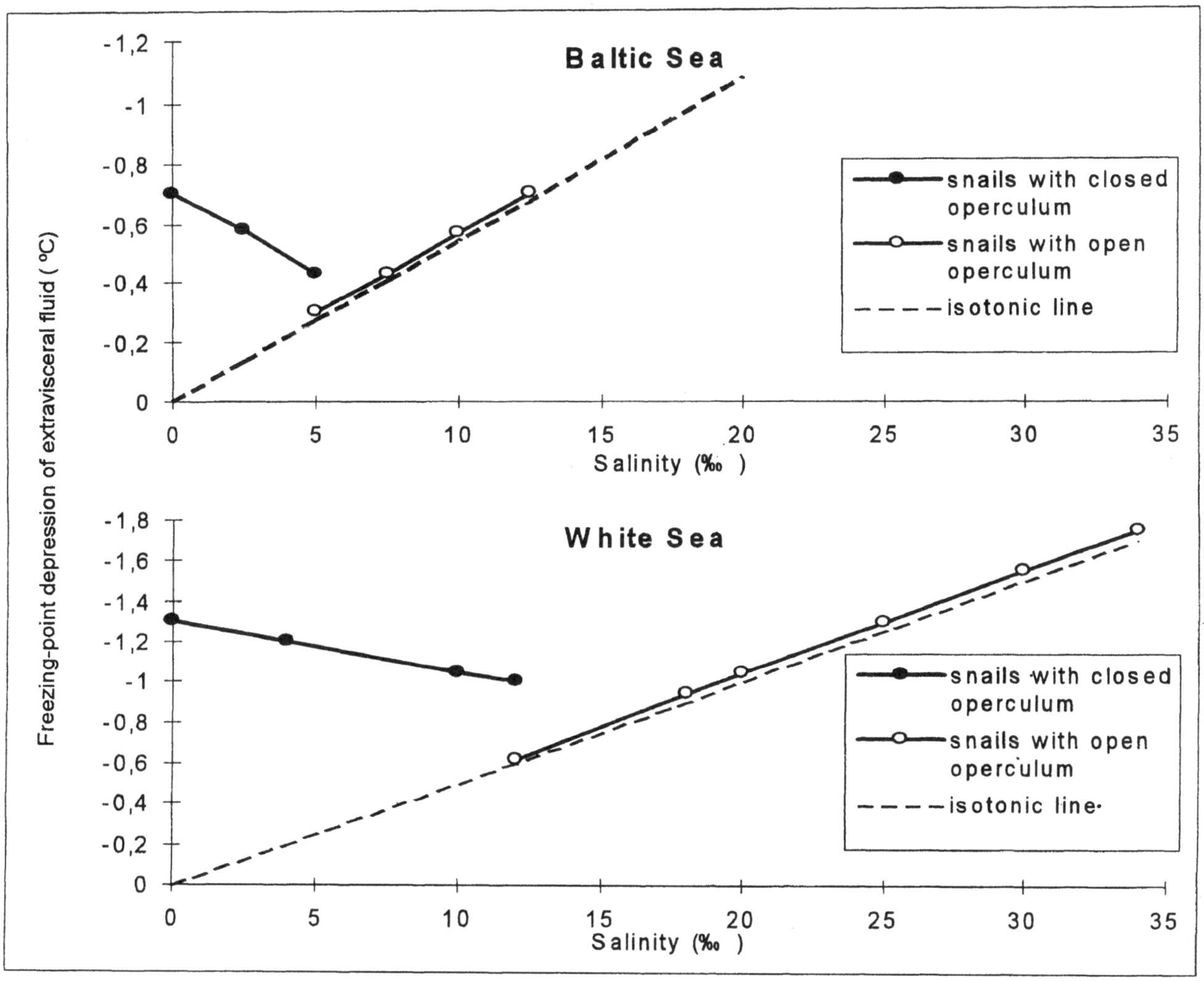

Figure 2. Changes in freezing-point depression of extravisceral fluid in *Littorina littorea* from the Baltic and White Seas at different salinities (after Klekowski, 1963 & Berger, 1976).

The functional basis of osmotic tolerance

The most common reaction of marine molluscs to salinity change is the decrease of functional activity. For example, when salinity is changed the respiration rate decreases and is kept at a comparatively low level, normally no longer than during 1 to 2 days. Later, when acclimation occurs, the intensity of respiration is restored to the original level or close to it. The rate and extent of functional activity normalisation depends on the magnitude of the external influence and the adaptive capacity of species. Thus, when salinity decreases from 25‰ (normal level in the White Sea) to 14‰, the respiration of intertidal White Sea periwinkles is at first inhibited. An increase of the oxygen consumption occurs after 1 (*Littorina saxatilis*) to 6 (*L. obtusata*) days. It is fully restored only in the more euryhaline *L. saxatilis*. The rate of oxygen consumption in the other species remains below the control level even after 16 to 22 days of acclimation to lowered salinity (Berger, 1986).

The ability of molluscs to restore the functional activity under long-term exposure to water of changed salinity is also manifested in locomotion and byssus production rate (Berger, 1986; Berger et al., 1985). All these data together demonstrate the considerable adaptive capacity of marine molluscs, by which they compensate the disturbances of functional activity during acclimation.

The transition of molluscs back to normal conditions after being completely acclimated to a changed salinity (de-adaptation, de-acclimation) results in a sharp increase of the functional activity, such as oxygen consumption, movement rate (Figure 4) and so

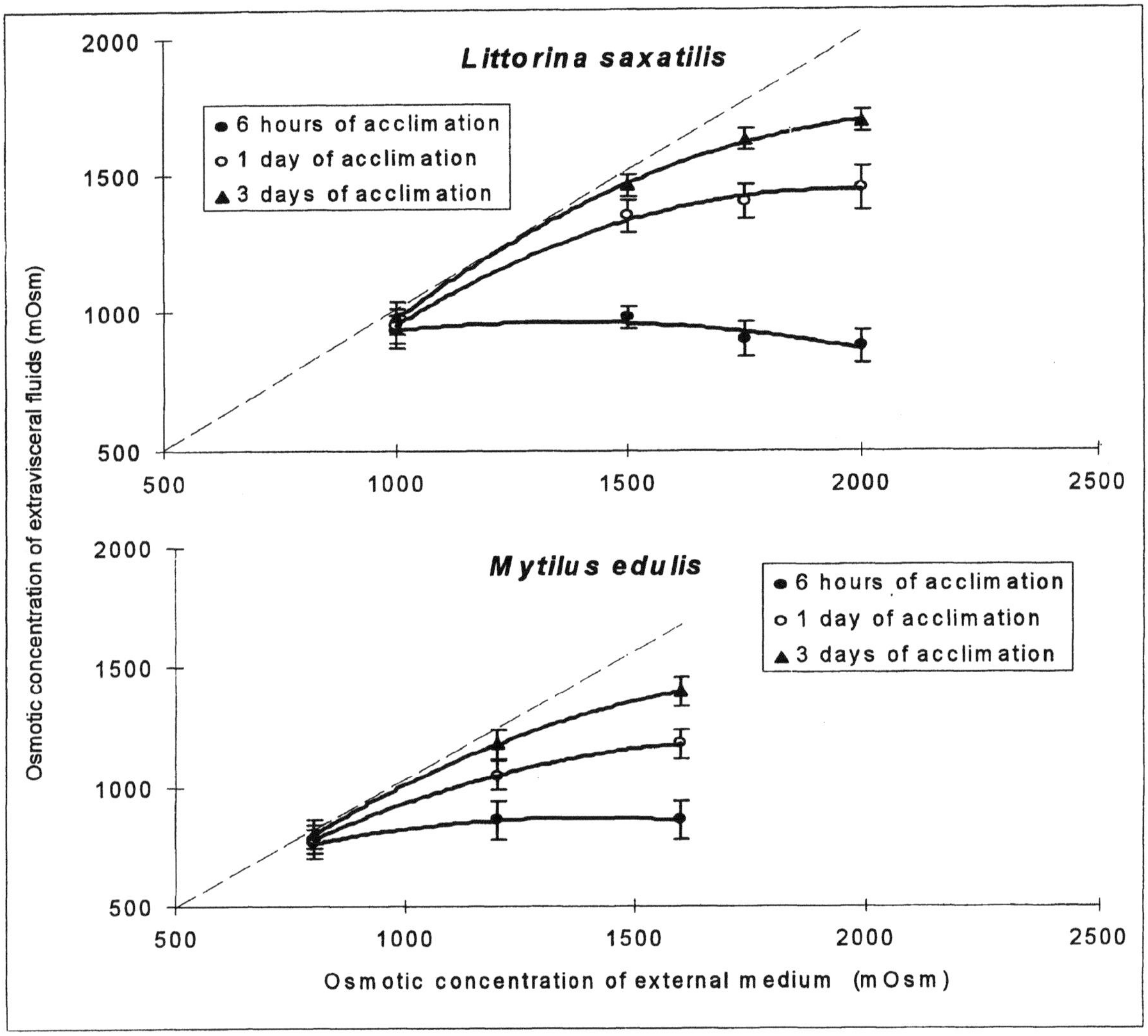

Figure 3. Osmotic concentration of extravisceral fluids of *Littorina saxatilis* and *Mytilus edulis* in water of high salinity (from Berger, 1989). In this figure and in subsequent figures the 95% confidence intervals are indicated by vertical lines.

on. Such a situation is more or less typical for de-adaptation processes of molluscs (Berger, 1976, 1986; Berger & Sergievskii, 1990). Probably, adaptive mechanisms for the restoration of functional activity are rather inert and normally do not switch off before 24 to 48 hours.

Thus, changes in the functional activity during acclimation of molluscs to altered salinity and subsequent de-acclimation, reveal a specific alternation of transitional processes and 'overshoot' reactions. Keeping in mind such dynamics of transitional processes we can suggest the existence of some regulatory systems providing the stabilisation of functional activity in molluscs within the zone of osmotic tolerance.

Cellular mechanisms: the role of amino acids, RNA and protein synthesis

Different aspects of various cellular mechanisms involved in the adaptation of molluscs, and other osmoconformers, to salinity changes have been demonstrated and discussed by a number of researchers (Kreps, 1929; Krogh, 1939; Savvateev, 1952; Schlieper, 1960; Zhirmunskiy,1962; Florkin & Schoffeniels, 1969; Gilles, 1979; Pierce, 1982, 1994; Bishop et al., 1994). Most research focused on the role of free amino acids in cell volume and iso-osmotic regulation. However, plastic metabolism of cells during salinity acclimation practically escaped the attention of scientists.

The role of cell synthetic activity during salinity acclimations of different marine molluscs (*Litto-*

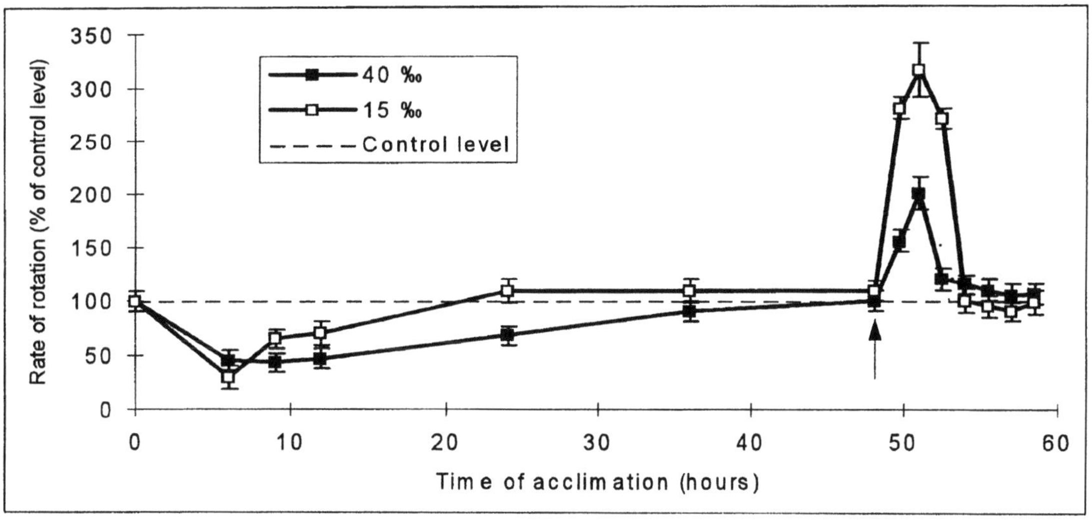

Figure 4. Changes of locomotion rate of *Epheria vincta* veligers during acclimation to 40‰ and 15‰ salinity and after de-acclimation (from Berger, 1986). The arrow indicates the moment of return to normal salinity (26‰).

rina littorea, Mytilus edulis, Crenomytilus grayana, Crassostrea gigas, Ruditapes japonica, Nuttalia olivacea, Coryphella rufibranchialis) has been studied using labeled precursors of protein and RNA synthesis (Berger & Kharazova, 1971, 1977; Kharazova & Berger, 1974; Gurina, 1975; Kharazova & Rostova, 1976; Kharazova et al., 1981, 1983; Kharazova, 1987, 1994). At early stages of acclimation significant changes in protein and RNA synthesis (bulk synthetic processes, amplitude of hourly oscillations of protein synthesis, etc.) occur in molluscan tissues. Electron microscopy studies revealed that at the same time the changes in synthetic activity of mussel gill cells are accompanied by the swelling of the cell and membrane organelles and de-condensation of perinuclear chromatin (Korolkova & Kharazova, 1994). During long-term acclimation (days, weeks) the synthetic activity returns to normal.

Different tissues display the same, above mentioned, sequence of changes (Figure 5). Differences observed in the development of the synthetic processes during salinity acclimation are connected with ecological peculiarities of the animals, type of cells and tissues and the extent of salinity changes. Similar data on alterations in synthetic processes were obtained in experiments with other invertebrates: ascidians (Kharazova et al., 1989), polychaetes (Lvova & Kulakovsky, 1979), scyphomedusae (Lukanin, 1976; Black & Bloom, 1984). Thus, cells of marine molluscs and other forms are capable of restoring the normal (or close to it) level of RNA and protein synthesis during adaptation to low salinity.

The above mentioned dynamics of synthetic processes during salinity acclimation are also shown in isolated tissues (Kharazova, 1994; Kharazova et al., 1981, 1983), which indicates the considerable autonomy of those cellular adaptive mechanisms. The data obtained by treating molluscs and other invertebrates and their isolated cells with various inhibitors of RNA and protein synthesis provide additional proof of the important role of the plastic metabolism of cells during the adaptation of molluscs to salinity changes (Berger et al., 1970; Berger & Lukanin, 1972; Lukanin & Khlebovich, 1979). It was shown that animals and their cells lost the capacity for salinity adaptations after inhibitor treatment.

These and other analogous facts support the idea that the ability of molluscs, and apparently of other osmoconformers, for phenotypic salinity adaptations is based on the capacity of their cells to change the RNA and protein synthesis. Acclimation to altered salinity is probably connected not only with quantitative variations in RNA and protein synthesis, but also with qualitative changes involving *de novo* synthesis. Experiments were undertaken (Berger et al., 1975) to verify this hypothesis. Multiple molecular forms of different enzymes taken from muscles and hepatopancreas of *Littorina littorea* acclimated to low salinity were investigated. After acclimation the spectrum of nonspecific esterases was considerably modified. An altered intensity of histochemical staining of some zones, emerging

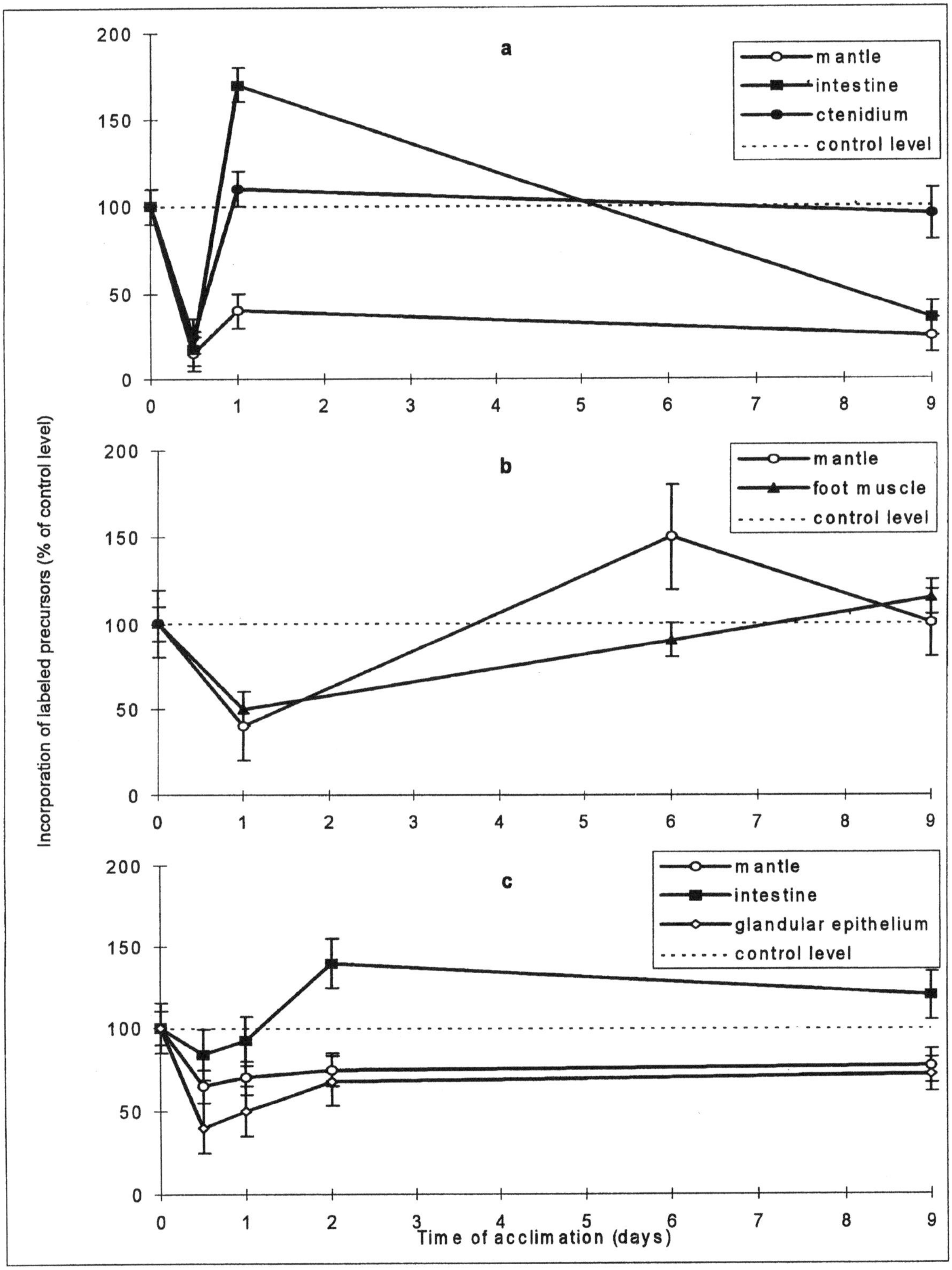

Figure 5. Changes of synthetic activity in the tissues of *Littorina littorea* during acclimation of the molluscs to lowered (14‰) salinity (after Berger & Kharazova, 1971 and Kharazova & Berger, 1974): incorporation of ^{35}S-methionine (a), ^{3}H-glycine (b) and ^{3}H-uridine (c).

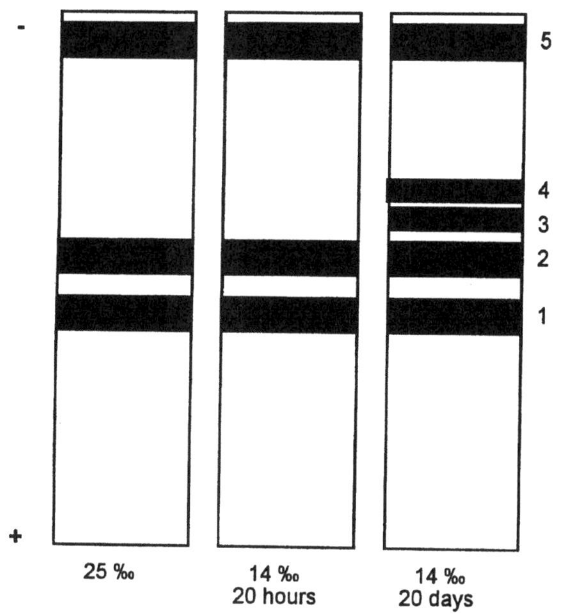

Figure 6. Pattern of multiple molecular forms of lactate dehydrogenases in hepatopancreas of *Littorina littorea* before and after exposure to low salinity (from Berger et al., 1975).

of new components and dissappearance of others, were detected. In control animals and periwinkles briefly exposed to low salinity three lactate dehydrogenase (LDH) fractions in the hepatopancreas were revealed (Figure 6). Prolonged acclimation (20 days) caused the appearance of two additional zones which were absent after a short period (20 hours) of exposure to lowered salinity. By means of such alterations in enzyme pattern the molluscs could better cope with new salinity conditions. There is some histochemical evidence that different enzymes of mussel gills vary in their salinity resistance, which can be increased by preliminary acclimation (Kharazova, 1994).

Changes in the balance between fractions in the spectrum of enzymes and in the activity of certain fractions during acclimation of molluscs to lowered salinity may occur due to de-repression and/or repression of loci responsible for the synthesis of the corresponding enzyme forms. This idea is supported by the following facts.

Changes in the environmental concentration of sodium and potassium ions result in the activation or inactivation of some loci of polythene chromosomes in insects (Kroeger, 1967; Lezzi, 1970). Actinomycin D and cycloheximid suppress the induction of new esterase forms in haemocytes and middle gut of insects, caused by shifting of the ratio of uni- and bivalent cations in the incubation medium (Marek & Kroeger, 1974, 1976). Moreover, recent findings on mammalian cells unequivocally demonstrate osmotically induced gene transcription (Ferraris & Garcia-Perez, 1996).

The role of inorganic ions

In spite of numerous investigations demonstrating the participation of free amino acids in intracellular iso-osmotic and volume regulation (see above), there are some facts which make us doubt the exclusive role of free amino acids in cell volume stabilisation and salinity adaptation. Evidence on marked alterations of inorganic ion content in cells and tissues of various animals caused by environmental salinity changes began to accumulate after the work of Potts (1958). It was shown that a significant decrease in sodium and chloride ion concentration in muscles occurs after acclimation of *Mytilus edulis* to water of low salinity. Similar investigations were carried out on other invertebrates (Freel, 1978; Gilles, 1979). However, these reports did not evaluate the role of electrolyte components in iso-osmotic intracellular regulation. The solution of this problem has arisen from experimental works undertaken to check the idea about the leading role of inorganic ions in osmotic and volume cell regulation. Experiments were carried out on adductor muscles of *Mytilus edulis,* foot muscles and hepatopancreas of *Littorina littorea, L. saxatilis* and *L. obtusata* from the White Sea.

Incubation of isolated adductor muscles for 24 hours in water of different salinity caused considerable variations of the cell volume (Natochin & Berger, 1979). The intracellular water content was equal to 650 ml kg^{-1} wet weight at 26.8‰ (normal salinity for the White Sea). It increased to 750 ml kg^{-1} wet weight at 8‰ and fell to 540 ml kg^{-1} wet weight at 50‰. At the same time oppositely directed changes in intercellular space were observed. Apparently, under conditions of varying salinity, cells of the investigated molluscs are capable of partial, but far from complete, stabilisation of their volume (Natochin et al., 1979). At a 15-fold increase of the salinity and a parallel increase of the sodium content in the environment, a rapid rise of the sodium concentration occurs in muscles as well as in muscles cells (Figure 7). At the same time the potassium content (per dry weight of muscles) increased only 1.7 fold. Analysis of the intracellular potassium concentration showed that its changes are markedly smaller than those of sodium. The potassium content in cells varies depending on their shrivelling

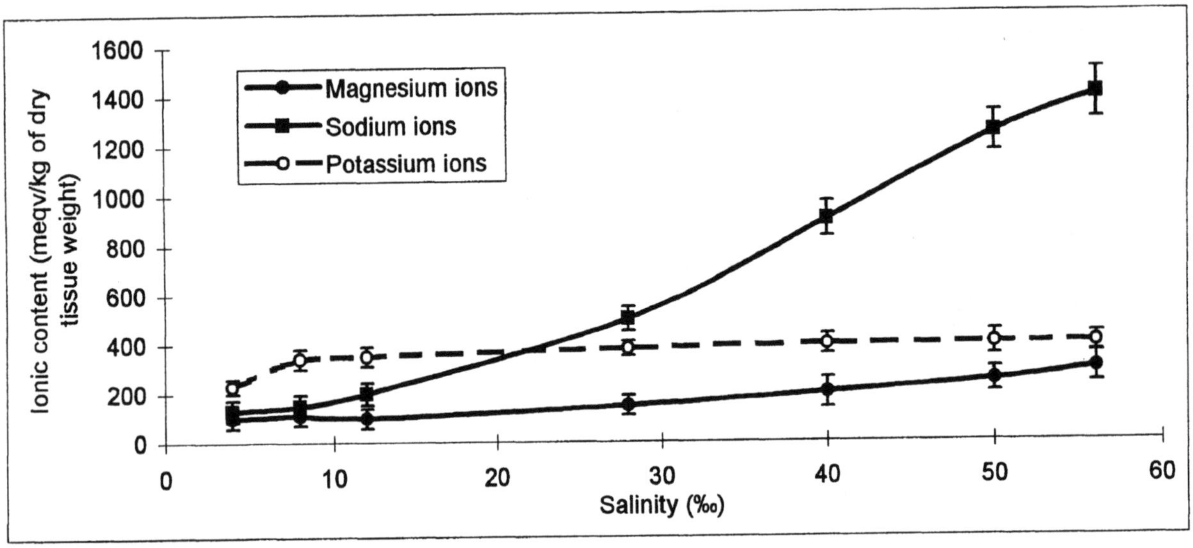

Figure 7. Ionic content (meqv/kg dry tissue weight) in the adductor muscle of *Mytilus edulis* after 24 hours acclimation to different salinities (from Natochin et al., 1979).

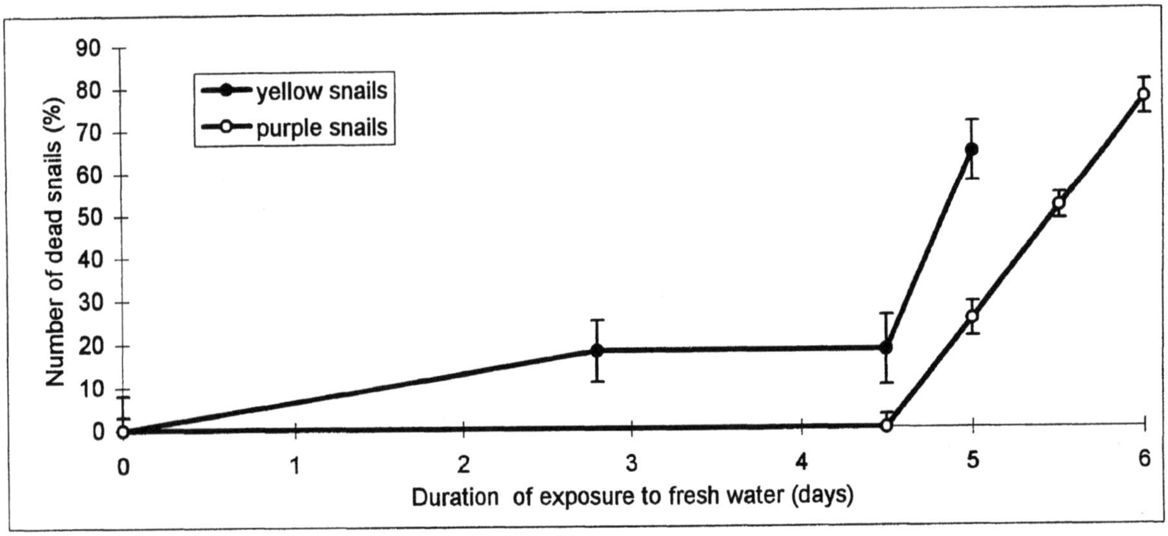

Figure 8. Resistance of different colour morphs of *Littorina obtusata* to fresh water (from Sergievskii & Berger, 1984).

or swelling, being mainly a function of cell volume. In other words, the total content of this ion alters slightly. It is just dissolved in a different volume of cell water.

Maintaining the cellular potassium concentration independently of other ions provides stabilization of the transmembrane K_i/K_e gradient. At the same time the Na_i/K_i ratio varies considerably. When the salinity is altered from 8 to 50‰ it increases from 0.29 to 1.30. In a salinity of 4 to 10‰ the sodium content in the muscle cells of mussels is maintained at the lowest stable level. Probably, the decrease of the intracellular sodium concentration initiates the osmoregulatory reflex and hermetization of the mantle cavity, protecting cells from the influence of extreme low salinities (Natochin, 1966). Exposure to higher salinity leads to a proportional increase of the intracellular sodium content and parallel changes in the chloride ion concentration in the cell (Kuzmina, 1982).

These data suggest that sodium and the corresponding anion (chloride) play an important role in osmotic cell volume regulation of various molluscs when environmental salinity is altered. Salinity being increased, sodium and chloride ions probably diffuse into the cell, and at a lowered salinity these ions are active-

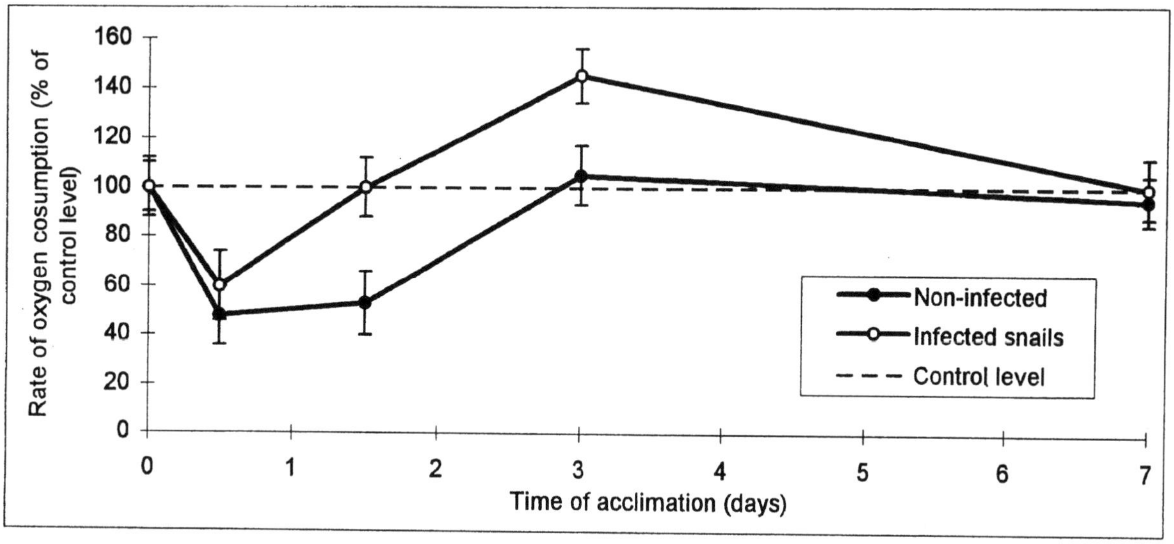

Figure 9. Changes of respiration rate of infected and non-infected *Hydrobia ulvae* during acclimation to low (14‰) salinity (from Berger, 1976).

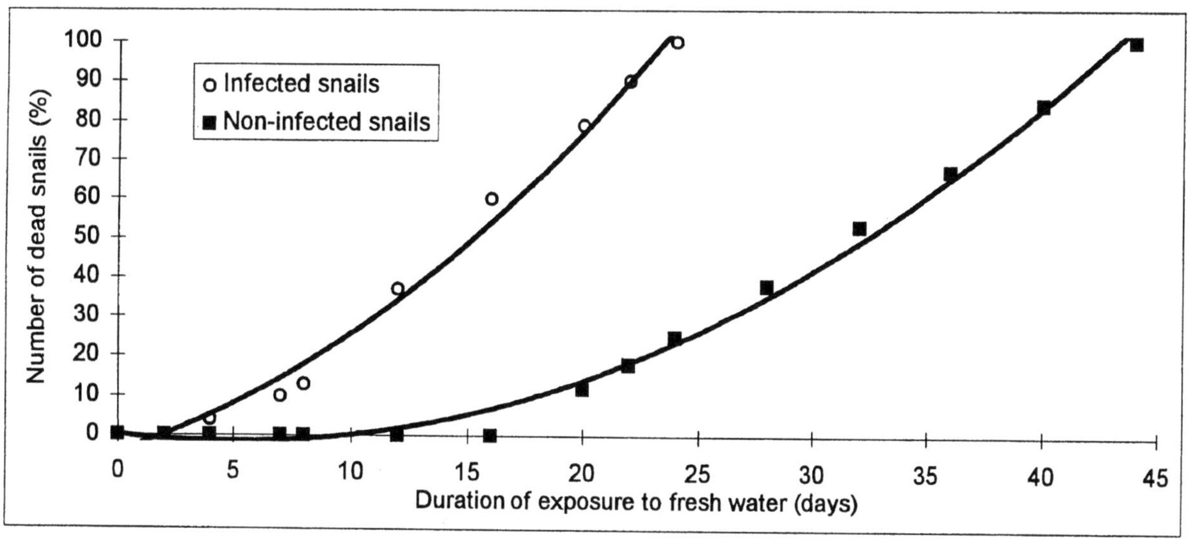

Figure 10. Resistance to fresh water of infected and non-infected *Hydrobia ulvae* (from Berger, 1976).

ly removed. This hypothesis was supported by results of inhibitory experiments (Natochin et al., 1979). The inhibitors used have different effects on the ionic composition in adductor muscles of mussels. Their influence depends on the environmental salinity. Ethakrine acid, known to block sodium and chloride ion transport (Lavrova & Natochin, 1981), causes an increase in the sodium content and swelling of the adductor muscle at low salinity (10‰). Under such conditions the inhibitor blocks the removal of sodium and chloride ions from the cell and, therefore, suppresses adaptation to low salinity. At 26.8 and 40‰, ethakrine acid has a weaker effect on the ion composition and hydratation of the muscles. In contrast to this, ouabain (an inhibitor of Na, K-ATP-ase) reduces the potassium concentration and increases the sodium content in muscles. This effect is independent of environmental salinity.

The ion balance of muscle cells during adaptation to various salinities could not be achieved by virtue of the Na/K-pump alone, removing sodium and accumulating potassium. As it is clear from the data obtained, the concentration of both ions drops at low salinity and increases at high salinity. Therefore, the effective ion regulation in molluscan cells can be provided only by cooperative action of two pumps – the Na/K-pump and Na,Cl-pump, independent of potassium transport. Moreover, the participation in cell volume regulation of a pump independent of potassium, insensitive to ouabain but inhibited by ethakrine acid, was demonstrated in the renal cells of vertebrates (Natochin, 1976). Thus, all these data demonstrate that, beside some nitrogen-bearing compounds of low molecular weight, sodium ions with accompanying anions (primarily chloride) are mainly responsible for the osmotic regulation in molluscan cells.

Salinity adaptations: a link with morphological polymorphism

The above mentioned, two relatively independent systems of adaptation of marine molluscs to extreme and moderate changes of environmental salinity determine different adaptive potentials of these animals. Realization of the latter depends not only on the efficiency of the mentioned systems but also on the specific ecological situation in certain habitat, interspecies (populational polymorphism) and intraspecies (e.g. parasite-host) interrelations and other factors. Investigation of the functional significance of and mechanisms determining populational polymorphism is one of the central problems of population research. Classical objects for such investigations are land Pulmonata: *Cepaea, Portula, Bradybaena*, etc. Marine molluscs are less studied in this respect, in spite of their well developed polymorphism. In the study on population structure of *Littorina obtusata* in the White Sea, the parametric system of shell colour polymorphism was worked out based on the analysis of pigment type, chromatophore functional activity and others (Sergievskii, 1983).

Comparison of *L. obtusata* from the White Sea with different shell colour phenotypes allowed demonstration of the existence of significant variation in adaptive reactions of snails to salinity changes (Sergievskii & Berger, 1983, 1984; Berger & Sergievskii, 1986, 1990). The investigated phenotypes were divided into groups according to their salt loss rate. The purple individuals, with a minimal rate of salt loss which corresponds to a maximal effect of mantle cavity hermetization, died in fresh water much more slowly than snails of other phenotypes (Figure 8). Comparison of specimens with two shell colour phenotypes, purple and yellow, differing significantly in the degree of shell hermetization and resistance to fresh water, showed that they reacted oppositely to moderate salinity changes. The snails with purple shells, more resistant to fresh water, were more sensitive to a moderate lowering of salinity. These differences were also observed in the study of respiration rate of the snails in diluted (16‰) sea water (Berger & Sergievski, 1986). Oxygen consumption was suppressed to a greater extent in purple molluscs. Thus, in relation to moderate (not lethal) salinity changes the adaptive potentials of the purple snails are lower than those of the yellow ones.

The influence of parasites

Marine molluscs and other animals are infected with different parasites, most often various trematodes, which cause considerable changes in the host. In the context of our problem it was essential to evaluate the influence of trematode infection on the salinity adaptations of the host. Infected molluscs are known to be less resistant to anoxia, desiccation, freezing and high temperature (Oliver & Brand, 1953; Vernberg & Vernberg, 1963; Jensen et al., 1996). But, there are few data on the changes in salinity resistance and tolerance of molluscs induced by parasites.

The respiration rate of infected and uninfected mud snails *Hydrobia ulvae* from the White Sea during acclimation to low salinity was investigated (Berger, 1976). Acclimation to low salinity was more effective in the snails infected by parthenogenetic stages of trematodes (Figure 9). So, these parasites apparently do not suppress the adaptive recovery of the energy metabolism of the host. Moreover, the comparison showed this process to be quicker and more effective in infected *H. ulvae*. Similar results were obtained later in experiments with the White Sea periwinkles *Littorina saxatilis* (Berger, 1986). Thus, adaptive modifications of salinity tolerance in molluscs are not suppressed by trematodes. A different situation was observed for the influence of parasites on the resistance of the host. The period of survival of uninfected *H. ulvae* in fresh water was twice that of infected ones (Figure 10). Similar results were obtained in experiments with *L. saxatilis*.

It has been already mentioned that mollusc resistance to extreme low salinity depends on the degree of mantle cavity hermetization. Therefore, less resistant infected molluscs should be expected to differ from the uninfected ones also by less efficient isolating reactions. Actually, the rate of salt loss in infected *H. ulvae* specimens was 2 to 3 times higher than in the uninfected ones (Berger, 1986).

These data suggest that parasites may suppress the resistance of the mollusc's-hosts to abiotic factors. At the same time they practically do not effect the host's capacity for adaptive changes in salinity tolerance during acclimation to altered salinity. This is additional evidence to support the idea that adaptations of molluscs to moderate (tolerance level) and extreme (resistance level) salinities are based on principally different mechanisms.

Acknowledgments

This work was supported by the Russian Foundation of Basic Research (grant number 96-04-48394 and 96-04-48965) and INTAS (project 94-391). We would like to express our gratitude to Dr A. Gorbushin and Dr V. Stefanov for their assistance in the preparation of the article. We also highly appreciate the kind support of Dr H. Hummel during our work on the manuscript.

References

Aliakrinskaya, I. O., 1972. Biochemical adaptations of aquatic molluscs to inhabiting air. Zool. Zhurn. 51: 1630–1636 [in Russian].

Beliaev, G. M., 1951. Osmotic pressure of cavity fluid of aquatic invertebrates in waters of different salinity. Trudy Vsesouz. Gidrobiol. ob-va. 3: 92–139 [in Russian].

Beliaev, G. M., 1957. Physiological peculiarities of representatives of the same species in water of different salinity. Trudy Vsesoyuzn. Gidrobiol. ob-va. 8: 321–353 [in Russian].

Berger, V. J., 1976. On adaptations of some littoral White Sea molluscs to salinity changes. In Khlebovich, V. V. & V. J. Berger (eds), Solenostnye adaptatsii vodnykh organizmov. Zooligical Institute, Leningrad: 69–111 [in Russian].

Berger, V. J., 1986. Adaptation of marine molluscs to environmental salinity changes. Nauka, Leningrad: 214 pp [in Russian].

Berger, V. J., 1989. On the adaptation of molluscs to increased salinity. Biologiia moria. 2: 30–35 [in Russian].

Berger, V. J. & A. D. Kharazova, 1971. Investigation of substantial changes in protein synthesis during adaptation to lowered salinity of the environment in some White Sea snails. Tsitologia 13: 1299–1303 [in Russian].

Berger, V. J. & A. D. Kharazova, 1977. The influence of low salinity on RNA passage from the nucleus to the cytoplasm of ctenidial cells of the snail *Littorina littorea*. Tsitologia 19: 233–35 [in Russian].

Berger, V. J., B. N. Letunov, G. V. Vshevtsov & O. L. Saranchova, 1985. Morpho-functional and ecological aspects of byssus formation in mussels *Mytilus edulis* L. In Berger, V. J. & L. N. Seravin (eds), Ekologija obrastanija v Belom more. Zoological Institute, Leningrad: 67–76 [in Russian].

Berger, V. J. & V. V. Lukanin, 1972. Inhibition of the capacity to salinity acclimation in *Aurelia aurita* (L.) larvae by actinomycin D. Dokl. Akad. nauk SSSR. 202: 205–207 [in Russian].

Berger, V. J., V. V. Lukanin & V. V.Khlebovich, 1970. Effect of actinomycin D on the capacity to salinity acclimation in larvae of the jellyfish *Aurelia aurita* and the mollusc *Epheria vincta*. Zhurn. Evolyuts. Biochim. Fiziol. 6: 636–638 [in Russian].

Berger, V. J., A. D. Naumov & A. I. Babkov, 1995. Quantity and diversity dependence of marine benthos on environmental salinity. Biologiia moria. 21: 45–50 [in Russian].

Berger, V. J., A. N. Pachomov & A. G. Mukhlenov, 1975. Investigation of esterase and lactatedehydrogenase isozymes spectra during adaptation of molluscs *Littorina littorea* to environmental salinity changes. Zhurn. Obsch. Biologii. 36: 579–584 [in Russian].

Berger, V. J. & S. O. Sergievskii, 1986. Differences in adaptive reactions on salinity changes of individuals of *Littorina obtusata* with different shell colour. Biologia moria. 1: 36–41 [in Russian].

Berger, V. J. & S. O. Sergievskii, 1990. Evolution of salinity adaptations in marine molluscs. Zhurn. Evolyuts. Biochim. Fiziol. 26: 462–468 [in Russian].

Bishop, S. H., D. E. Greenwalt, M. A. Kapper, K. T. Paynter & L. L. Ellis, 1994. Metabolic regulation of proline, glycine and alanine accumulation as intracellular osmolytes in ribbed mussel gill tissue. J. exp. Zool. 268: 151–161.

Black, R. E. & L. Bloom, 1984. Heat-shock proteins in *Aurelia* (Cnidaria, Scyphozoa). J. exp. Zool. 230: 303–307.

Davenport, J., 1979. Is *Mytilus edulis* a short term osmoregulator? Comp. Biochem. Physiol. 64: 91–95.

Davenport, J., 1981. The opening response of mussels (*Mytilus edulis* L.) exposed to rising sea-water concentrations. J. mar. biol. Ass. UK 61: 667–668.

Ferraris, J. D. & A. Garcia-Perez, 1996. Osmoregulatory gene expression and implications for evolutionary studies: Strategies in identification of the osmotic response element (ORE). In Ferraris, J. D. & R. Stephen (eds), Molecular Zoology: Advances, Strategies and Protocols. John Wiley & Sons, New York: 313–326.

Florkin, M. & E. Schoffeniels, 1969. Molecular Approaches to Ecology. John Wiley & Sons, New York, 203 pp.

Freel, R. W., 1978. Patterns of water solute regulation in the muscle fibres of osmoconforming marine decapod crustaceans. J. exp. Biol. 72: 107–126.

Freeman, R. F. & F. H. Rigler, 1957. The responses of *Scrobicularia plana* (Da Costa) to osmotic pressure changes. J. mar. biol. Ass. UK 36: 553–567.

Fretter, V. & A. Graham, 1962. British prosobranch molluscs. Proc. r. Soc., London. 144: 3–755.

Gilles, R., 1972. Osmoregulations of three molluscs: *Acanthochitona discrepans* (Brown), *Glycymeris glycymeris* (L.) and *Mytilus edulis* (L.). Biol. Bull. 142: 25–35.

Gilles, R., 1979. Mechanism of osmoregulation in animals. Wiley Interscience, New York, 667 pp.

Ginetsinskiy, A. G., 1963. Physiological mechanisms of water–salt balance. Nauka, Moscow-Leningrad, 276 pp [in Russian].

Golikov, A. N. & O. G. Kusakin, 1978. Shell gastropod molluscs from the intertidal zone of USSR seas. Nauka, Leningrad, 256 pp [in Russian].

Gurina, V. I., 1975. Investigation of RNA and protein synthesis in epithelial tissues of molluscs during adaptation to environmental salinity changes. Tsitologia 17: 298–303 [in Russian].

Hedgpeth, J. W., 1967. Ecological aspects of Laguna Madre, hypersaline estuary. In Lauff, G. H. (ed.), Estuaries. A.A.A.S., Washington: 408–419.

Jensen, K. T., G. Latama & K. N. Mouritsen, 1996. The effect of larval trematode on the survival rates of two species of mud snails (Hydrobiidae) experimentally exposed to dessication, freezing and anoxia. Helgolander wiss. Meeresunters. 50: 327–335.

Kharazova, A. D., 1987.The role of plastic metabolism in adaptations of hydrobionts to abiotic factors of the environment. Trudy zoologicheskogo instituta AN SSSR. 160: 59–84 [in Russian].

Kharazova, A. D., 1994. Protein and RNA metabolism in the tissues of marine molluscs at changes of environmental salinity. Izvestia Akad. Nauk. 4: 561–565 [in Russian].

Kharazova, A. D. & V. J. Berger, 1974. Changes of RNA synthesis in tissues of the mollusc *Littorina littorea* (L.) at decreasing water salinity. Tsitologia 16: 241–243 [in Russian].

Kharazova, A. D., V. J. Berger, V. I. Fateeva, L. M. Yaroslavtseva & P. V. Yaroslavtsev, 1981. Influence of salinity on the dynamics of protein synthesis in isolated gills of *Crenomytilus grayana*. Biologiia moria 6: 55–60 [in Russian].

Kharazova, A. D., V. J. Berger, V. I. Fateeva, L. M. Yaroslavtseva & P. V. Yaroslavtsev, 1983. On the correlation of organismic and cellular reactions at adaptation of mussels to environmental salinity changes. Dokl. AN SSSR. 269: 245–247 [in Russian].

Kharazova, A. D., N. V. Nechaeva & V. I. Fateeva, 1989. Circahouralian rhythms of protein synthesis in the tissues of some invertebrates. Tsitologia 31: 601–614 [in Russian].

Kharazova, A. D. & V. V. Rostova, 1976. Investigation of protein and RNA synthesis changes in tissues of White Sea mollusc *Coryphella rufibranchialis* at lowered salinity. In Khlebovich, V. V. & V. J. Berger (eds), Solenostnye adaptatsii vodnykh organizmov. Zoological Institute, Leningrad: 142–155 [in Russian].

Khlebovich, V. V., 1962. Pecularities of aquatic fauna composition in dependence of environmental salinity. Zhurn. Obsch. Biologii. 23: 90–97 [in Russian].

Khlebovich, V. V., 1974. Critical salinity of biological processes. Nauka, Leningrad, 230 pp [in Russian].

Khlebovich, V. V. & A. P. Kondratenkov, 1973. Stepwise acclimation – a method for estimating the potential euryhalinity of the gastropod *Hydrobia ulvae*. Mar. Biol. 18: 6–8.

Klekowski, R. Z., 1963. The influence of low salinity and dessication on the survival, osmoregulation and water balance of *Littorina littorea*. Polsk. arch. hydrobiol. 11: 241–250.

Korolkova, E. D. & A. D. Kharazova, 1994. EM investigation of mussel gill epithelium at a lowering of salinity. Tsitologia 36: 69–75 [in Russian].

Kreps, E. M., 1929. Investigations of gas exchange of *Balanus crenatus* at different salt concentrations of the environment. Trudy Murman. Biol. Stantsii. 3: 1–32 [in Russian].

Kroeger, H., 1967. Hormones, ion balance and gene activity in dipteran chromosomes. Mem. Soc. Endocrinol. 15: 55–56.

Krogh, A., 1939. Osmotic regulation in aquatic animals. University Press, Cambridge, 242 pp.

Kuzmina, O. Y., 1982. Role of intracellular inorganic ions in the adaptation of some poikilosmotic animals to environmental salinity changes. Ph.D. dissertation. Zoological Institute, Leningrad, 172 pp [in Russian].

Lavrova, E. A. & Y. V. Natochin, 1981. Inhibition of chloride permeability and sodium transport in frog skin by merkuzal and etakrinic acid. Biofizika 26: 651–656 [in Russian].

Lezzi, M., 1970. Differential gene activation in isolated chromosomes. Int. Revue Cytol. 29: 127–168.

Lukanin, V. V., 1976. Study of adaptive reactions of White Sea scyphomedusae *Aurelia aurita* (L.) on environmental salinity changes. In Khlebovich, V. V. & V. J. Berger (eds), Solenostnye adaptatsii vodnykh organizmov. Zoological Institute, Leningrad: 28–58 [in Russian].

Lukanin, V. V. & V. V. Khlebovich, 1979. Effect of inhibitors of protein synthesis on metamorphosis and adaptive capacity of scyphomedusae *Aurelia aurita* (L.) during water freshening. Ontogenez. 10: 80–83 [in Russian].

Lvova, T. G. & E. E. Kulakovsky, 1979. Investigation on protein and RNA synthesis in the tissues of the polychaete *Micronephthys minuta* at changes of environmental salinity. Tsitologia 21: 1356–1360.

Marek, M. & H. Kroeger, 1974. Influence of Na/Mg on the pattern of esterases in explanted *Galleria melonella* midgut. Comp. Biochem. Physiol. 47B: 503–506.

Marek, M. & H. Kroeger, 1976. Influence on Na, K, Mg and cooling on proteosynthesis in hemocytes of *Galleria melonella*. Comp. Biochem. Physiol. 53B: 45–47.

Natochin, Y. V., 1966. Reaction of mussels on separate changes of osmotic concentration and salinity in the environment. Zhurn. Obsch. Biologii. 27: 473–479 [in Russian].

Natochin, Y. V., 1976. The kidney: regulation of ionic balance. Nauka, Leningrad, 286 pp [in Russian].

Natochin, Y. V. & V. J. Berger, 1979. Ionic content of molluscs cells – evolutionary and ecological aspects. Zhurn. Evolyuts. Biokhim. Fiziol. 15: 295–302 [in Russian].

Natochin, Y. V., V. J. Berger, V. V. Khlebovich, E. A. Lavrova & O. Y. Mikhailova, 1979. The participation of electrolytes in adaptation mechanisms of intertidal mollusc cells to altered salinity. Comp. Biochem. Physiol. 63A.: 115–119.

Oliver, L. T. & M. Brand, 1953. The influence of lack of oxygen on *Schistosoma mansoni* cercariae and infected *Austrolorbus glabratus*. Exp. Parasitol. 12: 339–366.

Pierce, S. K., 1982. Invertebrate cell volume control mechanism: a coordinated use of intracellular amino acid and inorganic ions as osmotic solute. Biol. Bull. 169: 405–419.

Pierce, S. K, 1994. Osmolyte permeability in molluscan red cells is regulated by Ca^{2+} and membrane protein phosphorylation: the present perspective. J. exp. Zool. 268: 166–170.

Potts, W. T. W., 1958. The inorganic and amino acid composition of some lamellibranch muscles. J. exp. Biol. 53: 749–764.

Potts, W. T. W. & G. Parry, 1964. Osmotic and ionic regulations in animals. Pergamon Press, Oxford, 412 pp.

Savvateev, V. B., 1952. On the physiology of adaptations in *Balanus* to salinity oscillations. Zool. Zhurn. 31: 801–805 [in Russian].

Schlieper, C., 1960. Genotypische und phaenotypische Temperatur und Salzgehalts Adaptationen bei merinen Bodenvertebraten der Nord und Ostsee. Kieler Meeresforsch. 16: 180–185.

Sergievskii, S. O., 1983. Shell-colour polymorphism: paramethric systems. In Mollyuski, sistematika, ekologiia i zakonomernosti raspredeleniia. Leningrad: 52–54 [in Russian].

Sergievskii, S. O. & V. J. Berger, 1983. Population-physiological analysis of shell-colour polymorphism of *Littorina obtusata* (Gastropoda: Prosobranchia). In Likharev, I. M. (ed.), Mollyuski, sistematika, ekologiia i zakonomernosti raspredeleniia. Nauka, Leningrad: 55–56 [in Russian].

Sergievskii, S. O. & V. J. Berger, 1984. Physiological differences of principal shell-colour phenotypes of the gastropod mollusc *Littorina obtusata*. Biol. moria. 2: 36–44 [in Russian].

Todd, M. E., 1964. Osmotic balance in *Littorina littorea, L. litoralis* and *L. saxatilis* (Litorinidae). Physiol. Zool. 37: 33–44.

Vasilieva, V. F., A. G. Ginetsinskiy, M. G. Zaks & M. M. Sokolova, 1960. Two types of adaptation of poikilosmotic marine animals to hypodynamic environment. In Voprosy tsitologii i obscheiy fiziologii. Publishing house of USSR Academy of Sciences, Moscow-Leningrad: 50–60 [in Russian].

Vernberg, W. B. & F. J. Vernberg, 1963. Influence of parasitism on thermal resistance of the mud-flat snail, *Nassarius obsoleta*. Soc. exp. Parasitol. 14: 330–332.

Zhirmunskiy, A. V., 1962. The reaction of ciliary epithelia cells of mussels and sea anemones to reduced salinity. Zhurn. obsch. biologii. 23: 119–126 [in Russian].

Hydrobiologia **355**: 127–138, 1997.

A. D. Naumov, H. Hummel, A. A. Sukhotin & J. S. Ryland (eds), Interactions and Adaptation Strategies of Marine Organisms.

Sensitivity to stress in the bivalve *Macoma balthica* from the most northern (Arctic) to the most southern (French) populations: low sensitivity in Arctic populations because of genetic adaptations?

Herman Hummel[1], Roelof Bogaards[1], Tatiana Bek[2], Lennard Polishchuk[2], Claude Amiard-Triquet[3], Guy Bachelet[4], Michel Desprez[5], Peter Strelkov[6], Alex Sukhotin[6], Andrei Naumov[6], Salve Dahle[7], Stanislav Denisenko[8], Michael Gantsevich[2], Kirill Sokolov[2] & Lein de Wolf[1]

[1]*Centre for Estuarine and Coastal Ecology, Netherlands Institute of Ecology, Vierstraat 28, 4401 EA Yerseke, The Netherlands*

[2]*White Sea Biological Station Poyakonda, Moscow State University, Biology faculty, Vorobjovy Gory, Moscow 119899, Russia*

[3]*Université de Nantes, Faculté de Pharmacie, Laboratoire d'Ecotoxicologie, 1, Rue Gaston-Veil, 44035 Nantes Cedex, France*

[4]*Université de Bordeaux I, Laboratoire d'Océanographie biologique, 2, Rue du Professeur Jolyet, 33120 Arcachon, France*

[5]*GEMEL Picardie, Station d'Etudes en Baie de Somme (SEBS), 115, Quai Jeanne d'Arc, 80230 Saint Valéry sur Somme, France*

[6] *White Sea Biological Station, Zoological Institute, Russian Academy of Sciences, Universitetskaya nab. 1, St. Petersburg 199034, Russia*

[7]*Akvaplan-Niva, Strandtorget 2B, 9001 Tromso, Norway*

[8]*Murmansk Marine Biological Institute, Vladimirskaya Str. 17, Murmansk 183010, Russia*

Key words: Arctic, adaptation, copper, distribution limit, genetics, geographic cline, *Macoma balthica*, stress sensitivity, survival in air

Abstract

The stress sensitivity, determined in copper exposure experiments and in survival in air tests, and the genetic structure, measured by means of isoenzyme electrophoresis, were assessed in populations of the Baltic clam *Macoma balthica* (L.) from its southern to its northern distribution limit, in order to test the hypotheses that near the distribution limit the clams would be more stress sensitive and would have a lower genetic variability. The populations in west and north Europe show a strong genetic resemblance. The populations in the sub-Arctic White Sea are genetically slightly different, and show a low stress sensitivity. The populations in the Arctic Pechora Sea are genetically very distant from the other populations, and show the lowest stress sensitivity. Near the southern distribution limit, in agreement with the hypotheses, genetic variability is low and stress sensitivity high. On the other hand, in contrast to expectation, near the northern distribution limit, in the populations of the Pechora Sea, the genetic variability was higher, thus not reduced, and the stress sensitivity was low compared to all other populations. Yet, it remains a question if such is due to gradual physiological acclimatization (and ongoing differential selection) or to genetic adaptation.

Introduction

The sensitivity to stress in animal populations is assumed to be higher near the limits of a species' distribution (Hoffmann & Parsons, 1991; Hummel et al., 1996a). Because marginal populations are living at the limits of their adaptation capacities, they show a poor performance with regard to such parameters as growth, fitness or stress sensitivity, and reduced genetic variation (Conover, 1978; Hoffman & Parsons, 1991; Hummel et al., 1995, 1996a). Although in temperate areas many studies are carried out on stress sensitivity of marine and estuarine organisms, there is a strong lack on such studies in (sub-)Arctic specimens (Chapman, 1993).

In the Baltic clam *Macoma balthica* (L.), which is in NW Europe often a dominant species in the coastal benthic zone, a poor performance was found near the southern distribution limit (Hummel et al., 1995, 1996a, b, 1997). A decrease was found for growth, condition (low weight per volume), reserve constituents (low glycogen level), free amino acids (high taurine/glycine ratio), genetic structure (loss of variation at leucine aminopeptidase), and stress sensitivity (high mortality at low levels of copper) (Beukema & Meehan, 1985; Hummel et al., 1995, 1996a, b). Near the northern distribution limit of *M. balthica* no such studies were available.

Therefore, we assessed the stress sensitivity of the Baltic clam near the northern distribution limit. Our hypothesis was to find in the north a high stress sensitivity. To this end, clams were collected from populations in the White Sea and Pechora Sea and their sensitivity measured in copper exposure experiments or in 'survival in air' tests. In order to test whether the populations studied are genetically similar, and thus can be compared, the genetic structure was measured in the populations by means of electrophoretic isoenzyme analysis. It was hypothesized that towards the north low levels of genetic variation would be found, because near the geographic limits, a species' physiological acclimatization limits will be met and consequently differential selection will take place (Hoffmann & Parsons, 1991). The results were compared with similar analyses in France and the Netherlands (Hummel et al., 1995, 1996a).

Material and methods

Baltic clams, *M. balthica*, were sampled during 1991 to 1995 in populations from the most southern distribution limit in south-west France (Gironde, France; Bachelet, 1980) up to its most north-eastern distribution limit in the Arctic Pechora Sea (Khaypudyr Bay; Denisenko, Naumov: own observations) (Table 1; Figure 1). The stations in the Pechora Sea (Khaypudyr and Pechora) are in all winters at least 6 months (average 8 months) covered by ice (Figure 1) (Treshnikov, 1985). The water near the stations in the sub-Arctic White Sea (including Mezen, Table 1) is annually on average 4 months covered by ice, but in rare years not at all. The other stations (Murmansk, Tromso and the stations in the Netherlands and France) are all year free of ice, because of the warm Gulf Stream.

For genetic analyses 40 to 80 animals were dissected, and the total soft tissues taken and frozen in liquid nitrogen. The genetic structure of the animals was examined using electrophoretic isoenzyme analysis of seven loci according to Menken (1982) and Hummel et al. (1995): glucosephosphate isomerase (Gpi, E.C. 5.3.1.9), isocitrate dehydrogenase I and II (Idh, E.C. 1.1.1.42), leucine aminopeptidase (Lap, E.C. 3.4.11.1), malate dehydrogenase (Mdh, E.C. 1.1.1.37), phosphogluconate dehydrogenase (Pgd, E.C. 1.1.1.44), and phosphoglucomutase (Pgm, E.C. 5.4.2.2). The tissues were homogenized individually for a few seconds in maximally 0.2 ml of gel buffer. Electrophoresis was carried out in horizontal 12% starch gels (50% Sigma, 50% Connaught) at a temperature of 0 °C. The buffer systems used were Tris – citric acid gel buffer (8 and 3 mm resp.; pH 6.7) and Tris – citric acid electrode buffer (0.223 and 0.086 m resp.; pH 6.3). The electrophoresis was performed for 5 hours with a constant current of 100 mA. Staining procedures used Bush B Tris – hydrochloric acid (0.102 m; pH 8.4) according to Menken (1982).

The fastest allele was called A, the slower B, C, and so on. The data were analyzed and statistically tested for allele frequencies, heterozygosity, conformance to Hardy-Weinberg equilibrium (fixation index F_{IS}), coefficient of gene differentiation (fixation index F_{ST}) and genetic identities (standard genetic identity according to Nei, 1975) by the Biosys computer programme (Swofford & Selander, 1981). Differences in allele frequencies and heterozygote frequencies of different groups were tested with the χ^2 analysis (Sokal & Rohlf, 1995). The *F*-statistics F_{IS} and F_{ST} are defined according to Nei (1977), and have properties similar to that of Wright's (1965) definition. F_{IS} measures within a population the deviation of genotype frequencies from Hardy-Weinberg proportions, and the null hypothesis $F_{IS} = 0$ was tested for significance with $\chi^2 = NF_{IS}^2\ (b-1)$ and $b(b-1)/2$ degrees of freedom (*N* is number of specimens analyzed in the subpopulations, *b* number of alleles) (Li, 1955). F_{ST} measures the degree of genetic differentiation of subpopulations and was tested for significance with $\chi^2 = 2NF_{ST}\ (b-1)$ and $df = (b-1)(n-1)$ (*n* is number of subpopulations) (Workman & Niswander, 1970). The statistics were performed with Bonferroni correction, at a crit-

Table 1. Sampling stations (and abbreviations), cluster groups (W+N Eur.= West and North Europe), sampling dates and numbers of animals for genetic analyses (*N*).

Station	Code / Cluster	Country	Sampling date	(*N*)
Gironde	Gi / W+N Eur.	France	June 1991	40
Loire	Lo / W+N Eur.	France	June 1991	80
Baie des Veys	BdV / W+N Eur.	France	June 1991	40
Seine	Se / W+N Eur.	France	June 1991	80
Somme	So / W+N Eur.	France	June 1991	80
Baalhoek	Bh / W+N Eur.	Netherlands	June 1991	40
Paulina	Pa / W+N Eur.	Netherlands	June 1991	80
Tromso	Tr / W+N Eur.	Norway	September 1993	80
Dalnie Zelentsi	DZ / W+N Eur.	Russia	September 1993	80
Mezen	Me / W+N Eur.	Russia	July 1995	80
Luvenga	Lu / White Sea	Russia	July 1991	
Oil Station	Oi / White Sea	Russia	July 1991	
Kovda	Ko / White Sea	Russia	July 1991	
Poyakonda	Po / White Sea	Russia	August 1994 + January 1995	80
Kislaja Guba	KG / White Sea	Russia	August 1994	80
Pechora	Pe / Pechora Sea	Russia	July 1992 + July 1995	80
Khaypudyr	Kh / Pechora Sea	Russia	July 1995	80

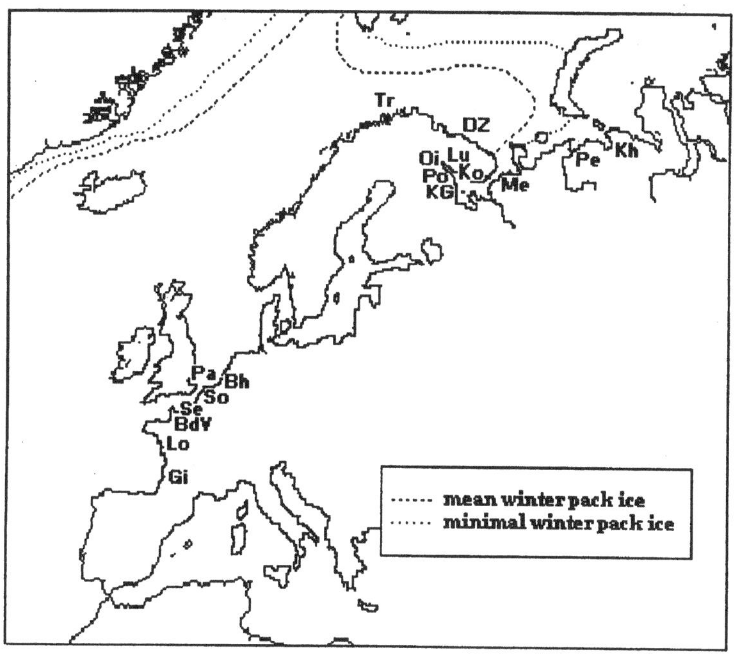

Figure 1. Location of the sampling stations (mean and minimal boundaries of ice cover in winter abstracted from Treshnikov, 1985; explanation of abbreviations of the stations in Table 1)

ical probability level of $\alpha' = 0.05\ x^{-1}$ (x is number of repetitions of the same test, i.e. 8; Sokal & Rohlf, 1995).

In order to test the survival in air, 100 specimens of each population were placed on wet filter paper on top of a 1 to 2 cm layer of wet sand, in such a way that the animals were exposed to air with a high humidity. The mortality was followed daily; those not reacting on piercing with a needle were judged to be dead.

Table 2. Allele frequencies in *Macoma balthica* of the different clustered groups and measures of genetic variability calculated on basis of all individual populations (stations of each group are given in table 1; (*N*) = number of specimens; abbreviations of isoenzymes in Materials and Methods; H_o = observed heterozygosity, H_e = expected heterozygosity, F_{IS} = conformance to Hardy-Weinberg equilibrium, F_{ST} = gene differentiation; Bonferroni correction for tests on significance of F_{IS} and F_{ST} $\alpha' = \alpha/8$; *: $p<0.05$, **: $p<0.01$)

Locus	Population			F_{ST}	F_{IS}
	W+N Europe	White Sea	Pechora	based on individual populations	
MDH (*N*)	677	160	160		
A	0.003	0.000	0.022	0.016	
B	0.983	1.000	0.978	0.020	
C	0.013	0.000	0.000	0.029	
Average				0.022	0.034
PGD (*N*)	677	160	160		
A	0.035	0.066	0.569	0.451 **	
B	0.951	0.856	0.366	0.451 **	
C	0.013	0.078	0.016	0.071	
D	0.001	0.000	0.050	0.059	
Average				0.407	0.271
IDH1 (*N*)	677	160	160		
A	0.023	0.009	0.056	0.017	
B	0.942	0.747	0.750	0.121	
C	0.035	0.244	0.194	0.132	
D	0.001	0.000	0.000	0.006	
Average				0.112	0.295
IDH2 (*N*)	677	160	160		
A	0.001	0.000	0.016	0.024	
B	0.018	0.025	0.056	0.010	
C	0.818	0.881	0.850	0.018	
D	0.017	0.022	0.019	0.031	
E	0.143	0.072	0.047	0.006	
F	0.003	0.000	0.013	0.016	
Average				0.019	0.136
GPI (*N*)	677	160	160		
A	0.013	0.003	0.003	0.016	
B	0.152	0.225	0.456	0.071	
C	0.002	0.003	0.013	0.013	
D	0.362	0.287	0.069	0.075	
E	0.006	0.000	0.016	0.014	
F	0.444	0.475	0.444	0.023	
G	0.021	0.006	0.000	0.016	
Average				0.051	0.025
PGM (*N*)	494	160	160		
B	0.047	0.013	0.025	0.028	
C	0.421	0.244	0.100	0.127 *	
D	0.008	0.041	0.003	0.018	
E	0.351	0.131	0.184	0.067	
F	0.015	0.344	0.022	0.235 **	
G	0.121	0.116	0.547	0.211 **	
H	0.028	0.112	0.087	0.041	
I	0.008	0.000	0.031	0.049	
Average				0.122	0.107

Table 2. Continued

Locus	Population W+N Europe	White Sea	Pechora	F_{ST} based on individual populations	F_{IS}
LAP (*N*)	643	160	160		
A′	0.003	0.000	0.041	0.032	
A	0.083	0.025	0.009	0.032	
B	0.565	0.278	0.125	0.145 **	
C	0.262	0.200	0.147	0.038	
D	0.061	0.287	0.475	0.202 **	
E	0.014	0.147	0.144	0.140 **	
G	0.011	0.059	0.059	0.047	
H	0.000	0.003	0.000	0.006	
Average				0.113	0.089
Averages					
Alleles	5.7	4.4	5.1		
H_o	0.320	0.362	0.375		
H_e	0.354	0.437	0.458		
F_{IS}	0.096	0.172	0.181		0.095
F_{ST}				0.116	

In the copper exposure experiments animals were kept in duplicate sets of aquaria with aerated water of 30 ppt salinity at a temperature of 10 to 12 °C (Avg 11, SD 1.5). Aquaria without sediment and containing groups of 50 clams were used for each experimental condition. Filtered (5 μm) water, was pumped from the Somme (in experiments with French and Dutch populations) or White Sea near Poyakonda (in experiments with White Sea populations). The water was changed three times a week, and copper concentrations adjusted to nominal concentrations of 0, 25 or 37.5 and 75 ppb Cu, respectively. Due to adsorption onto the walls of the aquaria and uptake of copper, the average copper concentration in the water became lower (the average concentration is approximately 50% of the nominal concentration; Hummel et al., 1995). Food, *Phaeodactylum tricornutum*, was added only 6 hours before changing the water. The mortality was followed daily. If 50% of the animals in an aquarium died the experiment was stopped.

Results

Genetic structure

The locus Mdh was monomorphic, according to the 0.95 criterion (Table 2). The other loci had a higher allelic variability, with two to four major alleles (Table 2). The average number of alleles was lowest in the southern populations and highest in the most north-eastern (Figure 2). The average heterozygosity across the seven loci ranged from 0.26 to 0.45 (Figure 2), with higher values going from south to north-eastern populations (Figure 2). A consistent heterozygote deficiency occurred (Figure 2); yet, the connected deviations from Hardy-Weinberg equilibrium (F_{IS}) were not significant (Table 2). Such a trend of a heterozygote deficiency is a common, but not yet understood phenomenon, in marine invertebrates (Berger, 1983; Singh & Green, 1984; Zouros, 1987; Zouros & Mallet, 1989; Gaffney, 1994).

The cluster-analysis on basis of the genetic identity between populations showed three major groups (Figure 3). Most distant is the group with populations of the Pechora Sea, with a genetic identity of only 0.78 to the other populations. The second group, consisting of populations in the White Sea, had a higher similarity (genetic identity of 0.93) to West and North European populations than to the Pechora Sea populations (Figure 3). The genetic identities between stations from west and north Europe were higher than 0.97 (Figure 3).

The low similarity (genetic identity) between populations was also apparent from the high genetic differentiation (F_{ST}), which was highly significant at 3 loci (Table 2). The differentiation coincided clearly with geographic changes (clines) in allele frequencies (Fig-

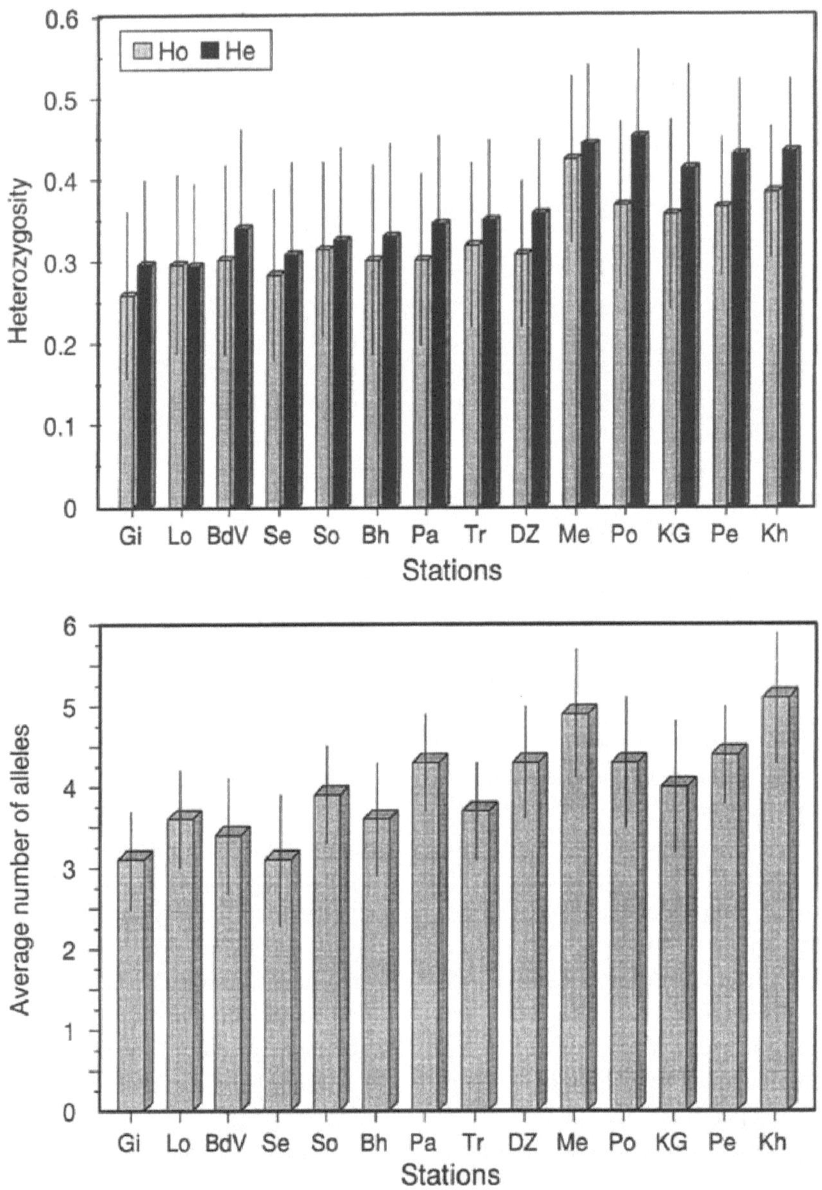

Figure 2. The heterozygosity (observed (Ho) and expected (He)) and average number of alleles (and the standard errors) for the 7 loci analyzed in populations from the northern (Gironde) to the southern (Khaypudyr) distribution limit of the Baltic clam (codes of stations in Table 1).

ure 4). Several gradual clines (Gpi, Idh1, Lap, Pgm) and one sharp cline, with a strong change in the Pechora Sea (Pgd), were found.

Stress sensitivity

The mortality due to copper exposure was higher in the west European populations than in the White Sea populations at all concentrations tested (Figure 5). Especially at an exposure of 75 ppb Cu the differences are strong: in the White Sea populations hardly any mortality was found during 4 weeks, whereas in the west European populations 50% mortality is reached in 2 to 3 weeks.

The survival in air period was shortest in the west European populations, longer in the White Sea populations, and longest in the Pechora Sea populations (Figure 5).

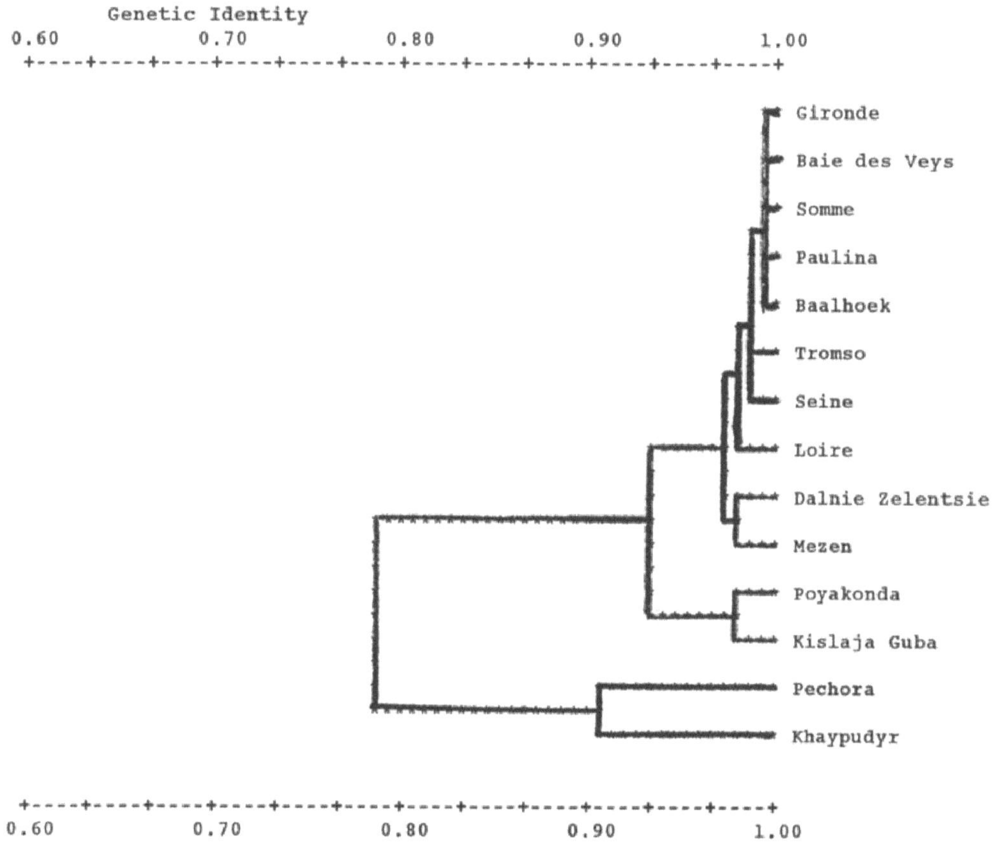

Figure 3. Genetic similarity between populations of Baltic clams measured as the standard genetic identity according to Nei (1975) (cluster analysis using the unweighted pair group method UPGM; codes of stations in Table 1).

Discussion

The genetic structure and variability of *M. balthica* in this study was similar to that reported in previous studies (Väinölä & Varvio, 1989; Green et al., 1983; Singh & Green, 1984; Meehan, 1985; Nilsson, 1985; Hummel et al., 1995), and was for the West and North European and White Sea populations within the range described for the East Atlantic group (Väinölä & Varvio, 1989; Hummel et al., 1995). Combined with the present data, the distribution of the East Atlantic group comprises populations from the southernmost distribution limit of the species in south-west France up to the Russian sub-Arctic White Sea. The data on the Pechora Sea populations are different from reported in earlier investigations.

The genetic identities between the populations in West and North Europe and the White Sea were all above 0.9, indicating according to Avise (1974) and Thorpe (1983) that they belong to the same (sub)species. Yet, the Pechora Sea populations, with a genetic identity of 0.78 to the other populations, form clearly a different group. Only a genetic identity below 0.75 would have indicated that populations are from different species (Avise, 1974; Thorpe, 1983). Thus, the populations in the Pechora Sea do apparently not belong to another species, but at least seem to present a subspecies (race) different from the other European populations.

For other bivalves, with high gene flow, the average gene differentiation (F_{ST}) amounts from 0.01 to 0.03 for populations at geographic distances of hundreds of kilometers, and 0.04 to 0.06 at distances of thousands of kilometers (Skibinski et al., 1983; Dillon & Manzi, 1992; Sarver et al., 1992; Grant et al., 1992; Saavedra et al., 1993; Hummel et al., 1994, 1995). In populations with limited gene flow a much higher F_{ST} can be found, e.g. 0.19 as found for *C. glaucum* (Hummel et al., 1994). Therefore, between the *M. balthica* populations studied, including the Pechora Sea, with an average F_{ST} of 0.10, gene flow is obviously hampered but not completely excluded. Gene flow between the

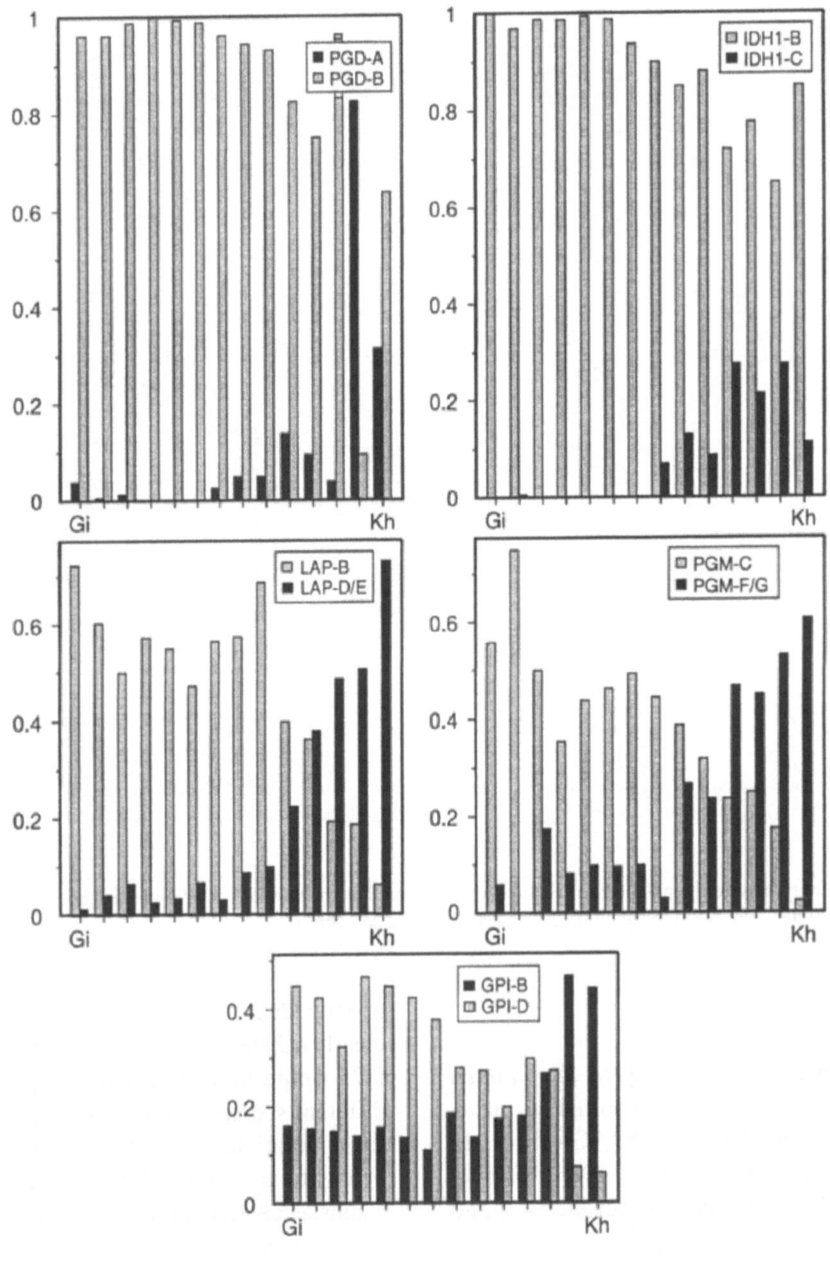

Figure 4. Geographic clines in allele frequencies.

Pechora Sea and the other teritories can be hampered (or blocked) by the minimally 6 months ice cover per year (Treshnikov, 1985) and the in general low Arctic temperatures, whereas the populations of West and North Europe are connected to the (warm) Gulf-stream. Gene flow to and from the White Sea may be hampered by an undeep rise at the entrance of the White Sea and an on average 4 months ice-cover in winter. Yet, evidently some water of the Gulf-stream enters regularly the White Sea, resulting in populations more related to the 'Gulf-stream' populations than to the nearby Pechora Sea populations.

Geographic clines in allele frequencies within a species, especially in a north-south direction, as found by us for most loci, are a common phenomenon, and are mostly related to temperature and salinity (Koehn et al., 1976, 1980b; Endler, 1977; Theisen, 1978; Buroker, 1983; Burton, 1983; Rose, 1984; Hoffman, 1985; Dil-

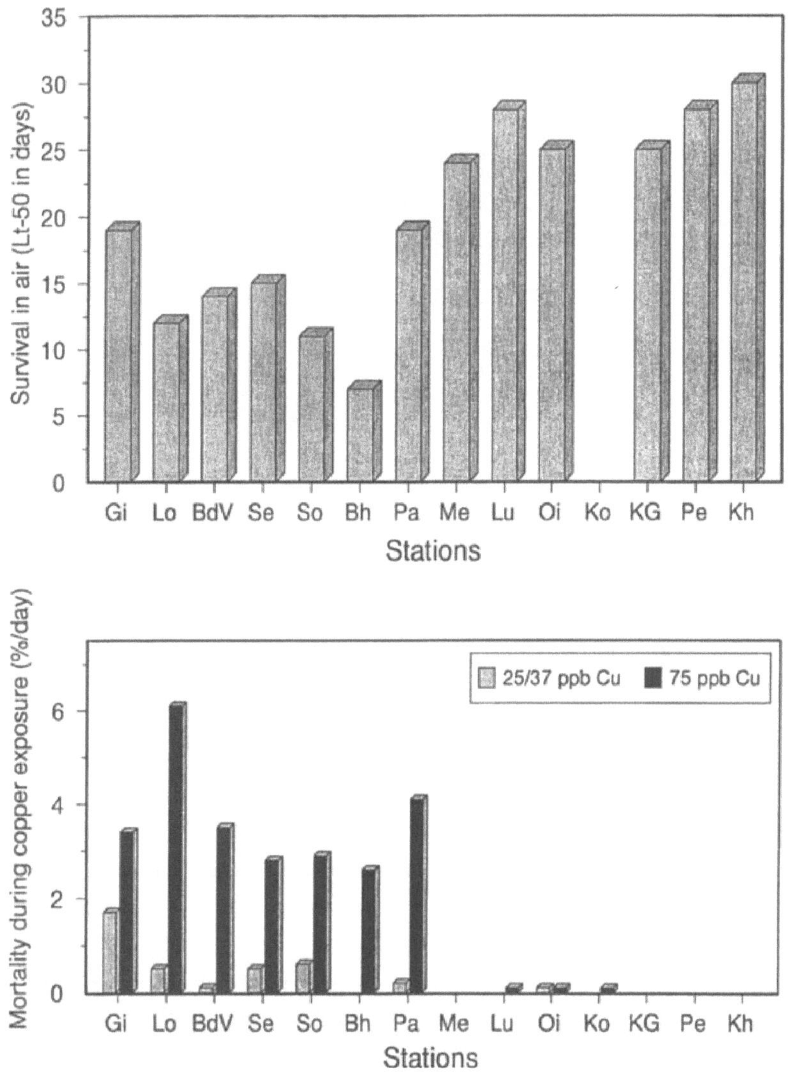

Figure 5. The survival in air (Lt-50; time (days) to reach 50% mortality) and the mortality during coper exposure (% per day during maximally 4 weeks exposure) in Baltic clams (for Me, KG, Pe and Kh no data on copper exposure available).

lon & Manzi, 1992; Hummel et al., 1995). The direct cause of such clines is not clearly known, although for Lap in *Mytilus* and Gpi in *Metridium* it has been shown that differential activities of allozymes are coupled to temperature and salinity (Koehn et al., 1976, 1980a; Koehn & Siebenaller, 1981; Hoffmann, 1985). The differential enzyme activities may lead to genotype dependent differential selection, most probably during juvenile stages (Levinton & Lassen, 1978; Hilbish, 1985). Some clines might also be caused by introgression of races or subspecies (Levinton & Lassen, 1978; Theisen, 1978; Beaumont, 1982; Koehn et al., 1984; Väinölä & Varvio, 1989). Similarly, it can not be excluded that some clines in the Baltic clams studied by us are due to the mixing of two races, which would then be the north-west European race and the Pechora Sea race. However, the mostly gradual clines in allele frequencies support the view that towards the northern or southern limits, stepwise stronger physiological limits, and thus acclimatization limits, are reached, resulting subsequently in gradual differential selection of specific genotypes. This process may have reached so far in the Pechora Sea, that a seemingly different race became apparent. Translocation experiments from, e.g., North Europe or White Sea to the Pechora Sea, and *vice versa*, may elucidate whether in the Pechora Sea the genotype of the populations is the result of differential selection because of reaching

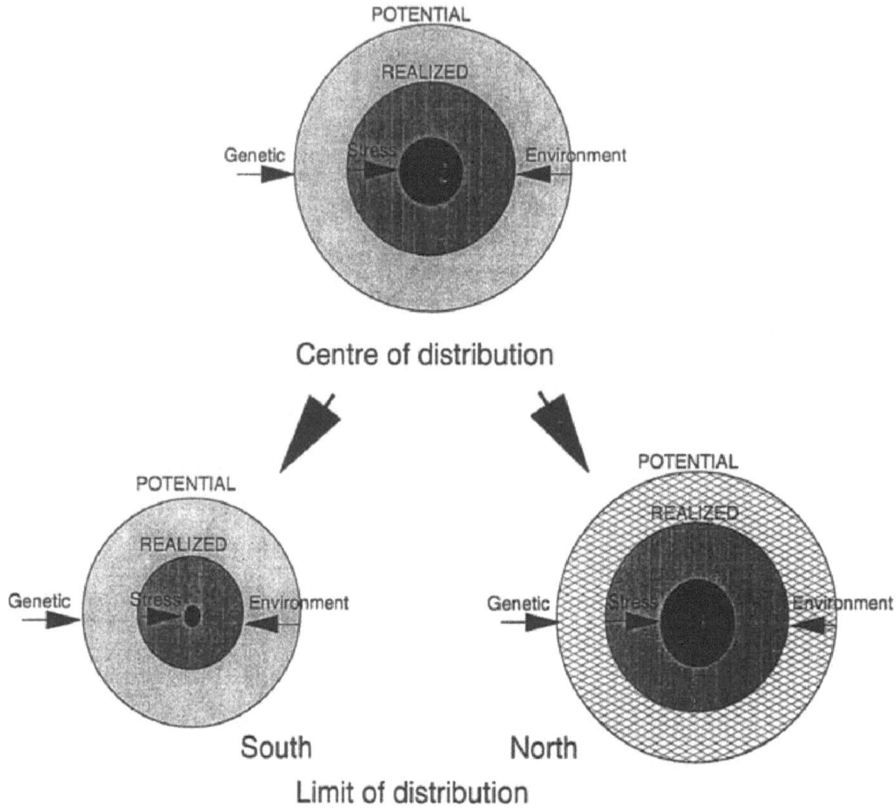

Figure 6. The performance capacity of the Baltic clam at the centre, and southern and northern limit of its distribution (adapted from a scheme by Schreck, 1981).

physiological acclimatization limits, or that the populations are really genetically adapted. Moreover, in the first case (limit of acclimatization) the phenomenon of 'fluctuating asymmetry' (Hoffmann & Parsons, 1991) would be prominent in the Pechora Sea populations, whereas if the second case (genetic adaptation) is true this phenomenon would be negligible. Studies on both the effects of translocation and on 'fluctuating asymmetry' are foreseen in the coming years to solve these issues.

In contrast to expectation we found a gradual increase, and not a decrease, of genetic variability (average number of alleles, heterozygosity) when going to the north-eastern distribution limit in the Pechora Sea. Only towards the southern limit were the lowest values found. A gradual increase of the genetic variability towards the north-eastern limit might be related to the north-eastern direction of the Gulf Stream currents. Because Baltic clams have a 3 wk pelagic larval stage, new genetic (allelic) variants may appear finally also in areas north-east from where they originated, but not in southern areas. Moreover, due to differential selection, specific variants may become extinct in the southern areas, whereas in north-eastern areas such extinct variants may be replenished from southern territories.

From the combination of the higher genetic variability and the lower stress sensitivity of the White Sea and Pechora Sea populations it can be concluded that towards the northern limits the Baltic clam shows a better performance. A similar conclusion was drawn from an analysis of shell-growth in clams: the largest and oldest specimens can be found in the Pechora Sea populations (Hummel et al., 1997). As indicated before, the performance of clams near the southern distribution limit was poor, especially in the Gironde (Hummel et al., 1996a). Such a reduction of the performance may be due to (1) reduction of the genetic potential because of a lower genetic variability, (2) physiological limits are reached and acclimatization is hampered, and (3) consequently, additional stress has a stronger impact (Figure 6) (Schreck, 1981; Hoffmann & Parsons, 1991). On the other hand, in parallel to the above, it has to be concluded that near the northern distribution

Table 3. Mortality of *M. balthica* during survival in air tests and during copper exposure experiments for. All tests at average temperatures of 10 degC. Dutch and French stations tested in May 1991 (from Hummel *et al.*, 1995, 1996b), White Sea stations in August 1991 (original data), Mezen and Pechora Sea stations in July 1995 (from Strelkov & Hummel, 1996).

	Survival in air LT-50 in days	Mortality by copper exposure % day^{-1} (4 wks exposure)		
		0 ppb	37.5 ppb	75 ppb
Gironde	19	0.0	1.7	3.4
Loire	12	0.1	0.5	6.1
Baie des Veys	14	0.1	0.1	3.5
Seine	15	0.1	0.5	2.8
Somme	11	0.0	0.6	2.9
Baalhoek	7	0.0		2.6
Paulina	19	0.0	0.2	4.1
Mezen	24			
Luvenga	28	0.0		0.1
Oil Station	25	0.1		0.1
Kovda		0.1		0.1
Kislaja Guba	25	0.1		0.0
Pechora	28			
Khaypudyr	30			

limit (1) the genetic potential is not reduced or even larger, (2) the limits for the physiological parameters measured are not reached, and (3) stress sensitivity is relatively very low (Figure 6). Yet, it remains for the Pechora Sea populations a question if such is due to gradual and ongoing differential selection of those specimens with genotypes (from basically the same gene-pool as in north-west Europe) that can physiologically acclimatize, or to genetic adaptation (of a different race).

Consequently, because of their different set of ecophysiological and toxicological reaction-norms, in order to predict eventual changes in Arctic systems by human impact, it will be necessary to study the endemic populations. Extrapolations from populations in temperate and sub-Arctic regions are most probably not valid.

Acknowledgments

This study was made possible by the continuous support of UNESCO-COMAR, Paris, several grants from the Netherlands Marine Research Foundation (SOZ, presently Geosciences Foundation GOA), a major grant from the International Association for the promotion of cooperation with scientists from the Independent States of the former Soviet Union (INTAS project 94-391), an INTAS related fund from the Netherlands Organization for Scientific Research (NWO 047.03.009), and grants for mg and KS from the Netherlands Commission for UNESCO (CN3013). NIOO-CEMO communication nr 2287.

References

Avise, J. C., 1974. Systematic value of electrophoretic data. Syst. Zool. 23: 465–481.

Bachelet, G., 1980. Growth and recruitment of the tellinid bivalve *Macoma balthica* at the southern limit of its geographical distribution, the Gironde estuary (SW France). Mar. Biol. 59: 105–117.

Beaumont, A. R., 1982. Geographic variation in allele frequencies at three loci in *Chlamys opercularis* from Norway to the Brittany coast. J. mar. biol. Ass. U.K. 62: 243–261.

Berger, E. M., 1983. Population genetics of marine gastropods and bivalves. In Russell-Hunter, W. D. (ed.), The Mollusca, Ecology. Academic Press, Orlando 6: 563–596.

Beukema, J. J. & B. W. Meehan, 1985. Latitudinal variation in linear growth and other shell charateristics of *Macoma balthica*. Mar. Biol. 90: 27–33.

Buroker, N. E., 1983. Population genetics of the American oyster *Crassostrea virginica* along the Atlantic coast and the Gulf of Mexico. Mar. Biol. 75: 99–112.

Burton, R. S., 1983. Protein polymorphism and genetic differentiation of marine invertebrate populations (Review). Mar. Biol. Lett. 4: 193–206.

Chapman, P. M., 1993. Are Arctic marine invertebrates relatively insensitive to metals. Envir. Toxicol. Chem. 12: 611–613.

Conover, R. J., 1978. Transformation of organic matter. In Kinne, O. (ed.), Marine Ecology, Volume IV, Dynamics. John Wiley & Sons, Chichester: 221–499.

Dillon, R. T. & J. J. Manzi, 1992. Population genetics of the hard clam, *Mercenaria mercenaria*, at the Northern limit of its range. Can. J. Fish. aquat. Sci. 49: 2574–2578.

Endler, J. A., 1977. Geographic Variation, Speciation, and Clines. Princeton University Press, Princeton, 246 pp.

Gaffney, P. M., 1994. Heterosis and heterozygote deficiencies in marine bivalves: More light? In Beaumont, A. R. (ed.), Genetics and Evolution of Aquatic Organisms. Chapman & Hall, London: 146–153.

Gilbert, M. A., 1973. Growth rate, longevity and maximum size of *Macoma balthica* (L.). Biol. Bull. 145: 119–126.

Grant, W. S., A. C. Schneider, R. W. Leslie & M. I. Cherry, 1992. Population genetics of the brown mussel *Perna perna* in southern Africa. J. exp. mar. Biol. Ecol. 165: 45–58.

Green, R. H., S. M. Singh, B. Hicks & J. McCuaig, 1983. An arctic intertidal population of *Macoma balthica* (Mollusca, Pelecypoda): genotypic and phenotypic components of population structure. Can J. Fish. aquat. Sci. 40: 1360–1371.

Hilbish, T. J., 1985. Demographic and temporal structure of an allele frequency cline in the mussel *Mytilus edulis*. Mar. Biol. 86: 163–172.

Hoffmann, R. J., 1985. Thermal adaptation and the properties of phosphoglucose isomerase allozymes from a sea anemone. In Gibbs, P. E. (ed.), Proceedings Nineteenth European Marine Biology Symposium. Cambridge University Press, Cambridge: 505–514.

Hoffmann, A. A. & P. A. Parsons, 1991. Evolutionary Genetics and Environmental Stress. Oxford University Press, New York: 284 pp.

Hummel, H., C. Amiard-Triquet, G. Bachelet, M. Desprez, J. Marchand, B. Sylvand, J. C. Amiard, H. Rybarczyk, R. H. Bogaards, J. Sinke, Y. de Wit & L. de Wolf, 1996a. Sensitivity to stress of the estuarine bivalve *Macoma balthica* from areas between the Netherlands and its southern limits (Gironde). J. Sea Res. 35: 315–321.

Hummel, H., C. Amiard-Triquet, G. Bachelet, M. Desprez, J. Marchand, B. Sylvand, J. C. Amiard, H. Rybarczyk, R. H. Bogaards, J. Sinke, Y. de Wit & L. de Wolf, 1996b. Free amino acids as a biochemical indicator of stress in the estuarine bivalve *Macoma balthica*. Sci. Tot. Envir. 188: 233–241.

Hummel, H., P. Bijok & R. H. Bogaards, 1996c. Effects of tidal zonation on size and genetic traits of *Mytilus edulis* (L.) and *Macoma balthica* (L.). Pol. Arch. Hydrobiol. 43: 431–445.

Hummel, H., R. H. Bogaards, C. Amiard-Triquet, G. Bachelet, M. Desprez, J. Marchand, B. Sylvand, Y. de Wit & L. de Wolf, 1995. Uniform variation in genetic traits of a marine bivalve related to starvation, pollution and geographic clines. J. exp. mar. Biol. Ecol. 191: 133–150.

Hummel, H., R. H. Bogaards, T. Bek, L. Polishchuk, K. Sokolov, C. Amiard-Triquet, G. Bachelet, M. Desprez, A. Naumov, P. Strelkov, S. Dahle, S. Denisenko, M. Gantsevich & L. de Wolf, 1997. Growth in the bivalve *Macoma balthica* from its northern to its southern distribution limit: A discontinuity in north-Europe because of genetic adaptations in Arctic populations? Comp. Physiol. Biochem. (submitted).

Hummel, H., M. Wolowicz & R. H. Bogaards, 1994. Genetic variability and relationships for populations of *Cerastoderma edule* and of the *C. glaucum* complex. Neth. J. Sea Res. 33: 81–89.

Koehn, R. K., B. L. Bayne, M. N. Moore & J. F. Siebenaller, 1980a. Salinity related physiological and genetic differences between populations of *Mytilus edulis*. Biol. J. Linn. Soc. 14: 319–334.

Koehn, R. K., J. G. Hall, D. J. Innes & A. J. Zera, 1984. Genetic differentiation of *Mytilus edulis* in eastern North America. Mar. Biol. 79: 117–126.

Koehn, R. K., R. Milkman & J. B. Mitton, 1976. Population genetics of marine pelecypods. IV. Selection, migration and genetic differentiation in the blue mussel *Mytilus edulis*. Evolution 30: 2–32.

Koehn, R. K., R. I. E. Newell & F. Immermann, 1980b. Maintenance of an aminopeptidase allele frequency cline by natural selection. Proc. Natl. Acad. Sci. USA 77: 5385–5389.

Koehn, R. K. & J. F. Siebenaller, 1981. Biochemical studies of aminopeptidase polymorphism in *Mytilus edulis*. II. Dependence of reaction rate on physical factors and enzyme concentration. Biochem. Gen. 19: 1143–1162.

Levinton, J. S. & H. H. Lassen, 1978. Selection, ecology and evolutionary adjustment within bivalve mollusc populations. Phil. Trans. r. Soc. London, Ser. B, 284: 403–415.

Li, C. C., 1955. Population Genetics. University of Chicago Press, Chicago, 366 pp.

Meehan, B. W., 1985. Genetic comparison of *Macoma balthica* (Bivalvia, Tellinidae) from the eastern and western North Atlantic Ocean. Mar. Ecol. Prog. Ser. 22: 69–76.

Menken, S. B. J., 1982. Biochemical genetics and systematics of small ermine moths (Lepidoptera, Yponomeutidae). Zeitschrift für Zoologischen Systematik und Evolutionsforschung 20: 131–143.

Nei, M., 1975. Molecular Population Genetics and Evolution. North Holland Publ. Co., Amsterdam, 288 pp.

Nei, M., 1977. F-statistics and analysis of gene diversity in subdivided populations. Ann. hum. Gen. 41: 225–233.

Nilsson, J., 1985. Allozyme variation of *Macoma balthica* (L.) in the Bothnian Sea. Heredita 102: 277–280.

Rose, R. L., 1984. Genetic variation in the oyster, *Crassostrea virginica* (Gmelin), in relation to environmental variation. Estuaries 7: 128–132.

Saavedra, C., C. Zapata, A. Guerra & G. Alvarez, 1993. Allozyme variation in European populations of the oyster *Ostrea edulis*. Mar. Biol. 115: 85–95.

Sarver, S. K., M. C. Landrum & D. W. Foltz, 1992. Genetics and taxonomy of ribbed mussels (*Geukensia* spp.). Mar. Biol. 113: 385–390.

Schreck, C. B., 1981. Stress and compensation in Teleostean fishes: Response to social and physical factors. In Pickering, A. D. (ed.), Stress and Fish. Academic Press, London: 295–321.

Singh, S. M. & R. H. Green, 1984. Excess of allozyme homozygosity in marine molluscs and its possible biological significance. Malacologia 25: 569–581.

Skibinski, D. O. F., J. A. Beardmore & T. F. Cross, 1983. Aspects of the population genetics of *Mytilus* (Mytilidae; Mollusca) in the British isles. Biol. J. Linn. Soc. 19: 137–183.

Sokal, R. R. & F. J. Rohlf, 1995. Biometry. Freeman & Co, New York, 887 pp.

Strelkov, P. & H. Hummel, 1996. Survival in air for Arctic bivalves. In Hummel, H. (ed.), Biodiversity and Adaptation Strategies of Arctic Coastal Marine Benthos. Interim report INTAS project 94-391, NIOO-CEMO, Yerseke.

Swofford, D. L. & R. B. Selander, 1981. Biosys-1: a Fortran program for the comprehensive analysis of electrophoretic data in population genetics and systematics. J. Heredity 72: 281–283.

Theisen, B. F., 1978. Allozyme clines and evidence of strong selection in three loci in *Mytilus edulis* L. (Bivalvia) from Danish waters. Ophelia 17: 135–142.

Thorpe, J. P., 1983. Enzyme variation, genetic distance and evolutionary divergence in relation to levels of taxonomic separation. In Oxford, G. S. & D. Rollinson (eds), Protein Polymorphism: Adaptive and Taxonomic Significance. Academic Press, London: 131–152.

Treshnikov, A. F., 1985. Atlas of the Arctic. GYGK (Soviet Ministry of Geodesy and Cartography), Moscow, 204 pp.

Väinölä, R. & S.-L. Varvio, 1989. Biosystematics of *Macoma balthica* in northwestern Europe. In Ryland, J. S. & P. A. Tyler (eds), Reproduction, Genetics and Distributions of Marine Organisms. Olsen & Olsen, Fredensborg: 309–316.

Workman, P. L. & J. D. Niswander, 1970. Population studies on southwestern Indian tribes. II. Local genetic differentiation in the Papago. Am. J. Hum. Gen. 22: 24–49.

Wright, S., 1965. The interpretation of population structure by F-statistics with special regard to systems of mating. Evolution 19: 395–420.

Zouros, E., 1987. On the relation between heterozygosity and heterosis: An evaluation of the evidence from marine mollusks. In Rattazi, M. C., J. G. Scabdalios & G. S. Whitt (eds), Isozymes: Current Topics in Biological and Medical Research, Genetics, Development, and Evolution. Liss Inc., New York 15: 255–270.

Zouros, E. & A. L. Mallet, 1989. Genetic explanations of the growth/heterozygosity correlation in marine mollusks. In Ryland, J. S. & P. A. Tyler (eds), Reproduction, Genetics and Distribution of Marine Organisms. Olsen & Olsen, Fredensborg: 317–324.

Hydrobiologia **355:** 139–144, 1997.
A. D. Naumov, H. Hummel, A. A. Sukhotin & J. S. Ryland (eds), Interactions and Adaptation Strategies of Marine Organisms.

Defenses against oxidative stress in the Antarctic scallop *Adamussium colbecki* and effects of acute exposure to metals

F. Regoli[1], M. Nigro[2], E. Bertoli[1], G. Principato[1] & E. Orlando[2]
[1]*Istituto di Biologia e Genetica, Università di Ancona, Via Ranieri, Monte D'Ago, I-60100 Ancona, Italy*
[2]*Dipartimento di Biomedicina Sperimentale, Università di Pisa, Via A. Volta 4, I-56100 Pisa, Italy*

Key words: molluscs, extreme environment, oxidative stress, heavy metals, biomarkers

Abstract

Since a general pathway of toxicity induced by pollutants is the enhancement of reactive oxygen species, biochemical responses associated with variations in the antioxidant cellular system have been often proposed as biomarkers of pollutant-mediated toxicity associated with oxidative stress. Antarctic organisms live in an extreme environment characterized by low water temperature, high level of dissolved oxygen, presence of ice and strong seasonal changes in light intensity and availability of food, conditions which could influence both the formation of reactive oxygen species and the mechanisms for their removal. In this respect and considering the utility of this as a key species for monitoring marine Antarctic environment it was of interest to investigate the antioxidant defense system of the scallop *Adamussium colbecki.*

The parameters examined in the digestive gland of the scallop were the concentration of glutathione and the activity of several glutathione dependent and antioxidant enzymes (glyoxalase I and II, glutathione S-transferases, glutathione peroxidases, glutathione reductase, catalase, superoxide dismutase). Very high levels of catalase suggest a possible adaptation to Antarctic extreme conditions, while the high activities of glutathione S-transferases are more probably related to the feeding behavior of Pectinids. Enzymes from *Adamussium colbecki* generally appeared to be active at low temperatures but, with a few exceptions, their activities increased with rising temperature. Exposure of *A. colbecki* to sublethal concentrations of Cu or Hg resulted in a significant reduction in the levels of total glutathione and in the activity of catalase and glutathione S-transferases. Antioxidant responses of *A. colbecki* could represent a useful tool in assessing the biological impact of environmental pollutants in the Antarctic ecosystems.

Introduction

Marine bivalves are commonly used in temperate waters as suitable indicators of environmental pollution (Phillips, 1980). These organisms can concentrate contaminants within the tissues, so providing a time-integrated measurement of their bioavailability. Moreover, in environmental disturbance assessment, the integration of chemical data with biological responses (the so called biomarkers) is strongly recommended in order to assess effects of pollutants on the organisms (Bayne et al., 1988).

A similar approach is urgently also needed for Antarctic ecosystems, where the increase of human activities makes necessary the development of such integrated biomonitoring programs. In fact, an effective biological monitoring would 'allow the assessment of the impacts of ongoing activities and facilitate the early detection of the possible unforeseen impacts' (Protocol on Environmental Protection to the Antarctic Treaty, 1991).

Biomarkers can be investigated at different levels of biological organization and, at the biochemical one, many studies are based on the enhancement of reactive oxygen species as a general pathway of toxicity induced by pollutants and associated with oxidative stress (Winston, 1991). In this respect, antioxidant cellular responses to pollutants are well documented for several temperate molluscs (Winston & Di Giulio, 1991), while data are still limited for Antarctic invertebrates. Also considering the extreme Antarctic environmental conditions (higher oxygen solubility at low water temperature and seasonality in light intensity and food availability) it was of interest to make a preliminary characterization of the cellular antioxidant system in the scallop *Adamussium colbecki* which is considered a key species for biomonitoring pollution in Antarctic ecosystems (Berkman & Nigro, 1992).

The possible biochemical adaptation to cold seawater has been considered by evaluating the enzymatic responses at different temperatures.

The utility of antioxidant responses as biomarkers of oxidative stress induced by pollutants has also been examined in scallops exposed to metals under different laboratory conditions.

Materials and methods

Sampling and laboratory exposures

Specimens of *Adamussium colbecki* were collected by Scuba diving in Terra Nova Bay (Ross Sea) near the Italian Antarctic Base. For preliminary biochemical characterization, digestive glands were dissected from 20 specimens, grouped in 5 pools and maintained in liquid nitrogen till processing for analyses.

Exposures to metals (Cu and Hg at 20 and 5 $\mu g\ l^{-1}$ respectively) were carried out at the Italian Antarctic Base during the Austral summer 1995–96. Scallops were acclimatized without sediments for some days in running filtered seawater and they were not fed during the experiments. Seawater (1 l ind^{-1}) was changed and redosed daily and the temperature maintained at 0 ± 1 °C. At each sampling time 30 specimens (from control, Cu and Hg exposed groups) were collected and digestive glands dissected with the same procedures previously described.

Biochemical analyses

Sample preparation was carried out at 4 °C and detailed procedures have been described elsewhere (Regoli et al., 1997). Total glutathione was assayed by the enzymatic method of Akerboom & Sies (1981). Glyoxalase I (EC 4.4.1.5) was measured by the increase in absorbance due to the formation of S-D-lactoylglutathione from the hemimercaptal adduct of methylglyoxal (MG) and reduced glutathione. Glyoxalase II (EC 3.1.2.6) activity was followed by the reaction of 5,5′-dithio-bis-2-nitrobenzoic acid (DTNB) with GSH formed from S-D-lactoylglutathione (LSG). Glutathione S-transferase (GST) (EC 2.5.1.18) activities were determined, according to Habig & Jakoby (1981), using 1-chloro-2,4-dinitrobenzene (CDNB) as substrate. Glutathione reductase (EC 1.6.4.2.) was assayed following the decrease in absorbance at 340 nm due to the oxidation of NADPH in the presence of GSSG (Ramos-Martinez et al., 1983). Glutathione peroxidase (GPx) activities were measured in a coupled enzyme system where the formed GSSG is converted to its reduced form by glutathione reductase (Lawrence & Burk, 1976); H_2O_2 or cumene hydroperoxide were used as substrates (respectively for the Se-dependent, EC 1.11.1.9, and the sum of Se-dependent and Se-independent, EC 2.5.1.18, activities). The rate of blank reaction was subtracted from the total rate. Catalase (EC 1.11.1.6) activity was measured, according to Greenwald (1985), by the decrease in absorbance at 240 nm due to H_2O_2 consumption. Superoxide dismutase (SOD) (EC 1.15.1.1) activity was determined by its ability to inhibit the reduction of cytochrome c by $O_2^{\cdot -}$ generated by the xanthine oxidase/hypoxanthine system (McCord & Fridovich, 1969). One unit of SOD has been calculated as the amount of enzyme inhibiting by 50% the reduction of cytochrome c. Protein concentration was determined according to Lowry et al. (1951) by using Bovine Serum Albumin (BSA) as standard.

Results

Table 1 reports the levels of glutathione and of antioxidant enzymes measured at different temperatures between 0 ° and 34 °C, in the digestive gland of *Adamussium colbecki*. Enzymatic activities generally increased at higher temperatures but with some considerable differences among the enzymes. In this respect, increasing activities of glyoxalase I and glutathione S-transferase were observed up to 19 °C, while similar or even reduced values were found at higher temperatures. A similar trend was exhibited by glutathione peroxidases (especially with cumene hydroperoxide) with maximum enzymatic activity at 15 °C. On the other hand, the increase of glyoxalase II and glutathione reductase was more constant up to 32 °C, while no clear effect of temperature was evident for catalase activity. Superoxide dismutase was measured only at 10 °C, since the blank reaction was greatly affected by temperature, making difficult the interpretation of differences.

Table 1. Levels of glutathione and of antioxidant enzymes measured at different temperatures in the digestive gland of *Adamussium colbecki*. Mean values ± standard deviations ($n = 5$). GST: glutathione S-transferases; G.R.: glutathione reductase; GPx (H_2O_2): glutathione peroxidases (with H_2O_2 as substrate); GPx (CHP): glutathione peroxidases (with cumene hydroperoxide as substrate); SOD: superoxide dismutase.

	0 °C	10 °C	19 °C	26 °C	34 °C
Glutathione[1]			0.84 ± 0.11		
GST[2]	4870 ± 310	5530 ± 410	7260 ± 530	6870 ± 570	5930 ± 370
Glyoxalase I[2]	303 ± 21.7	387 ± 37.2	459 ± 55.8	419 ± 35.7	407 ± 42.3
Glyoxalase II[2]	8.77 ± 2.58	15.6 ± 6.12	34.9 ± 6.80	40.0 ± 9.71	42.3 ± 11.4
G.R.[2]	15.8 ± 4.08	27.2 ± 5.14	37.1 ± 5.51	41.0 ± 8.37	56.3 ± 12.8
GPx (H_2O_2)[2]	7.78 ± 2.12		10.1 ± 1.78		
GPx (CHP)[2]	10.7 ± 1.66		17.9 ± 4.75		
Catalase[3]	555 ± 107	495 ± 66.5	526 ± 87.8	605 ± 110	562 ± 118
SOD[4]		12.9 ± 1.67			

[1]: μmol g^{-1} tissue; [2]: nmol min^{-1} mg^{-1} protein; [3]: μmol min^{-1} mg^{-1} protein; [4]: SOD Units mg^{-1} protein.

Among the variations of antioxidant responses in *A. colbecki* treated with copper or mercury, Figure 1 shows the levels of glutathione and the activities of catalase and glutathione S-transferases after 0, 3 and 6 days exposure. After 6 days exposure to Cu and Hg, the level of total glutathione was reduced to 65–70% compared to values of control organisms. Also catalase and glutathione S-transferases showed a general reduction of activity ranging from 25 to 35% after 3 days exposures. Moreover the biochemical variations induced by copper and mercury were very similar and no significant differences were obtained between these two metal-treated groups.

Discussion

From the obtained results, the levels of antioxidant defenses in *A. colbecki* did not show striking differences compared to other described molluscs (Livingstone, 1991, 1993; Regoli & Principato, 1995; Regoli et al., 1997) with the only exception of catalase and, to a lesser extent, of glutathione S-transferases. In fact, the levels of catalase exhibited by the Antarctic scallop were particularly high, with values approximately 1 order of magnitude greater than those reported for temperate bivalves, including some systematically related Mediterranean Pectinids (Regoli et al., 1997). Catalase has been indicated as the principal enzyme involved to protect against lipid peroxidation processes during adaptation to high oxygen concentrations (Chance et al., 1979; Barja de Quiroga et al., 1989) and, in this respect, high levels of catalase in *A. colbecki* could represent an important adaptation to the Antarctic extreme environment, considering the high oxygen solubility at low water temperatures. In addition, catalase was very active at 0 °C, but it was not particularly influenced by increasing the temperature: similar enzymatic responses to temperature have been obtained in many Antarctic fishes by Macdonald et al. (1987) who indicates as 'cold adaptation' the presence of enzymes very active at low temperatures but characterized by a reduced temperature sensitivity.

While the activity of catalase in *A. colbecki* could be related to pecularities of the environmental conditions, a similar hypothesis does not seem to be reasonable for the high levels of glutathione S-transferases. Even though much lower values have been reported for several temperate molluscs (Lee, 1988; Fitzpatrick & Sheehan, 1993; Regoli & Principato, 1995), activities of glutathione S-transferases similar to those exhibited by *A. colbecki*, have been obtained in other Mediterranean scallops (Regoli et al., 1997). In this respect, it could be hypothesized that high activities of glutathione S-transferases in the digestive gland are typical of Pectinids and probably related to some natural compounds in their diet, largely based on particulate matter associated to surface sediments.

The activities of the other antioxidant enzymes in the digestive gland of *A. colbecki* were quite similar to those described for other temperate molluscs, but generally more active at low temperatures (Livingstone, 1993; Regoli et al., 1997). Additionally, enzymes of the Antarctic scallop were often susceptible to heat denat-

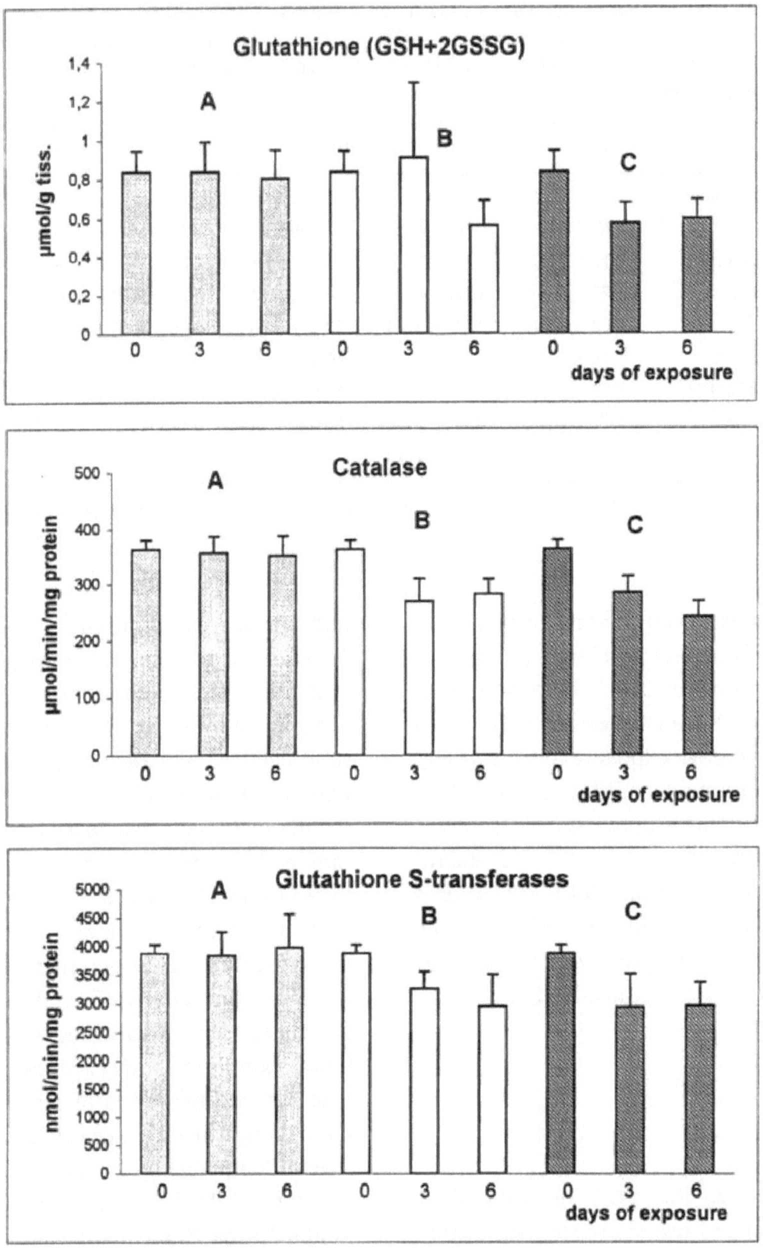

Figure 1. Variations of total glutathione, catalase and glutathione S-transferases in *Adamussium colbecki* exposed to metals (mean values ± standard deviations, $n = 5$). A: control scallops; B: scallops exposed to Cu (20 $\mu g\ l^{-1}$); C: scallops exposed to Hg (5 $\mu g\ l^{-1}$).

uration at relatively high temperature (over 25 °C), but the presented data do not allow us to hypothesize the presence of molecular adaptation involving protein structure.

The results obtained from laboratory exposures to metals demonstrated that the antioxidant system of *A. colbecki* is affected by heavy metals. In fact, copper and mercury, which are well known inducers of lipid peroxidation processes, induced in treated scallops biochemical alterations indicative of oxidative stress.

The levels of glutathione, one of the most important soluble antioxidant compounds, were significantly reduced after 6 days exposure to Cu and Hg. This effect could be due either to a direct binding of metals to the GSH molecule (yielding mixed disulfides) or to an enhanced oxidation of this thiol which, in such circumstances, would not be reconverted to GSH by glutathione reductase but actively excreted from the cell (Meister, 1989). Moreover, the results obtained in *A. colbecki* are in agreement with those previously reported by many authors indicating decreased levels of glutathione in molluscs from contaminated field and mesocosm experiments (Suteau et al., 1988; Viarengo et al., 1990; Regoli & Principato, 1995); in this respect the analysis of glutathione content is confirmed, also for the Antarctic environment, as a useful biomarker of contaminant-mediated oxidative stress.

Exposure to metals, also resulted in a significant decrease of catalase activity in digestive gland of *A. colbecki*. Since this enzyme catalyses the removal of hydrogen peroxide, the reduction of catalase activity could render *A. colbecki* more susceptible to toxicity induced by these reactive oxygen species. However, data on Mediterranean mussels (Regoli & Principato, 1995) indicated that decrease of catalase activity is a transitory response to contaminants during acute stress but not evident in chronically polluted organisms. Further investigations on the presence of similar compensatory mechanisms also in the Antarctic scallop would be important for a correct interpretation of data on catalase in assessing the environmental disturbance.

Variations observed on glutathione S-transferases after exposure to Cu and Hg are more difficult to discuss. Glutathione S-transferases comprise a multigene family encoding for several isoenzymes with different functions but often overlapping specificities for the commonly used substrate CDNB. These enzymes act as the catalysts of a very wide variety of conjugation reactions of glutathione with xenobiotic compounds, but some isoforms are also involved in protection against oxidative stress through the reduction of organic hydroperoxides. Since our experimental conditions did not resolve variations in the activities of the single isoforms, it is not possible, at the moment, to hypothesize specific effects of the mean decreased activity of glutathione S-transferases on the health condition of *A. colbecki* exposed to metals.

This preliminary characterization of antioxidant features of *Adamussium colbecki* and of the responses to metal exposures, seems to indicate that similar biomarkers could represent, also for the Antarctic environment, useful tools for monitoring contaminant-mediated oxidative stress.

References

Akerboom, T. P. M. & H. Sies, 1981. Assay of glutathione, glutathione disulfide, and glutathione mixed disulfides in biological samples. Meth. Enzym. 71: 373–382.

Barja de Quiroga, G., M. Lòpez-Torres & R. Pérez-Campo, 1989. Catalase is needed to avoid tissue peroxidation in *Rana pereri* in normoxia. Comp. Biochem. Physiol. 94C: 391–398.

Bayne, B. L., K. R. Clarke & J. S. Gray, 1988. Background and rationale to a practical workshop on biological effects of pollutants. Mar. Ecol. Prog. Ser. 46: 1–5.

Berkman, P. A. & M. Nigro, 1992. Trace metal concentrations in scallop around Antarctica: extending the Mussel Watch Program to the Southern Ocean. Mar. Pollut. Bull. 24: 322–323.

Chance, B., H. Sies & A. Boveris, 1979. Hydroperoxide metabolism in mammalian organs. Physiol. Rev. 59: 527–605.

Fitzpatrick, P. J. & D. Sheehan, 1993. Separation of multiple forms of glutathione S-transferases from the blue mussel, *Mytilus edulis*. Xenobiotica 23: 851–861.

Greenwald, R. A., 1985. Handbook of Methods for Oxygen Radical Research. CRC Press, Boca Raton, Florida, 276 pp.

Habig, W. H & W. B. Jakoby, 1981. Assays for differentiation of glutathione S-transferases. Meth. Enzym. 77: 398–405.

Lawrence, R. A. & R. F. Burk, 1976. Glutathione peroxidase activity in selenium-deficient rat liver. Biochem. Biophys. Res. Commun. 71: 952–958.

Lee, R. F., 1988. Glutathione S-transferases in marine invertebrates from Langesundfjord. Mar. Ecol. Prog. Ser. 46: 33–36.

Livingstone, D. R., 1991. Organic xenobiotic metabolism in marine invertebrates. In Gilles, R. (ed.), Advances in Comparative and Environmental Physiology. Springer-Verlag, Heidelberg, 7: 45–185.

Livingstone, D. R., 1993. Biotechnology and pollution monitoring. Use of molecular biomarkers in the aquatic environment. J. Chem. Technol. Biotechnol. 57: 195–211.

Lowry, O. H., N. J. Rosenbrough, A. L. Farr & R. J. Randall, 1951. Protein measurement with the folin phenol reagent. J. Biol. Chem. 193: 266–275.

MacDonald, J. A., J. C. Montgomery & R. M. G. Wells, 1987. Comparative physiology of Antarctic fish. Adv. Mar. Biol. 24: 321–388.

McCord, J. M. & I. Fridovich, 1969. Superoxide dismutase: an enzymatic function for erythrocuprein (hemocuprein). J. Biol. Chem. 244: 6049–6055.

Meister, A., 1989. On the biochemistry of glutathione. In Taniguchi, N., T. Higashi, S. Sakamoto & A. Meister (eds), Glutathione Centennial, Molecular Perspectives and Clinical Implications. Academic Press, San Diego, CA: 3–22.

Phillips, D. J. H., 1980. Quantitative aquatic biological indicators: their use to monitor trace metal and organochlorine pollution. Appl. Sci. Publ. Ltd, London, 181 pp.

Ramos-Martinez, J. I., T. R. Bartolomé & R. V. Pernas, 1983. Purification and properties of glutathione reductase from hepatopancreas of *Mytilus edulis* L. Comp. Biochem. Physiol. 75B: 689–692.

Regoli, F. & G. Principato, 1995. Glutathione, glutathione-dependent and antioxidant enzymes in mussel, *Mytilus galloprovincialis*, exposed to metals under field and laboratory conditions: implications for the use of biochemical biomarkers. Aquat. Toxicol. 31: 143–164.

Regoli, F., G. B. Principato, E. Bertoli, M. Nigro & E. Orlando, 1997. Biochemical characterization of the antioxidant system in the scallop *Adamussium colbecki*, a sentinel organism for monitoring the Antarctic environment. Polar Biol. 17: 251–258.

Suteau, P., M. Daubeze, M. L. Migaud & J. F. Narbonne, 1988. PAH-metabolizing enzymes in whole mussels as biochemical tests for chemical pollution monitoring. Mar. Ecol. Prog. Ser. 46: 45–49.

Viarengo, A., L. Canesi, M. Pertica, G. Poli, M. N. Moore & M. Orunesu, 1990. Heavy metal effects on lipid peroxidation in the tissues of *Mytilus galloprovincialis* Lam. Comp. Biochem. Physiol. 97C: 37–42.

Winston, G. W., 1991. Oxidants and antioxidants in aquatic animals. Comp. Biochem. Physiol. 100C: 173–176.

Winston, G. W. & R. T. Di Giulio, 1991. Prooxidant and antioxidant mechanisms in aquatic organisms. Aquat. Toxicol. 19: 137–161.

Hydrobiologia **355:** 145–151, 1997.

A. D. Naumov, H. Hummel, A. A. Sukhotin & J. S. Ryland (eds), Interactions and Adaptation Strategies of Marine Organisms.

A new species of *Hyalopomatus* (Serpulidae: Polychaeta) which lacks an operculum: is this an adaptation to low oxygen?

E. W. Knight-Jones[1], Phyllis Knight-Jones[1], P. G. Oliver[2] & A. S. Y. Mackie[2]
[1] *School of Biol. Sciences, Univ. of Wales, Swansea, SA2 8PP, U.K.*
[2]*National Museum and Galleries of Wales, Cardiff, CF1 3NP, U.K.*

Key words: Hyalopomatus, Protis, Protula, serpulid operculum, oxygen

Abstract

Hyalopomatus cancerum n.sp., epizoic on spider-crabs (*Encephaloides*) in a low-oxygen area of the Arabian Sea, differs from other species of the genus in lacking opercula. Larger serpulids in Indian Ocean 'Galathea' samples from great depths, tentatively referred to *Protis simplex* Ehlers, mostly bear vesicles on tips of pinnulate radioles. These are too small to occlude the tube mouth, so perhaps they and the larger vascularised vesicles of *Apomatus* may help in respiration. The BIOFAR Survey off the Faroe Islands showed that *Protula* (always non-operculate) and *Apomatus* are the main serpulid genera in the deeper channels. *Protis, Protula* and *Apomatus*, which are amongst the best-known of deep-sea serpulids, are like early postlarval stages of *Serpula* and *Hydroides* in lacking opercula, or in having thin-walled opercular vesicles on pinnulate stalks. They thus support the view that hypomorphy is somewhat characteristic of abyssal taxa.

Considering non-operculate serpulids of shallow seas, many species of *Spiraserpula* lack opercula, but secrete sharp ridges and spines on the inner walls of their tubes, which must deter or trap predatory tube-invaders; *Floriprotis* may be protected by coral, whilst *Salmacina, Paraprotula, Microprotula* and *Paraprotis dendrova* Uchida incubate their embryos, so have special respiratory needs.

Introduction

Cruise 211 of the 'Discovery' in October 1994 found dense populations of spider crabs, *Encephaloides armstrongi* Wood-Mason, at depths of about 700 m off the S coast of Oman, where the concentration of dissolved oxygen near the bottom was <0.1 ml l^{-1} (Gage, 1995). Many *Encephaloides* were collected from that zone by P. G. Oliver, because they bore serpulid tubes. The tubes contained *Hyalopomatus*, which agreed with previous descriptions of the genus (Zibrowius, 1969, 1977; Kupriyanova, 1993a) except that opercula were lacking.

Most genera and species of Serpulidae have been recorded, in greatest numbers, on east-facing coasts in shallow seas and warm latitudes (Knight-Jones & Knight-Jones, 1991). Their tubes are often invaded by predators such as anthurid isopods (Wägele, 1979; Pillai & ten Hove, 1994), although most of these shallow-water serpulids have opercula. The opercula are often found in fish stomachs, but are readily regenerated on stalks which lack pinnules. At great ocean depths few serpulid genera are found (Zibrowius, 1977) and prominent amongst these are *Protis, Protula* and *Apomatus*. *Protula* lacks opercula, *Apomatus* has vesicular opercula without any distal protective thickening, whilst *Protis* (reviewed briefly below) was regarded as lacking opercula until Kupriyanova's (1993b) revision.

This paper aims to answer the question posed in its title. It suggests an affirmative, as far as the new *Hyalopomatus* is concerned; the *Encephaloides* substratum seems similarly adapted, having very large branchial chambers (Griffin, 1974). This paper also studies 'Galathea' samples from great depths, kindly sent by Dr J. B. Kirkegaard, which raise the further question of what is meant by 'operculum'. Finally it considers vertical distributions down to moderate depths in cool latitudes, where oxygen minimum zones are not formed (Sverdrup et al., 1942), through a preliminary analysis of results from the BIOFAR Survey.

Chaetae were studied in polyvinyl-lactophenol mounts, using a ×100 objective for *Hyalopomatus* and ×40 for *Protis*.

Taxonomic and morphological descriptions

Genus Hyalopomatus Marenzeller

Includes *Hyalopomatopsis* Saint-Joseph and *Cystopomatus* Gravier.

Diagnosis. Operculum (if present) vesicular, borne on thin apinnulate peduncle; six thoracic chaetigers, 5 with uncini; thoracic membrane short, not extending to posterior half of thorax; uncini of both thorax and abdomen with anterior gouge followed by about 20 transverse rows of up to 4 very small teeth.

Hyalopomatus cancerum n.sp.
Material examined: worms from about 20 of the larger tubes encrusting *Encephaloides* from 'Discovery' Stn 211/12696*1, depth 685 m, 24 Oct.1994, National Museum of Wales, Cardiff. Holotype fixed out of tube NMW, Z1997.0151, paratypes NMW, Z1997.0152. Measurements and numbers given below were from holotype, followed (in brackets) by those from a paratype (Figure 1C).

Tubes round in section, up to 8 mm long, attached for most of length, often ascending anteriorly, external diameter at mouth up to 0.8 mm, without flanges, but with thickened base increasing area of attachment (Figure 1B). Thorax width up to 0.5 mm, lengths of crown 1.75 (2), thorax 1 (1.2) abdomen 1.7 (2), total lengths 4.4 (5.2) mm. Radioles about 8 or 9 pairs, with no operculum. Collar dorsally divided, but ventrally complete. Thoracic membrane extending half way down thorax (Figure 1A). Thorax with 6 fascicles and 5 tori on each side. Achaetigerous region as long as sum of 3 adjacent chaetigers, with gut lumen full of diatom frustules. Abdominal chaetigers about 22. Pygidium bilobed.

Thoracic notochaetae 7 or 8 per fascicle, about 4 mm in diameter basally and for 0.75 of length, geniculate, with striated border on tapering part distal to knee. Collar chaetae mostly with small teeth at knee (Figure 1E). Abdominal capillary chaetae 0.2 to 0.25 mm long, 2.5 mm wide basally, tapering gradually throughout length, occurring on only about 10 posterior-most chaetigers, one (occasionally two) per chaetiger on each side of body. No other chaetae found in abdomen, except uncini. Uncini similar on both thorax and abdomen, each with rasp-like distal edge, 20–26 mm long and 3 mm wide for most of length, narrowing anteriorly to gouge about 2 mm wide. Distal edge with about 22 transverse rows of fine teeth, 3 or 4 teeth in most rows, but paired or single teeth anteriorly (Figure 1F, J). In side view (Figure 1G) rows of teeth appear as continuous lines, at right angles to striations in basal plate of uncinus. Uncini in end view (Figure 1H) show wide distal edges closely packed, with 4 teeth per transverse row, but spaces for uncinal follicles between otherwise narrow uncinal plates.

Remarks. Specific name is genitive plural of Latin *cancer*. Some *Encephaloides* bore crowded serpulids (Figure 2), but we saw no evidence of scissiparity or brooding in any tube which we opened. Crabs were about 10 per square metre on a muddy bottom (Gage, 1995), perhaps offering the only substrata suitable for firm attachment. Distribution between crabs seemed non-random, indicating larval gregariousness during searching and settlement.

Three small *Hyalopomatus* from a stone at 378 m depth, 'Discovery' Stn. 211:12697*1, lack opercula but are juvenile, with short abdomens, so may not be the same species as those on *Encephaloides*. From 919 m depth a 'Discovery' trawl brought up coal and clinkers bearing another species of *Hyalopomatus*, which was operculate yet not much bigger than *H. cancerum*. Its tubes are sculptured like those of *Hyalopomatus variorugosus* Ben-Eliahu & Fiege (1996), but two individuals had the opercular peduncle annulated. These and juveniles of various *Hyalopomatus* species are being studied further, to see at what size opercula are first developed.

Genus Protis Ehlers 1887

Diagnosis as in Kupriyanova (1993b). Type species *Protis simplex* Ehlers, described from 1700 m depth in Gulf of Mexico.

Protis simplex *Ehlers 1887*
Material examined: from *Globigerina* ooze at three 'Galathea' stations (Bruun, 1957–1959), two near Madagascar, the other between Seychelles and Sri Lanka: Stn 234, 5°25′ S, depth 4800 m, one *Protis* lacking crown but with tube; Stn 235, 4°47′ S, 4810 m, at least two *Protis*; Stn 281, 3°38′ N, 3310 m, pieces of ca 19

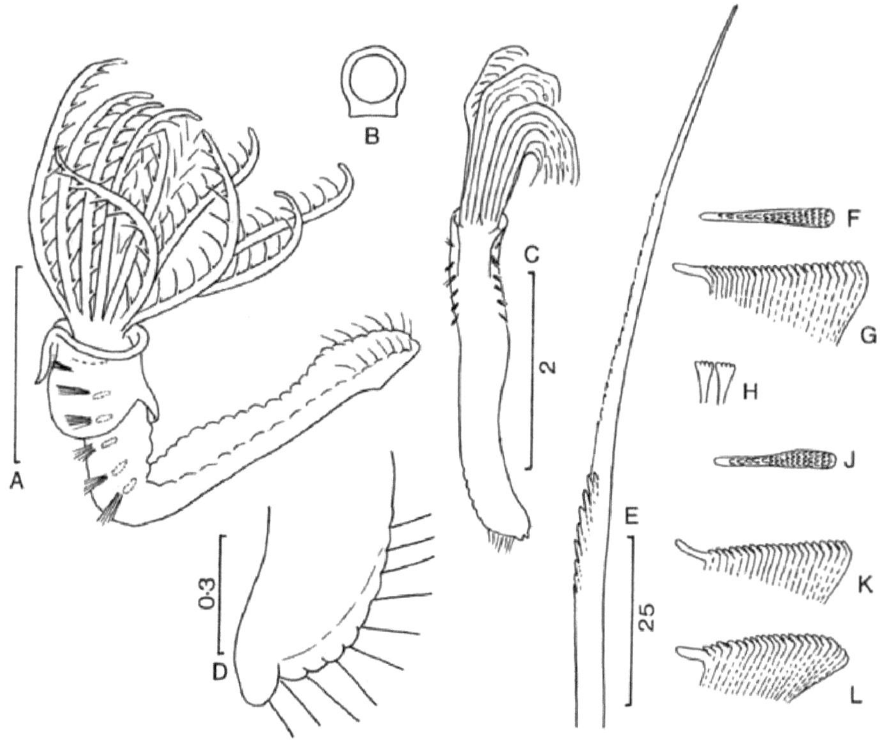

Figure 1. *Hyalopomatus cancerum* n.sp. A, type specimen from right side, with anterior margin of collar rolled back, before mounting in polyvinyl-lactophenol; B, cross-section tube and C, dorsal view body of paratype; D, posterior end of abdomen from right side; E, distal 0.3 of collar chaeta; F, face view and G, side view of thoracic uncini; H, end view of two abdominal uncini; J, face, and K, L, side views of abdominal uncini. All chaetae to same scale, in mm; other scales in mm, B as C.

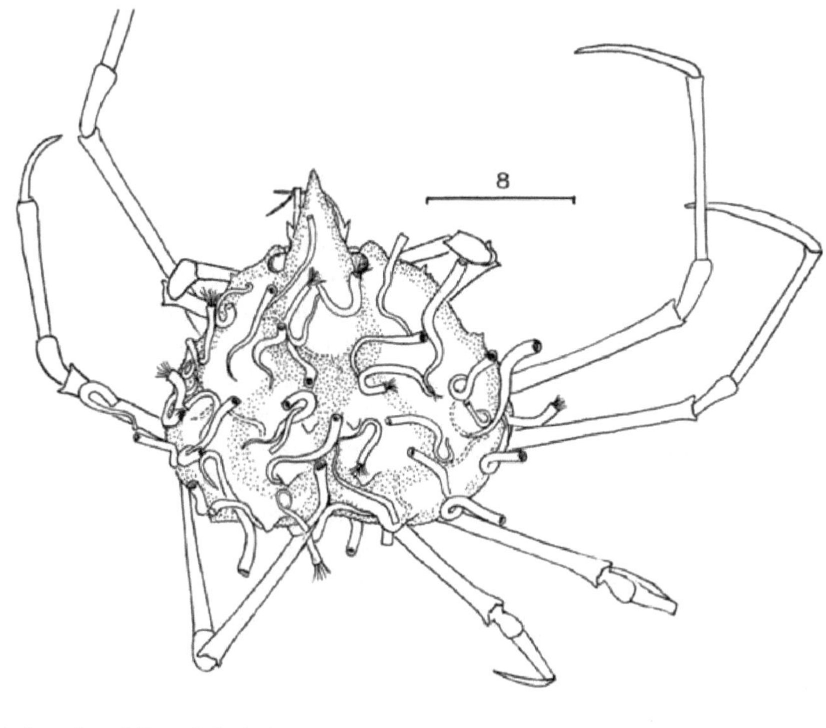

Figure 2. Dorsal view of small *Encephaloides* bearing many *Hyalopomatus* tubes, some with radioles protruding. Scale in mm.

Protis, probably broken during extraction from tubes, which were not included. Lengths (mm) of crown 10–15, thorax 6–7, abdomen ca 25, width of thorax up to 2.5, generally 2 or less, total length probably <47 mm. Tube from Stn 234 straight, with irregular growth rings and no sign of having been attached, in 3 pieces, total length 42 mm, external diameter 3.5 tapering to 1.5 mm, width of mouth 2.5, accompanied by thorax 2 mm wide.

All crowns had been fixed whilst within tubes, so tips of radioles had to be teased apart, with a needle and fine paint-brush, to search for opercular vesicles. Some were accidentally torn from the radioles which bore them, so six crowns were left untouched, for future investigators. Of ten others, searched carefully, three seemed to lack any opercular rudiments; one had a spherical vesicle (arrowed in Figure 3A); in each of two there was a transparent disc, 1 mm (Figure 3B) and 0.3 mm diameter respectively; three had two such discs each, diameters 0.4–0.6 mm; one had three 'opercula', a misshapen vesicle 0.9 mm diameter with two others ca 0.3 mm diameter. All were at the tips of pinnulate radioles. One vesicle (Figure 3A) seemed to be on the second dorsal radiole, right side, and the smallest discs seemed more ventral, but the length and twisting of crowns made this difficult to determine. Radioles 20–22 in each half of crown, arising from spiralling base nearly as long as worm's circumference, as figured by Wollebaek (1912) in *Protis arctica* (Hansen). In retracted crown one half-base folds over the other (Figure 3C). Collar long, folded, sometimes with midventral cleft (Figure 3C, D), but often entire ventrally, always cleft dorsally and forming dorsal margins of thoracic cloak (Figure 3A). Thoracic chaetigers 7. Ventral thorax bears (glandular?) patches, one patch per torus (Figure 3D). Faecal groove on right side, with very distinct anterior border. Achaetigerous region about as long as 2 or 3 thoracic chaetigers. Abdominal chaetigers about 140. Posterior 0.25 of abdomen fringed by long capillary chaetae. Postero-dorsal gland tapering along abdomen as yellowish mid-dorsal ridge. Collar chaetae 'fins' fringed by small teeth (Figure 3E). Thoracic uncini with single row of 5 or 6 'saw' teeth, including main fang (Figure 3F). Abdominal uncini 'rasps', with irregular teeth, about 3 in a row at widest part near crest (Figure 3G). Posterior uncini with pairs of teeth, small (from end of torus? Figure 3H).

Remarks. This material is referred to *Protis simplex* because the (glandular?) patches on the ventral thorax (Figure 3D) are like those figured and described by Ehlers (1887). 'Ehlers patches' shine conspicuously in bright reflected light, but have not been noted in any other species of *Protis*. We did not record them in Hansen's (1878) type material of *P. arctica*, in the Bergen Museum, and could not see them in material of *P. arctica* from NE Iceland, held in Moscow State University and described by Kupriyanova & Jirkov (1997). They are conspicuous, however, in type material of *Apomatus brownii* Pixell (1913, type of *Pixellgrana* Uchida, 1978), held in the Royal Museum of Scotland, Edinburgh. That is clearly a *Protis* as redefined by Kupriyanova (1993b) and perhaps a junior synonym of *Protis simplex*.

Zibrowius (1969) favoured Eliason's (1951) suggestion that *P. simplex* may be a junior synonym of *Protis arctica*, which has been recorded from Greenland to the equator at depths of 2000–5000 m in the Atlantic (Zibrowius, 1977) and Mediterranean (Ben-Eliahu and Fiege, 1996). This synonymy, however, seems doubtful (Kupriyanova & Jirkov, 1997). Considering other species, *Protis pacifica* Moore is like *P. arctica*, but with 6–8 teeth on thoracic uncini (Zibrowius, 1969); *Protis hydrothermica* ten Hove & Zibrowius (1986) has distinctive collar chaetae; and *Protis polyoperculata* Kupriyanova (1993b) has up to 6 opercular vesicles and lives at depths of >5000 m. in the Kurile Trench. That name *polyoperculata* is a good one, derived from comparative anatomy, but such small vesicles cannot close the tube-mouth effectively against invasion by predators.

Depth distribution of serpulid genera off Faroe Islands

The BIOFAR survey is still in progress, using various collecting gears. Results here are from identifications and counting prior to a Symposium convened at Tórshavn in 1991. Most serpulids were collected by triangular dredge at depths <700 m, but many by detritus sledge in both deep and shallow samples. Numbers on the left of Figure 4 show how many sampling stations yielded serpulids within each 100 m depth zone. Histogram widths indicate mean numbers of each genus per station per zone. Those widths were measured against the logarithmic scale shown, to make the vast preponderance of operculate serpulids in shallow water seem less overwhelming. In depths of 701–1006 m there were only 5 stations which yielded serpulids, all obtained by detritus sledge. These 5 samples were

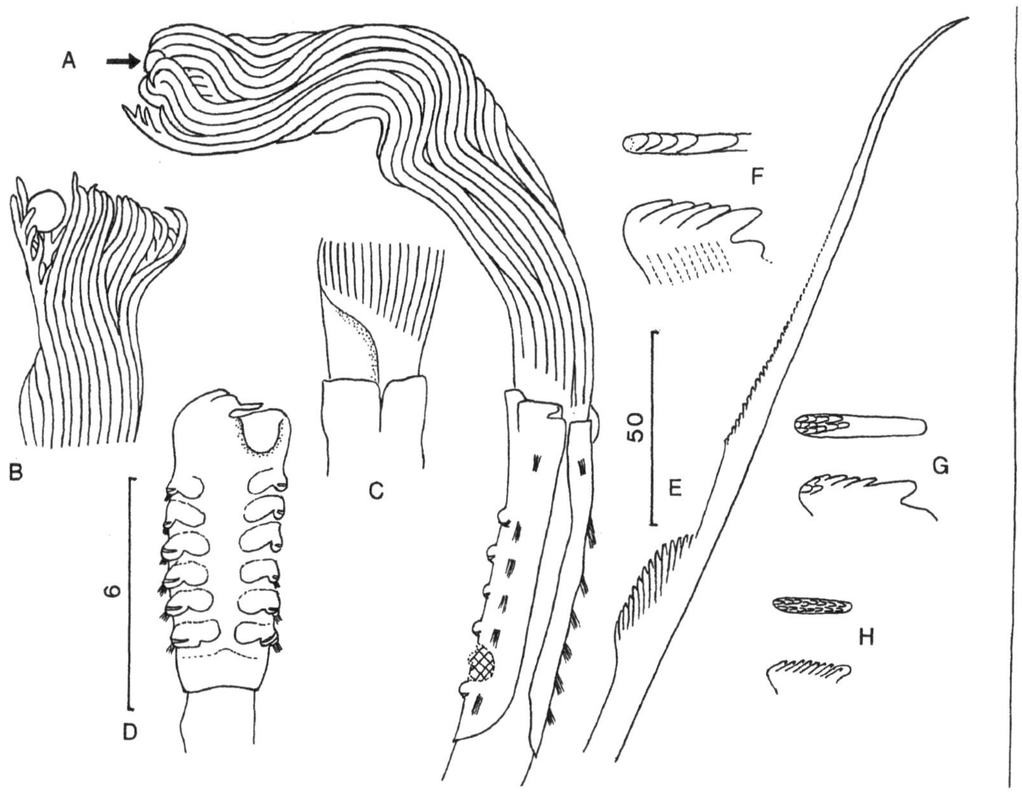

Figure 3. *Protis simplex.* A, crown and thorax, with arrow indicating an 'opercular' vesicle hidden amongst tips of radioles; B, distal end of another crown with larger, flattened vesicle; C, ventral view showing base of crown partly hidden by ventral lobes of collar; D, ventral view of thorax, showing rounded patch of thickened epithelium associated with each torus; E, distal end of collar chaeta; F, face and side views of thoracic uncini; G, face and side views of uncini from mid-abdomen; H, ditto from posterior end. All chaetae to same scale, in mm; other drawings as D, in mm.

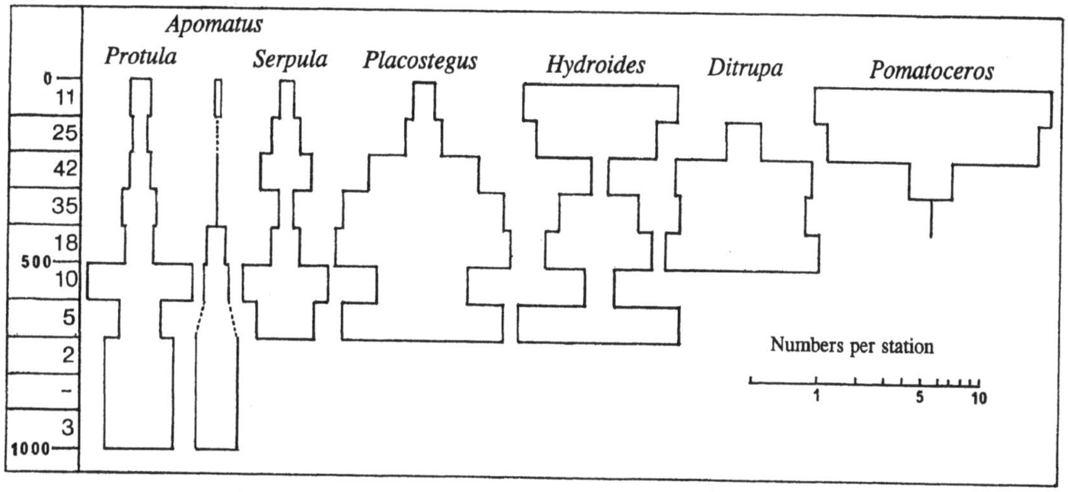

Figure 4. Depth distributions of serpulid genera off Faroe Islands. Histograms indicate, by width against logarithmic scale, mean numbers of individuals per BIOFAR station within each 100 m depth zone.

pooled (in Figure 4) to smooth over a sampling gap in the 801–900 m depth zone.

Counting the numbers of individuals representing each genus, in samples from gradually increasing depths, shows that *Protula*, always non-operculate, is the genus best represented in deeper channels. About 33% of all *Protula* found so far were from depths >500 m, these figures for operculate genera being 30% of *Apomatus*, 23% of *Serpula*, 10% of *Placostegus* and 9% of *Hydroides*, with no *Ditrupa* below 500 m and no *Pomatoceros* below 400 m. *Filograna*, numerous at <300 m, scarce below 500 m and not found below 600 m, sometimes had and sometimes lacked opercula.

Discussion

Perhaps development of *Filograna* opercula is related to cycles of fission and sexual reproduction, for embryos which are brooded within tubes need oxygen. Considering non-operculate serpulids of shallow seas, *Salmacina, Paraprotula* Uchida, *Microprotula* Uchida and *Paraprotis dendrova* Uchida incubate their embryos, so have special respiratory needs; *Floriprotis* Uchida (1978) may be protected by the coral with which it is associated, whilst the spines in tubes of *Spiraserpula* (see Pillai & ten Hove, 1994) may be as good as opercula for combatting predatory invaders.

Although most *Paraprotis pulchra* Imajima (1979) were without opercula, some had a vesicle as big as the tube mouth. The similar operculum of *Apomatus similis* has a vascular network, with the blood tinged green (Marion & Bobretzky, 1875), probably by chlorocruorin (Fox, 1933), so it may indeed close the tube yet help in some exchange of dissolved gases. The presence or absence of opercular vesicles, with or without the horny or calcareous coverings of true opercula, must depend on the balance between conflicting needs for protection and respiration. The answer to the question in our title is likely to be 'yes', as far as *Hyalopomatus cancerum* is concerned, but it lives in an oxygen minimum zone, where predators are probably scarce.

Off the Faroe Islands (latitude 62° N) concentrations of dissolved oxygen were near saturation at all depths of the survey (Appendix in Westerberg, 1990). The decline in serpulid diversity below 700 m should probably be attributed to lower temperatures and increased detritus. Most serpulids (but not *Ditrupa*) attach their tubes to hard substrata, which must be scarcer at great depths. Tubes of *Protula* and *Apomatus* from deeper samples were mostly unattached. These two genera are difficult to separate. *Apomatus* is indeed often included within *Protula* (Zibrowius, 1973), but here it was convenient to record as *Apomatus* individuals which retained their (easily shed) opercular vesicles.

To judge from their thoracic uncini, which somewhat resemble Figures 1F & G, *Apomatus* and *Protula* are more closely related to *Hyalopomatus* than to *Protis*, which has uncini like Figure 3F. In describing *Protis brownii* (Pixell, 1913) as a species of *Apomatus* Pixell was misled by a vesicle on the end of a pinnulate radiole. In sharing this character *Protis* and *Apomatus* may well be regarded as neotenous, and *Protula* still more so, considering how opercula develop in at least two other genera (Sentz-Braconnot, 1964). Soon after metamorphosis young *Serpula* and *Hydroides* develop a terminal vesicle on one of the second dorsal radioles, which had earlier developed pinnules. The vesicle differentiates to form the first operculum. Only subsequent replacement opercula have stalks without pinnules.

Studies of deep-sea Brachiopoda often met difficulties of the kind that misled Pixell, so it became convenient to call such resemblances homoeomorphy (Muir-Wood, 1960). The resemblances often involved underdevelopment of adult organs and systems, with some clear examples of neoteny in several major taxa of deep-sea animals, in addition to the brachiopods. The terms hypomorphy and paedomorphy were introduced (Zezina, 1994) and could well be applied to the pinnulate opercula of *Apomatus*, the vesicles of some *Protis* and the lack of even rudimentary opercula in other *Protis* and *Protula*. These three genera are prominent amongst the few serpulids found at great depths (Zibrowius, 1977). There the oxygen is generally about 0.5 of saturation (Sverdrup et al., 1942), so lack of opercular protection in *Protula* and *Protis* probably reflects scarcity of predators rather than respiration difficulties. Abyssal hypomorphy may also be associated with low temperatures and poor feeding (Zezina, 1994).

Acknowledgments

In preparing this paper we have been greatly helped by Drs Nechama Ben-Eliahu, Simon Creasey, I. A. Jirkov, J. B. Kirkegaard, Elena Kupriyanova, T. G. Pillai, A. V. Rzhavsky, Harry ten Hove, J. W. Wägele, Olga Zezina, Helmut Zibrowius and three referees.

References

Ben-Eliahu, M. N. & D. Fiege, 1996. Serpulid tube-worms (Annelida: Polychaeta) of the central and eastern Mediterranean, with particular reference to the Levant Basin. Senckenbergiana mar. 28: 1–51.

Bruun, A. F., 1957–59. General introduction to the reports and list of deep-sea stations. Galathea Rep. 1: 7–48.

Ehlers, E., 1887. Reports on dredgings in the Gulf of Mexico by the U.S. Coast Survey Steamer 'Blake'. XXXI, Report on the Annelids. Mem. Mus. comp. Zool. Harv. 15: 1–355.

Eliason, A., 1951. Polychaeta. Rep. Swed. Deep Sea Exped. 2, Zoology 11: 131–148.

Fox, H. M., 1933. The blood circulation of animals possessing chlorocruorin. Proc. r. Soc. Lond. B 112: 479–495.

Gage, J. D., 1995. A cruise to investigate the deep-sea sediment community and processes in the Arabian Sea. Deep-Sea Newsletter 23: 23–27.

Griffin, D. J. G., 1974. Spider crabs (Crustacea: Brachyura: Majidae) from the International Indian Ocean Expedition. Smiths. Contr. Zool. 182: 1–35.

Hansen, G. A., 1878. Annelider fra den norske Nordhavsexpedition i 1876. Nyt Mag. Naturvid. 24: 1–17.

Imajima, M., 1979. Serpulidae collected around Cape Shionomisaki, Kii Peninsula. Mem. natn. Sci. Mus. Tokyo 12: 159–183.

Knight-Jones, P. & E. W. Knight-Jones, 1991. Ecology and distribution of Serpuloidea (Polychaetae) round South America. Ophelia 5: 569–586.

Kupriyanova, E. K., 1993a. Deep-water Serpulidae from the Kurile-Kamchatka Trench 1. Genus *Hyalopomatus*. Zool. Zhurn. 72: 145–152 (in Russian).

Kupriyanova, E. K., 1993b. Deep-water Serpulidae from the Kurile-Kamchatka Trench 2. Genera *Bathyditrupa*, *Bathyvermilia* and *Protis*. Zool. Zhurn. 72: 21–28 (in Russian).

Kupriyanova, E. K. & I. A. Jirkov, 1997. Serpulidae (Annelida, Polychaeta) of the Arctic Ocean. Sarsia 81: in press.

Marion, A. F. & N. Bobretzky, 1875. Étude des annélides du Golfe de Marseille. Ann. Sci. nat. 6: 1–106.

Muir-Wood, M., 1960. Homoeomorphy in recent Brachiopoda: *Abyssothyris* and *Neorhynchia*. Ann. Mag. nat. Hist. 13: 521–528.

Pillai, T. G. & H. A. ten Hove, 1994. On recent species of *Spiraserpula* Regenhardt, 1961, a serpulid polychaete genus hitherto known only from Cretaceous and Tertiary fossils. Bull. nat. Hist. Mus. Lond. 60: 39–104.

Pixell, H. L. M., 1913. Polychaeta of the families Serpulidae and Sabellidae, collected by the Scottish National Antarctic Expedition. Trans. r. Soc. Edinb. 49: 347–358.

Sentz-Braconnot, E., 1964. Sur le développement des Serpulidae *Hydroides norvegica* (Gunnerus) et *Serpula concharum* Langerhans. Cah. Biol. mar. 5: 385–389.

Sverdrup, H. U., M. W. Johnson & R. H. Fleming, 1942. The Oceans. Prentice-Hall, New York, 1087 pp.

Ten Hove, H. A. & H. Zibrowius, 1986. *Laminatubus alvini* gen. et sp. n. and *Protis hydrothermica* sp. n. (Polychaeta Serpulidae) from the hydrothermal communities in the eastern Pacific. Zool. Scr. 15: 21–31.

Uchida, H., 1978. Serpulid tube worms from Japan with the systematic review of the group. Bull. mar. Park Res. Stn. 2: 1–98.

Wägele, J. W., 1979. Morphologische Studien om *Eisothistos* mit Beschreibung von drei neuen Arten (Crustacea, Isopoda, Anthuridea). Mitt. Zool. Mus. Univ. Kiel 1: 1–19.

Westerberg, H., 1990. Benthic temperature in the Faroe area. Rep. Dep. Oceanography Univ. Gothenburg 51: 1–19.

Wollebaek, A., 1912. Nordeuropaeiske Annulata Polychaeta 1, Ammocharidae, Amphictenidae, Ampharetidae, Terebellidae og Serpulidae. Skr. VidenskSelsk. Christiana 18: 1–144.

Zezina, O. N., 1994. Deep-sea brachiopods. Their peculiarities in morphology and evolution. Sarsia 79: 59–64.

Zibrowius, H. W., 1969. Review of some little known genera of Serpulidae. Smiths. Contr. Zool. 42: 1–22.

Zibrowius, H. W., 1973. Serpulidae (Annelida Polychaeta) des côtes ouest de L'Afrique et des archipels voisins. Ann. Mus. r. Afr. c. Zool. 207: 1–98.

Zibrowius, H. W., 1977. Review of Serpulidae from depths exceeding 2000 metres. In Reish, D. J. & K. Fauchald (eds), Essays on Polychaetous Annelids in Memory of Dr Olga Hartman. Allan Hancock Foundation, Los Angeles: 289–306.

Hydrobiologia **355:** 153–158, 1997.

A. D. Naumov, H. Hummel, A. A. Sukhotin & J. S. Ryland (eds), Interactions and Adaptation Strategies of Marine Organisms.

Adaptation capabilities of marine modular organisms

N. N. Marfenin
Dept. Invertebrate Zoology, Biological Faculty, Moscow State University, Moscow 119899, Russia

Key words: ecology, modular organisms, coloniality, growth strategy

Abstract

Marine modular organisms such as hydroids and corals are plastic in their responses to continuously changing environments. Morphogenetic limitations are less important for modular animals and plants, than for unitary ones. Although each module varies relatively little, modular organisms are characterized by an extremely broad plasticity of shape. Sessile colonial animals grow into a heterogenous environment and so each modular organism has its own often unique shape. The mechanism of modular body plasticity and adaptation to the environment is based on cyclical morphogenesis through replication of modules. Plasticity of shape is achieved not only by colonial growth, but during unfavorable periods also by body reduction due to module reabsorbtion.

Introduction

Colonial patterns are still a subject of scientific discussion and investigation. In his famous comparative anatomy treatise on invertebrates, Beklemishev (1969) summarized a long period of attention to the problem of coloniality, in which general principles concerning the evolution of life and especially the origin of the Metazoa, were elaborated. The distinguishing features of this period were the special attention paid to homologies among zooids and to the determination of anatomical prototypes and their possible transformations during evolution. Since 1970 a new approach to the problem of coloniality has arisen as comparative anatomy was supplanted by general system analysis. Parallelisms in the architecture and ecology of different taxonomic groups such as plants, fungi and colonial animals have been proposed instead of homologies among initially solitary individuals. New terms such as 'module', 'modular pattern', 'modular growth' and 'modular strategy' overcome taxonomic and morphological differences between organisms (Fry, 1979; Harper & Bell, 1979; Rosen, 1979; Chapman & Stebbing, 1980; Chapman, 1981; Grasshoff, 1981; Begon et al., 1986; Ryland & Warner, 1986; Ellison & Harvell, 1989).

Modular organisms have some common ecological peculiarities (Ryland, 1981; Hughes & Cancino, 1985; Jackson, 1985) such as the following: relative indifference to mechanical damage (Mackie, 1963; Werner, 1979; Rosen, 1979), gradually increasing growth, conformity to ephemeral ecological niches (Chapman & Stebbing, 1980), exploration of patchily distributed resources (Buss, 1979), and preadaptation to relatively stable environments on a geographical scale (Jackson, 1977, 1985; Ryland, 1981). In the present report I try to explain the ecological significance of modular patterns among marine invertebrates, with examples from hydroids, by reviewing – further to the accounts of Boero (1984) and Gili & Hughes (1995) – the morphogenetical properties and the physiological integration of a colony.

Material and methods

The spatial adaptation of marine modular organisms was studied in living colonial hydroids of the families Clavidae, Bougainvilliidae, Campanulariidae and Sertulariidae. More detailed information about the material can be found in a series of publications by the author which are quoted in the text and contain original results.

Whole colonies were grown on natural and artificial substrata according to the method described in Burykin et al. (1984). Overall increases of all parts of the colonies within a standard period of time, and morphological changes of the colonies in the course of growth, were recorded with the help of a specially developed method of 'mapping' (Marfenin, 1980). The rate of growth was determined by recording stolon and stalk tip growth pulsations. The morphogenesis of hydroids was studied with time-lapse video recording. To record changes in the colony as it grew, a system of quantitative indices was used (Marfenin, 1977). Morphogenetic plasticity in corals was studied in the genus *Acropora* at the Great Barrier Reef in Australia (Heron Island Research Station) by means of direct observations on the compatibility between the body shape of colonies and environmental space features, such as the local topography of free spaces in the coral 'bushes'.

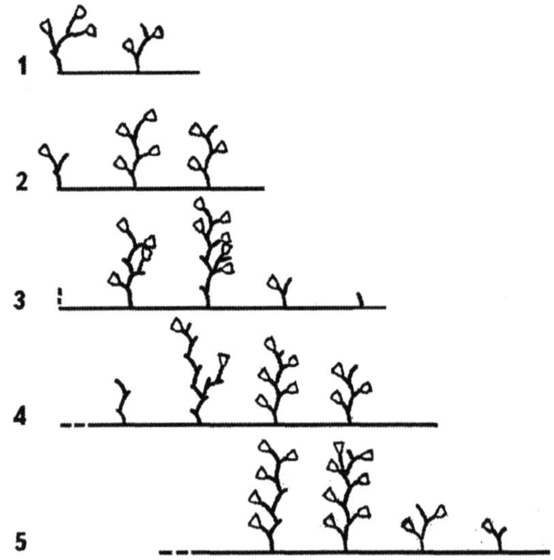

Figure 1. Obelia colony moving along a substrate: new stems arise from a growing stolon, while the oldest one is reabsorbed. Time intervals between 1, 2, etc. is 4 days.

Results

This account contains a summary of the ecological results of the author's earlier experimental research on the growth of colonies. During this work, data for the analysis of spatial adaptation of modular invertebrates such as hydroids and corals were also accumulated. The mechanism of spatial adaptation to the environment became clear only after the growth of various types of colonies, behaviour of hydranths, mechanisms of physiological integration of colonies and a number of other questions had been studied. Details of the experiments and primary statistical data are given in earlier publications by the author and his colleagues (Marfenin & Burykin, 1979; Burykin & Marfenin, 1983; Karlsen & Marfenin, 1984; Marfenin & Kosevich, 1984a, b; Marfenin, 1984, 1985a, b, c, 1993a, b).

Generalization of separate facts makes it possible to consider the assumed mechanisms of environmental space adaptation, which are realized only due to modular organization. The mechanisms determining the relationship between body shape and environmental space features are the following: (1) modifying the growth rate, (2) branching (or expansion of growth zones), (3) reabsorption of several parts of a colony.

In hydroid colonies the growth rate of separate parts can change quickly and unevenly with some parts more affected than others. For example, the rate of growth of the tips of upright stalks usually changes little before a new internode is formed, but in stolons it can vary considerably. Changes in the number of growing upright stalks corresponds proportionally to the daily food ration for the whole colony. If food is lacking after a new internode is formed, the tops of the oldest upright stalks stop growing. Young upright stalks, however, continue to grow (Marfenin, 1993b).

Stolons react to changes in food supply within 1 to 2 days at 15–18 °C. A hungry hydroid colony grows with straight stolons. Thus, the growth rate of a modular animal depends on ration. A colony slowly grows along a surface (Figure 1), changing its position on the substrate (Marfenin & Burykin, 1979; Kosevich, 1984).

Growth is accompanied by dedifferentiation of the oldest zooids (Thatcher, 1903; Huxley & De Beer, 1923) and of big fragments of coenosarc, which is a well known phenomenon (Crowell, 1953; Brock, 1974; Hale, 1973). Its ecological significance is still discussed (Hughes, 1987). From our observations and laboratory experiments we found that growth and dedifferentiation together could reconstruct a colony to a large extent according with the environmental situation, which is of important adaptive significance because of its direct effect on the modular pattern. The adaptive reaction depends on environmental factors such as food and water currents.

These features depend on the degree of physiological integration within an colony. Food given to the oldest parts of a colony is transported in the hydroplasm to the youngest parts (Marfenin, 1983a, 1985a). If food is abundant, all the upright stalks grow and, at a given temperature, the rate of stolon growth gradually increases up to a species specific maximum. Branching becomes intensive when the colony receives more food than the growing stalks and stolons can utilize. When stolons and stalks start to branch, the colony turns from stolonial growth without stalk branching, which is an extensive occupation of the substrate surface, to a more intensive exploitation of its resources due to an increase of the number of hydranths, and thus an increase of the food intake, per unit surface. The hydroid *Clava multicornis* normally is distinguished by its compact colonies. Nevertheless if it is maintained in a different habitat (Edwards & Harvey, 1975), or in experiments under long-term shortage of food, the colony can become linear in form (Marfenin, 1985c). It then shifts along the substrate. After an increase of the ration, stolons form a lot of short lateral branches, which grow twisted. As a result of the change in growth the colony remains stationary. This result can be used to explain the mosaic distribution of *C. multicornis* on *Ascophyllum nodosum* (Aldrich et al., 1980).

In more complicated colonies, like *Obelia longissima*, branching takes place in a certain regular order. At the periphery of the colony just behind the newest internodes of stolons and the branching starts earlier than in the old central areas. This means that the colonies first 'elaborate' the peripheral zone of contact with the environment. If there are sufficient food resources, branching also continues in the central part under the cover of the developed periphery.

Branching automatically accelerates the consumption of food. Consumption grows exponentially and must lead to an imbalance between the food available and the demand for food of the increasing number of growing stolons, stalks and new branches. Some branches then stop growing.

In poorly integrated species such as the creeping colonies of *Cordylophora inkermanica* (Marfenin, 1983b), the discrepancy between growth and the amount of received food causes an almost simultaneous rapid reabsorbtion of all the hydranths.

In colonies with a more complex structure, for example those of Campanulariidae or Sertulariidae, growth stops unevenly and asynchronously. Our experiments have shown that the cessation of growth is more abrupt in those sections of a colony (a radial stolon with stalks) which receive less food. We have found nevertheless that semidigested food is transferred from one section of the colony to another, which means that there is physiological integration (Marfenin, 1985a, 1993b). That the integration is incomplete is also a matter of adaptive significance, because not all parts of the colony are then affected simultaneously.

Under a shortage of food modular organisms decrease their body size many times. As in every organism during starvation periods, modular animals are spending their own internal resources by digesting the oldest cell material (Brock, 1970). Whereas the organs of an unitary organism become smaller during autolysis, which limits its tolerance to starvation, a modular body becomes smaller mainly through decreasing the number of identical modules, rather than reducing their size. Colonial animals stay alive, even when getting smaller several hundred fold, without damage to their viability.

In this way the adjustment of shape, size and fertility of a colony depends upon important ecological factors such as physical obstacles hindering growth, factors reducing the probability of catching prey at a certain site, or intensity of illumination (important for the growth of corals in symbiosis with zooxanthellae).

As a rule every colony has abundant growth zones during a favourable season, but nevertheless only some of them continue to grow. That is a way to modify the direction of growth and the intensity of a colony's expansion (Marfenin, 1973).

Discussion

There are many ecological advantages of the modular pattern among metazoan phyla. Modular organisms can easily sacrifice modules to predators while keeping their viability (Chapman & Stebbing, 1980). With an increase in colony size the risk of being killed is reduced.

Because of branching their growth is exponential in the absence of any exogenous limiting factor (Davis, 1971; Simkina & Turpaeva, 1980; Chapman & Stebbing, 1980; Marfenin, 1993b; Gili & Hughes, 1995). At the same time, unlike that of a nonmodular body, the surface area to volume ratio of a modular organism is rather stable (Jackson, 1977, 1979; Hughes & Cancino, 1985). A colony can function at any size, from one module to millions. Its degree of complexity does not change. Finally, modular organisms on average live far longer than unitary aclonal individuals (Jackson, 1985).

Our investigations of colonial hydroids focused on two main aspects. These are the morphogenetic basis of colonial structure and growth, and the role of physiological integration (via food transportation in the

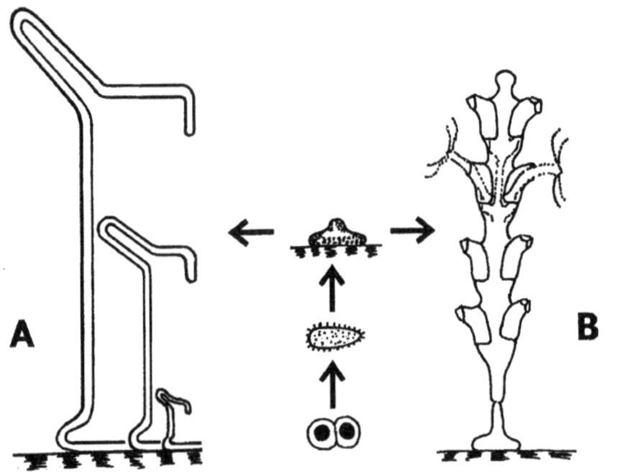

Figure 2. Two morphogenetic strategies: A – Growth of an unitary organism like a sea-anemone with long-time increase in body size; B – Growth by multiplication of modules, like the hydroid colony.

coenosarc) in spatial adaptation of modular organisms to a constantly changing environment.

The differences between unitary and modular types of organisms is determined by their morphogenetic strategies (Figure 2). In the first, such as a sea anemone, the process of individual development and growth is prolonged for the whole life cycle of the organism. In the second, the morphogenetic strategy is typically a repeated process of individual development realized as cyclical morphogenesis. It is easily demonstrated with colonial hydroids, in which modular growth is a clear example of a morphogenetic cycle. Cyclical morphogenesis differs from cloning or asexual reproduction. In asexual reproduction zooids are liberated from the mother organism, and each daughter zooid develops to a complete organism. In contrast to this, the replication of modules builds a colony because liberation of zooids is prevented. This is due to some morphogenetic mechanisms which limits longevity of zooids, or which holds back growth and reproduction. Owing to cyclical morphogenesis a modular organism possesses: (1) a low degree of morphological variability among modules; (2) a sequence of branches and of their orientation; (3) a definite plan for the colonial form.

As already noted a modular strategy permits easy variation of colony size and, within limits, does not restrict colony shape. A broad size range and plasticity of shape are both advantageous for sedentary organisms. They inhabit an environment characterized by an irregular distribution of several ecological factors. This heterogeneity has a more significant influence on sedentary organisms than on mobile species. The most important environmental parameters are: (1) water movements, (2) overgrowth effects, (3) sediment impact, (4) competition for space, (5) feeding probability (Boero, 1984; Gili & Hughes, 1995). Together these factors contribute to a highly heterogeneous environment.

A genetically determined stable body shape should be poorly compatible with immobility. Mobile animals are able to change their position and find a better place, which will be more compatible with their shape, size, and life strategy. But sessile animals use different means to adapt to their environment. Their life strategy could be called a 'progressive growing into' a local asymmetric space. Their shape becomes adequate for the peculiarities of an occupied space (Figure 3). Corals provide many good examples of the adaptation of their shape to the environment. Besides this, the shape of encrusting colonial ascidians and bryozoans depends on the substrate. The hydroids *Orthopyxis integra* and *Laomedea angulata* are good examples of successful colonial growth upon narrow leaves of *Zostera marina* (Burykin & Marfenin, 1983; Hughes et al., 1991b), as are the hydroids *Sertularia perpusilla* on the leaves of the seagrass *Posidonia oceanica* (Hughes et al., 1991a). The growth regulation of *Clava multicornis* colonies, described above, illustrates further the mechanism of spatial adaptation.

Conclusions

A remarkable variability in size and shape is a characteristic property of modular organisms as opposed to unitary ones. Cyclical morphogenesis is the basic process of colonial growth. Cyclical morphogenesis differs from cloning or asexual reproduction. It limits the growth of single modules. Owing to cyclical morphogenesis a modular organism possesses a regular construction, while the modular strategy permits variation of colony size and does not strongly restrict colonial shape. Due to modular growth and a proper physiological integration colonial organisms adapt easily to (changes in) complex environmental situations. The modular pattern is especially beneficial for sessile life, even if it is also found among pelagic invertebrates. Three mechanisms ensure the plasticity of colonial shape: (1) regulation of growth rate; (2) branching; (3) reabsorption of certain modules.

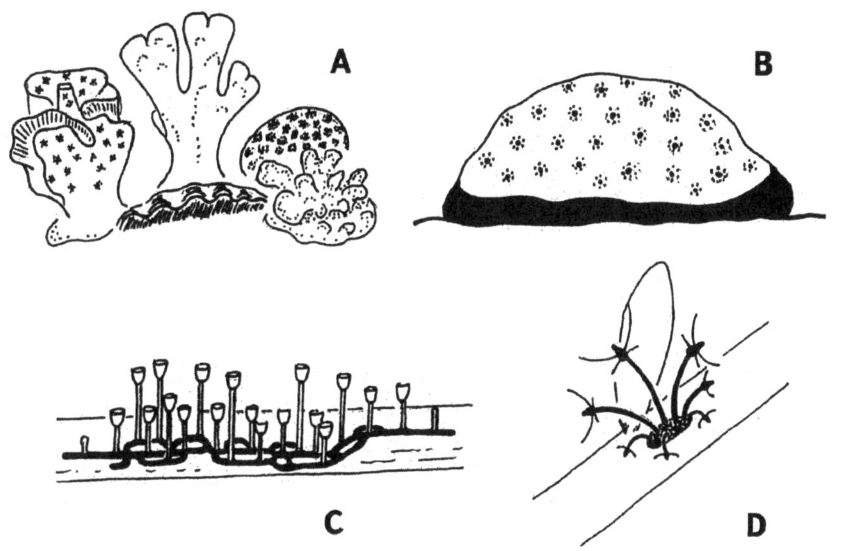

Figure 3. Four examples of progressive growth into local asymmetric space: A – Corals growing in a reef with complex morphology; B – Colonial ascidian encrusting a stone surface; C – Linear growth of the hydroid *Orthopyxis integra*: a colony on the leaves of *Zostera marina* (after Burykin & Marfenin, 1983); D – Compact colonies of the hydroid *Clava multicornis* occupying limited parts of *Ascophyllum nodosum*.

Acknowledgments

The current investigations have been supported by the Russian Foundation for Fundamental Research: Grant #95-04-12071Œ. I am grateful to Dr E. Robson for correction of the English version, and to Prof. F. Boero for valuable criticism of the manuscript.

References

Aldrich, J. C., W. Crowe, M. Fitzgerald, M. Murphy, C. McManus, B. Magennis & D. Murphy, 1980. Analysis of environmental gradients and patchiness in the distribution of the epiphytic marine hydroid *Clava squamata*. Mar. Ecol. Prog. Ser. 2: 293–301.

Begon, M., J. L. Harper & C. R. Townsend, 1986. Ecology: individuals, population and communities. Blackwell Sci. Publ., Boston, 851 pp.

Beklemishev, W. N., 1969. Principles of Comparative Anatomy of Invertebrates, 1: Promorphology. Oliver and Boyd, Edinburgh, 490 pp.

Boero, F., 1984. The ecology of marine hydroids and effects of environmental factors: a review. P.S.Z.N. I, Mar. Ecol. 5: 93–118.

Brock, M. A., 1970. Ultrastructural studies on the life cycle of a short-lived metazoan, *Campanularia flexuosa*. II. Structure of the old adult. J. ultrastruct. Res. 32: 118–141.

Brock, M. A., 1974. Growth, developmental, and longevity rhythms in *Campanularia flexuosa*. Am. Zool. 14: 757–771.

Burykin, J. B., N. N. Marfenin & A. G. Karlsen, 1984. Some experience on the cultivation of the marine colonial hydroid *Dynamena pumila* (L.) in a laboratory. Biologicheskie nauki (Biological Science): 102–106 (in Russian).

Burykin, J. B. & N. N. Marfenin, 1983. The growth and structure of the colony of a hydroid polyp *Campanularia platycarpa* (Hydrozoa, Campanulariidae). Zool. Zh. 62: 1417–1419.

Buss, L. W., 1979. Hebitat salection, directional growth and spatial refuges: Why colonial animals have more hiding places. In Larwood, G. & B. R. Rosen (eds), Biology and Systematics of Colonial Organisms. Systematics Association, Special Volume 11. Academic Press, London: 459–497.

Chapman, G. & A. R. D. Stebbing, 1980. The modular habit – a recurring strategy. In Tardent, P. & R. Tardent (eds), Developmental and Cellular Biology of Coelenterates. Elsevier, Amsterdam: 157–162.

Chapman, G., 1981. Individuality and modular organisms. Biol. J. Linn. Soc. 15: 177–183.

Crowell, S., 1953. The regression-replacement cycle of hydranths of *Obelia* and *Campanularia*. Physiol. Zool. 26: 319–327.

Davis, L. V., 1971. Growth and development of colonial hydroids. In Lenhoff, H. M., L. Muscatine & L. V. Davis (eds), Experimental Coelenterate Biology. Honolulu: 16–36.

Edwards, C. & S. M. Harvey, 1975. The hydroids *Clava multicornis* and *Clava squamata*. J. mar. biol. Ass. U.K. 55: 879–886.

Ellison, A. M. & C. D. Harvell, 1989. Size hierarchies in *Membranipora membranacea*: Do colonial animals follow the same rules as plants? Oikos 55: 349–355.

Fry, W. G., 1979. Taxonomy, the individual and the sponge. In Larwood, G. & B. R. Rosen (eds), Biology and Systematics of Colonial Organisms. Systematics Association, Special Volume 11. Academic Press, London: 39–80.

Gili, J.-M. & R. G. Hughes, 1995. The ecology of marine benthic hydroids. Oceanogr. mar. Biol. Ann. Rev. 33: 351–426.

Grasshoff, M., 1981. Polipen und Kolonien der Blumentiere (Anthozoa). 2. Die Achtstrahligen Korallen (Octocorallia). Natur und Museum 3: 29–45.

Hale, L. J., 1973. The pattern of growth of *Clythia johnstoni*. J. Embryol. exp. Morph. 29: 283–309.

Harper, J. L. & A. D. Bell, 1979. The population dynamics of growth form in organisms with modular constraction. In Anderson, R. M., H. D. Turner & L. R. Taylor (eds), Population Dynamics. Blackwell, Oxford: 29–52.

Hughes, R. G., 1987. The loss of hydranths of *Laomedea flexuosa* Alder and other hydroids, with reference to hydroid senescence. In Bouillon, J., F. Boero, F. Cicogna & P. F. S. Cornelius (eds), Modern Trends in the Systematics, Ecology and Evolution of Hydroids and Hydromedusae. Clarendon Press, Oxford: 171–184.

Hughes, R. N. & J. M. Cancino, 1985. An ecological overview of cloning in Metazoa. In Jackson, J. B. C., L. W. Buss & R. E. Cook (eds), Population Biology and Evolution of Clonal Organisms. Yale University Press, New Haven: 153–186.

Hughes, R. G., A. Garcia Rubies & J. M. Gili, 1991. The growth and degeneration of the hydroid *Sertularia perpusilla*, an obligate epiphyte of leaves of the seagrass *Posidonia oceanica*. In Williams, R. B., P. F. S. Cornelius, R. G. Hughes & E. A. Robson (eds), Coelenterate Biology: Recent Research on Cnidaria and Ctenophora. Hydrobiologia 216–217/Dev. Hydrobiol. 66: 211–214.

Hughes, R. G., S. Johnson & I. D. Smith, 1991. The growth patterns of some hydroids that are obligate epiphytes of seagrass leaves. In Williams, R. B., P. F. S. Cornelius, R. G. Hughes & E. A. Robson (eds), Coelenterate Biology: Recent Research on Cnidaria and Ctenophora. Hydrobiologia 216–217/Dev. Hydrobiol. 66: 205–210.

Huxley, J. S. & G. R. de Beer, 1923. Studies in dedifferentiation. IV. Resorption and differential inhibition in *Obelia* and *Campanularia*. Q. J. microsc. Sci., N. Ser. 67: 473–495.

Jackson, J. B. C., 1977. Competition on marine hard substrata: the adaptive significance of solitary and colonial strategies. Am. Nat. 118: 743–767.

Jackson, J. B. C., 1979. Morphological strategies of sessile animals. In Larwood, G. & B. R. Rosen (eds), Biology and Systematics of Colonial Organisms. Systematics Association, Special Volume 11. Academic Press, London: 499–555.

Jackson, J. B. C., 1985. Distribution and ecology of clonal and aclonal benthic invertebrates. In Jackson, J. B. C., L. W. Buss & R. E. Cook (eds), Population Biology and Evolution of Clonal Organisms. Yale University. Press, New Haven: 297–355.

Karlsen, A. G. & N. N. Marfenin, 1984. Hydroplasm movements in the colony of the hydroid *Dynamena pumila* (L.) and some other species. Jurnal obschey biologii (J. General Biology) 45: 670–680 (in Russian).

Kosevich, I. A., 1984. Moving of a colony of *Obelia loveni* (Allm.) (Hydrozoa, Thecaphora, Campanulariidae) on a substrate. In Actual problems of oceanology. Institute of Oceanology, Moscow: 92–93 (in Russian).

Mackie, G. O., 1963. Siphonophores, bud colonies, and superorganisms. In Dougherty E.C. (ed.). The Lower Metazoa, Comparative Biology and Physiology. University of California Press, California: 329–337.

Marfenin, N. N., 1973. Growth morphology in a colony of the hydroid polyp *Dynamena pumila* (Hydrozoa, Leptolida). Jurnal obshey biologii (J. General Biology) 34: 727–737 (in Russian).

Marfenin, N. N., 1977. Study of *Dynamena pumila* colonial integration using quantitative morphological criteria. Jurnal obshey biologii (J. General Biology) 38: 409–422 (in Russian).

Marfenin, N. N., 1980. Method of mapping the colonial structure in Hydrozoa, and its significance for the investigation of colonial parts. In Naumov, D. V., N. N. Marfenin & S. D. Stepanjants (eds), The Theoretical and Practical Importance of the Coelenterates. Zoological Institute, Leningrad: 66–69 (in Russian).

Marfenin, N. N., 1983a. Morphology of a colony and transport system in two species of the genus *Acropora* (Anthozoa). Zoologichesky jurnal (Zool. J.) 62: 5–13 (in Russian).

Marfenin, N. N., 1983b. The new species *Cordylophora* (Hydrozoa, Clavidae) from the Black sea. Zoologicheskii jurnal (Zool. J.) 62: 1732–1734 (in Russian).

Marfenin, N. N., 1984. A morpho-functional analysis of a temporary hydroid colony, *Moerisia maeotica* (Ostr., 1896) (Leptolida, Limnomedusae), taken as an example. Jurnal obschey biologii (J. General Biology) 45: 660–669 (in Russian).

Marfenin, N. N., 1985a. The functioning of the pulsatory-peristaltic type of a transport system in colonial hydroids. Jurnal obschey biologii (J. General Biology) 46: 153–164 (in Russian).

Marfenin, N. N., 1985b. A morpho-functional analysis of monopodial colony organization in hydroids (*Tubularia larynx* Ell. et Sol.), characterized by terminally arranged zooids. Izvestiy Akademii nauk SSSR, ser. biologicheskay (Transaction of the Academy of Science USSR, Ser. Biol.) 2: 238–247 (in Russian).

Marfenin, N. N., 1985c. The formation of compact stolonal colonies in hydroids, *Clava multicornis* (Leptolida, Athecata) taken as an example. Zoologicheskiy jurnal (Zool. J.) 64: 975–981 (in Russian).

Marfenin, N.N., 1993a. The Phenomenon of Coloniality. Moscow University Press, Moscow, 237 pp. (in Russian)

Marfenin, N. N., 1993b. Functional Morphology of Colonial Hydroids. Zoological Institute, St.-Petersburg, 151 pp. (in Russian)

Marfenin, N. N. & Y. B. Burykin, 1979. The growth of hydroid colonies *Dynamena pumila* (L.) as a function of the quantity of food. Vestnik Moskovskogo Universiteta, Seria Biologia (Rep. Moscow State Univ., Ser. Biol.) 1: 61–68 (in Russian).

Marfenin, N. N. & I. A. Kosevich, 1984a. Colonial morphology of the hydroid *Obelia loveni* (Allm.)(Campanulariidae). Vestnik Moscovskogo universiteta, Seria Biologia (Rep. Moscow State Univ., Ser. Biol.) 2: 37–46 (in Russian).

Marfenin, N. N. & I. A. Kosevich, 1984b. Biology of the hydroid *Obelia loveni* (Allm.): colony formation, behavior and life cycle of hydranths, reproduction. Vestnik Moscovskogo universiteta, Seria Biologia (Rep. Moscow State Univ., Ser. Biol.) 3: 16–24 (in Russian).

Rosen, B. R., 1979. Modules, members and communes: a post-script introduction to social organisms. In Larwood, G. & B. R. Rosen (eds), Biology and Systematics of Colonial Organisms. Systematics Association, Special Volume 11. Academic Press, London: 13–25.

Ryland, J. S., 1981. Colonies, growth and reproduction. In Larwood, G. P. & C. Nielsen. Recent and Fossil Bryozoa. Olsen & Olsen, Fredensborg, Denmark: 221–226.

Ryland, J. S. & G. F. Warner, 1986. Growth and form in modular animals: ideas on the size and arrangement of zooids. Phil. Trans. r. Soc., Lond. B 313: 53–76.

Simkina, R. G. & E. P. Turpaeva, 1980. Character of the hydroid *Perigonimus* ecology in the Azov Sea. In Naumov, D. V., N. N. Marfenin & S. D. Stepanjants (eds), The Theoretical and Practical Importance of the Coelenterates. Zoological Institute, Leningrad: 100–104 (in Russian).

Thatcher, H. F., 1903. Adsorption of the hydranth in hydroid polyps. Biol. Bull. 5: 295–303.

Werner, B., 1979. Coloniality in the Scyphozoa: Cnidaria. In Larwood, G. & B. R. Rosen (eds), Biology and Systematics of Colonial Organisms. Systematics Association, Special Volume 11. Academic Press, London: 81–103.

Hydrobiologia **355:** 159–165, 1997.
A. D. Naumov, H. Hummel, A. A. Sukhotin & J. S. Ryland (eds), Interactions and Adaptation Strategies of Marine Organisms.

Morphological convergence of resting stages of planktonic organisms: a review

Genuario Belmonte[1], Anna Miglietta[1], Fernando Rubino[2] & Ferdinando Boero[1]
[1]*Dipartimento di Biologia, Universita' di Lecce, 73100 Lecce, Italy*
[2]*Istituto Sperimentale Talassografico 'A. Cerruti', CNR, 74100 Taranto, Italy*

Key words: resting stages, convergence, adaptation, plankton

Abstract

In temperate seas, many plankters avoid unfavourable periods by producing resting stages which accumulate in the sediments to form biodiversity banks from which plankton communities are seasonally restored. Most resting stages have typical spiny coverings. This morphology is common across phyla, and even kingdoms, and favours flotation, passive transport, and sensory activity, also opposing both predation and burial into the sediments. Spiny coverings are considered a convergence allowing survival of resting forms.

Introduction

At temperate latitudes, many species of coastal plankton produce resting stages (usually encysted) to overcome unfavourable periods which recur with the season cycle (for a list of different taxa, see Dale, 1983; Montresor, 1992; Madhupratap et al., 1996). Such stages can survive for long periods on or in the sediments, a habitat different from that where the active stages live (Marcus et al., 1994; Hairston et al., 1995).

Both the abundance of resting stages in coastal sediments, and the number of species showing life cycles which go through a resting period, led to the consideration of this trait as heavily influencing plankton community dynamics (Boero, 1994). The presence of benthic resting stages of plankters represents a strong link, although still underevaluated, between pelagic and benthic compartments (Boero et al., 1996a). At present, the knowledge of the ecology of resting stage pools (or marine 'seed' banks) in the sediments is scant; nevertheless, their influence on some aspects of the benthos can be easily supposed (Belmonte et al., 1995).

The majority (80%) of known resting stages of planktonic species show sculptured and/or protective coverings (Table 1), generally in the form of projections (Figure 1). The presence of projections in planktonic organisms is usually justified by flotation enhancement, but we propose here that they represent an adaptation to passive existence (rest). In fact, the usually planktonic active stages of protists having spiny cysts are mostly unsculptured, and also the subitaneous eggs of metazoans with spiny resting eggs are smooth.

Flotation

Undoubtedly, projections and/or long appendages favour flotation of planktonic organisms even when they can swim (e.g. plutei, which have cilia, or zoeae, which have swimming legs). Projections are probably an important part of the floating device for those plankters which do not swim (e.g. Rhizopoda). Protistan cysts (e.g. those of Dinophyta, and Tintinnina), however, are spiny, differently from most corresponding active stages, which swim by using flagella or cilia. The spiny covering probably delays sinking because it increases the cyst bearing surface. In fact, the spiny egg of *Centropages ponticus* (a neritic calanoid) sinks with a rate 5 times slower than that of the smooth egg of *Calanus finmarchicus* (a pelagic species) (Sazhina, 1987). Consequently, *C. ponticus* eggs hatch (embryonic development requires 24–48 h) after 6–8 m of

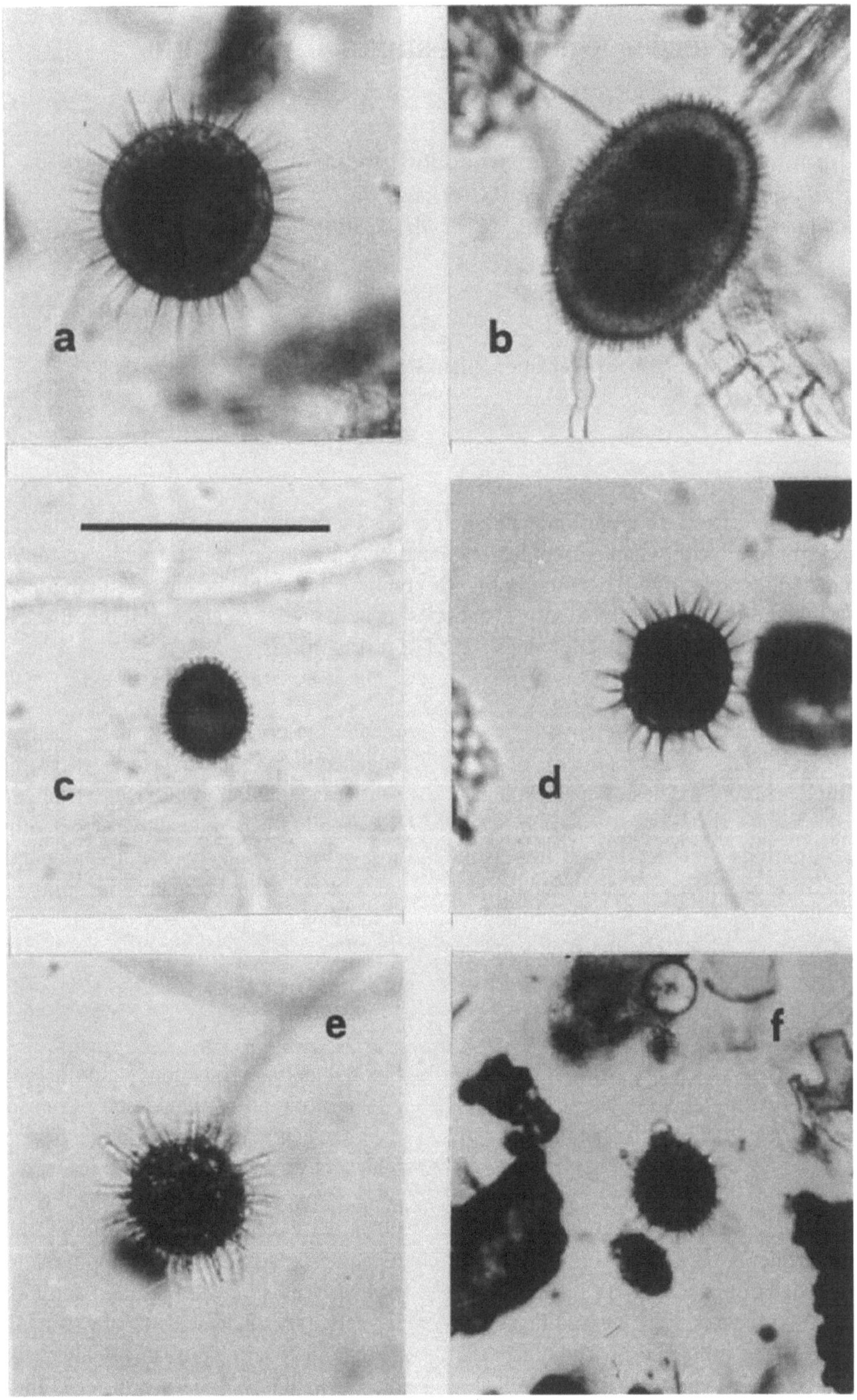

Figure 1. Some typical spiny resting stages from Italian coastal sediments. a. *Acartia* sp. (Copepoda); b. *Synchaeta* sp. (Rotifera); c. *Scrippsiella* sp. (Dinophyta); d. *Protoperidinium* sp. (Dinophyta); e. *Gonyaulax* sp. (Dinophyta); f. Tintinnina (Ciliophora). Scale bar, 100 μm.

Table 1. Typology of resting stage coverings among planktonic metazoans and protists.

Taxon	Resting stage covering		References
	sculptured	smooth	
Planktonic metazoans			
Anostraca	55	4	Cesar, 1989; Gilchrist, 1978; Mura, 1986; 1991; 1992; Mura et al., 1978.
Cladocera	2	5	Marcus, 1990; Viitasalo & Katajisto, 1994
Rotifera	1	4	Marcus, 1990; Viitasalo & Katajisto, 1994
Calanoida	12	3	Belmonte, 1992; 1996; Belmonte & Puce, 1994; Kasahara et al., 1974; Marcus, 1990; Santella & Ianora, 1992.
Planktonic protists			
Bacillariophyceae	6	1	Itakura et al., 1993; Marino et al., 1987, 1991; Riaux-Robin & Descolas-Gros, 1992.
Dinophyceae	82	28	Akselman, 1987; Blanco, 1989; Bolch & Hallegraeff, 1990 De Vernal et al., 1992; Drebes, 1981; Ellegaard et al., 1994; Hallegraeff, 1993; Harland, 1982; Ishikawa & Taniguchi, 1993; Larrazabal et al., 1990; Lewis, 1991; Lewis et al., 1984; Matsuoka, 1985a, 1985b, 1988; McMinn, 1991, 1992; Montresor, 1995; Montresor & Zingone, 1988; Montresor et al., 1993; 1994; Morey-Gaines & Ruse, 1980; Nehring, 1994; Reguera & Fraga, 1995; Reid, 1974, 1977; Wall & Dale, 1968.
Tintinnina	1	4	Paranjape, 1980; Reid, 1987; Reid & John, 1978
Total	*159*	*49*	

sinking, while *C. finmarchicus* eggs hatch after 40 m. This probably allows the nauplius of the neritic species to be born before the egg have reached the bottom.

In addition, spine length is directly correlated with water temperature (being, consequently, inversely correlated with water density). In cold areas, in fact, *Centropages typicus* has eggs with shorter spines than in warm areas (Gaudy, 1971).

However, spines can also promote sinking: the majority of bloom-forming diatoms, in fact, have simple or even barbed protuberances with which they become entangled, forming aggregations. This enhances their sinking rate, subtracting them from the nutrient depleted area where the bloom occurred (Smetacek, 1985).

The subitaneous eggs of the forms inhabiting the most confined, shallowest areas (e.g. *Synchaeta*, among rotifers, or *Acartia*, among copepods) have reduced or no spines. In fact, in these areas, even the spine covering cannot prevent the eggs from reaching the bottom (which is very close), and they cannot utilize currents (which are typically weak in such environments) for passive dispersal. As a result, the species typical of these areas do not consume energy and materials in spine building around subitaneous eggs. The same species, however, produce spiny resting eggs (e.g., for the Acartiidae, Belmonte, 1992, 1997; Belmonte & Puce, 1994) suggesting that this character probably favours more the passive existence in sediments than flotation in the water column.

Coastal areas, and confined ones overall, are characterized by high sedimentation rates. Resting stages programmed to rest for months can be buried by continuous sediment fall. Probably the spiny surface plays a role in the reactivation of buried resting stages. In fact, an episodic (seasonal) water turbulence in shallow waters can re-suspend fine-grained sediments, and the bearing surface given by projections should allow resting stages to remain longer than sediment grains in the water column, where they perceive seasonal stimuli better than in sediments. In fact, once re-activation has started, less than 24 h could be sufficient for hatching.

Passive defence

A spiny surface is also a defence for resting stages against predators and abiotic adversities. A long existence in the sediments probably exposes resting stages to predation. It is possible that resting stages have also chemical defences against microbes (bacteria and fungi), or against the digestive juices of sediment swallowers (as indirectly tested by Marcus, 1984, with the polychaete *Capitella capitata*), and spines could represent the morphological level of a complex anti-predator device set. Spines, in this framework, might represent an obstacle for chewers or drillers which do not ingest cysts with the sediments but attack them individually.

But spines could also be useful against abiotic adversities by keeping away from the surface the mineral particles of the sediment which could scrape and injure the resting stages. In fact, resting stages are present not only in surface sediments: those that do not hatch at the return of favourable conditions are buried by the continuous sediment fall. Viable resting eggs (which hatched in laboratory) have been found under 24 cm of sediment (Hairston et al., 1995).

Sensory activity and chemical exchanges

The spines of protistan cysts are sometimes hollow (Figure 2) and contain cellular substances. In this way they represent an increased cellular surface which probably plays a role in the information exchange between the resting organism and the environment. The chemistry of sediments, which varies seasonally in confined areas, could inform the resting stages about the best time for reactivation. In fact, oxygen concentration is probably the main factor responsible for the maintenance of diapause (see Uye et al., 1979; Grice & Marcus, 1981, for copepods; Anderson et al., 1987, for dinoflagellates), and the increased surface could better perceive vertical displacements of the redox horizon within the sediments.

In addition, a wide surface probably favours chemical exchange (e.g. respiratory gas diffusion) deriving from metabolic activity of resting organisms at least in the re-activation period which precedes hatching (e.g. Romano et al., 1996). Spines could form a microenvironment surrounding the resting stage even in the presence of fine grained sediments. The importance of a free space around the resting body has been proposed for *Artemia salina* diapause eggs (Gilchrist, 1978). In this case, a smooth outer cortex delimits a thick alveolar layer of the chorion (a true 'physiological gill') which represents a constant, outer environment for the embryo. The extreme variability of *Artemia* environments (from fresh- to salt water, to dry environments) probably justifies the existence of the outer cortex (a protection for the alveolar layer), but resting stages from marine areas do not need it.

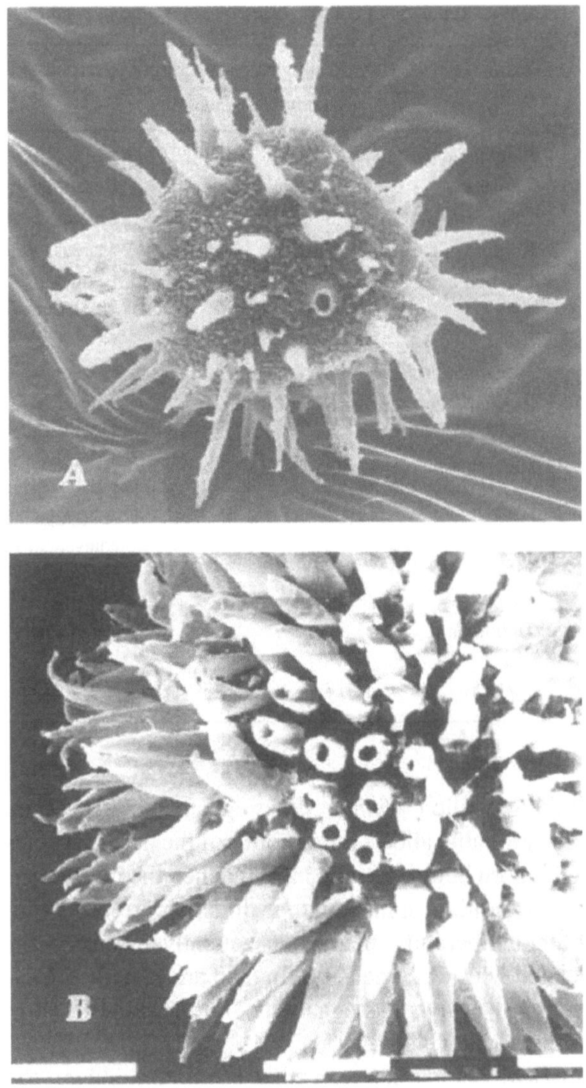

Figure 2. Dinoflagellate cysts with hollow spines (cavities visible in fractured spines). A. *Gonyaulax* sp., B. unidentified dinoflagellate. Scale bar, A = 20 μm; B = 10 μm.

Dispersal

With the increase of the bearing surface, the sculptured coverings of resting stages also favour dispersal by currents, as does delayed development. Furthermore, their projections allow resting stages to adhere to floating objects and swimming or even flying organisms (vectors), so being able to cross eco-geographical barriers for active stages with them. Many species of confined areas (e.g. lagoons, harbours, etc.) show wide geographical distribution, probably due to resting stage circulation which, in this way, maintains the genetic flux among apparently isolated populations (e.g. Carlton & Geller, 1993) also avoiding genetic drift.

Perspectives

Even if evolved to favour flotation, a spiny covering is a pre-adaptation to long passive survival on or in the sediments of coastal areas, where this morphology seems almost obligatory. This statement is supported by several lines of evidence:

- Among planktonic metazoans of confined areas (e.g. rotifers and copepods) the resting eggs are spiny, while the subitaneous ones are only lightly spiny, or lack spines. The spiny covering of resting stages is probably an adaptive constraint. Although its role can be only supposed at this stage of knowledge, we suggest that it was adopted by many planktonic phyla of two kingdoms (protists and metazoans).
- Sculptured surfaces characterize resting stages of non planktonic species (e.g. Porifera, Hydrozoa, Bryozoa, Tardigrada, Nematoda) or even of non aquatic species (e.g. angiosperms) further suggest that projections (generally spines) must have a key role in the survival of non motile stages, which does not involve merely flotation.

As remarked by Boero (1994) and Boero et al. (1996a), coastal plankton dynamics in temperate seas are characterized by an alternation of planktonic and benthic stages, even in species considered as 'holoplanktonic' such as calonoid copepods or diatoms. Resting stages, in this framework, play an essential role in explaining the seasonal changes in plankton composition. In spite of a huge body of knowledge on plankters' life cycles, highlighting the presence of resting stages in many species, plankton ecology is still explained mainly in terms of biogeochemical cycles, whereas biological cycles are not considered of ecological importance, and even the energy flow from plankton to benthos via cysts is totally unknown. This attitude must change, in spite of a perceived resistence in the scientific community, to recognize the role of biological cycles in one of the main processes of the biosphere: the cycles of coastal plankton. This resistence is probably linked to the disappearance of morphology and taxonomy from ecological studies, with the triumph of physico-chemical approaches to the study of ecosystem functioning. The expertise in zoology and botany, much widespread just a few decades ago, has disappeared from advanced countries (Boero et al., 1996b), leading to the lack of advocates for this type of knowledge in various advisory boards. We think that the burial of such disciplines has been premature and that it is time for them to come back to life after a period of rest due to adverse conditions. Besides morphology, furthermore, also physiology is to be put into an ecological framework. The discovery by Liang et al. (1997) of chaperon proteins allowing a reversible metabolic standstill in *Artemia* cysts under different environmental conditions demonstrates that we are just scratching the surface of an enormous reservoir of still unexplored perspectives of research.

References

Akselman, R., 1987. Quistes planctonicos de dinoficeas en areas de plataforma del Atlantico sudoccidental. I. Reporte taxonomico de la familia Peridiniaceae Ehrenberg. Boll. Inst. Oceanogr. S. Paulo 35: 17–32.

Anderson, D. M., C. D. Taylor & V. Armbrust, 1987. The effect of darkness and anaerobiosis on dinoflagellate cyst germination. Limnol. Oceanogr. 32: 340–351.

Belmonte, G., 1992. Diapause egg production in *Acartia (Paracartia) latisetosa* (Crustacea, Copepoda, Calanoida). Boll. Zool. 59: 363–366.

Belmonte, G., 1997. Resting eggs in the life cycle of *Acartia italica* and *A. adriatica* (Copepoda, Calanoida, Acartiidae). Crustaceana 70: 114–117.

Belmonte, G. & M. Puce, 1994. Morphological aspects of subitaneous and resting eggs from *Acartia josephinae* (Calanoida). Hydrobiologia 292/293: 131–135.

Belmonte, G., P. Castello, M. R. Piccinni, S. Quarta, F. Rubino, S. Geraci & F. Boero, 1995. Resting stages in marine sediments off the Italian coast. In Eleftheriou, A., A. D. Ansell & C. J. Smith (eds), Biology and Ecology of Shallow Coastal Waters. Olsen & Olsen Publ., Fredensborg: 53–58.

Blanco, J., 1989. Quistes de dinoflagelados de las costas de Galicia. II. Dinoflagelados peridinioides. Sci. Mar. 53: 797–812.

Boero, F., 1994. Fluctuations and variations in coastal marine environments. P.S.Z.N.I: Mar. Ecol. 15: 3–25.

Boero, F., G. Belmonte, G. Fanelli, S. Piraino & F. Rubino, 1996a. The continuity of living matter and the discontinuities of its constituents: do plankton and benthos really exists? Trends Ecol. Evol. 11: 177–180.

Boero, F., G. Belmonte, G. Fanelli, S. Piraino & F. Rubino, 1996b. Benthic pelagic uncoupling of carbon flow. Reply from Boero et al. Trends Ecol. Evol. 11: 471–472.

Bolch, C. J. & G. M. Hallegraeff, 1990. Dinoflagellate cysts in recent marine sediments from Tasmania, Australia. Bot. Mar. 33: 173–192.

Carlton, J. T. & J. B. Geller, 1993. Ecological roulette: the global transport of non indigenous marine organisms. Science 261: 78–82.

Cesar, I. I., 1989. Comparative study on the resting eggs of several anostracans (Crustacea). Key for the determination of the species based upon the egg structure and diameter. Stud. Neotrop. Fauna Envir. 24: 169–181.

Dale, B., 1983. Dinoflagellate resting cysts: benthic plankton. In Fryxel, G. A. (ed.), Survival Strategies of the Algae. Cambridge Univ. Press: 69–136.

De Vernal, A., L. Londeix, P. J. Mudie, R. Harland, M. T. Morzadec-Kerfourn, J. L. Turon & J. Wrenn, 1992. Quaternaey organic-walled dinoflagellate cysts of the North-Atlantic ocean and adjacent seas: ecostratigraphy and biostratigraphy. In Head, M. J. &

J. H. Wrenn (eds), Neogene and Quaternary Dinoflagellate Cysts and Architarchs. American Association of Stratigraphic Palynologist Foundation, Dallas: 289–328.

Drebes, G., 1981. Possible resting spores of *Dissodinium pseudolunula* (Dinophyta) and their relation to other taxa. Br. Phycol. J. 16: 207–215.

Ellegaard, M., N. F. Christensen & O. Mostrup, 1994. Dinoflagellate cysts from recent Danish marine sediments. Eur. J. Phycol. 29: 183–194.

Gaudy, R., 1971. Etude expérimentale de la ponte chez trois espèces de copepodes pélagiques (*Centropages typicus*, *Acartia clausi* et *Temora stylifera*). Mar. Biol. 9: 65–70.

Gilchrist, B. M., 1978. Scanning electron microscope studies of the egg shell in some Anostraca (Crustacea: Branchiopoda). Cell Tiss. Res. 193: 337–351.

Grice, G. D. & N. H. Marcus, 1981. Dormant eggs of marine copepods. Oceanogr. Mar. Biol. annu. Rev. 19: 125–140.

Hairston, N. J., R. A. van Brunt, C. M. Kearns & D. R. Engstrom, 1995. Age and survivorship of diapausing eggs in a sediment egg bank. Ecology 76: 1706–1711.

Hallegraeff, G. M., 1993. A review of harmful algal blooms and their apparent global increase. Phycologia 32: 79–99.

Harland, R., 1982. A review of recent and quaternary organic-walled dinoflagellate cysts of the genus *Protoperidinium*. Paleontology 25: 369–397.

Ishikawa, A. & A. Taniguchi, 1993. Some cysts of the genus *Scrippsiella* (Dinophyceae) newly found in Japanese waters. Bull. Plankton Soc. Japan 40: 1–7.

Itakura, S., M. Yamaguchi & I. Imai, 1993. Resting spore formation and germination of *Chaetoceros didymus* var. *protuberans* (Bacillariophyceae) in clonal culture. Nippon Suisan Gakkaishi 59: 807–813.

Kasahara, S., S.-I. Uye & T. Onbé, 1974. Calanoid copepod eggs in sea-bottom muds. Mar. Biol. 26: 167–171.

Larrazabal, M. E., P. Lassus, P. Maggi & M. Bardouil, 1990. Kystes modernes de dinoflagelle en baie Vilaine-Bretagne Sud (France). Cryptogamie Algol. 11: 171–185.

Lewis, J., 1991. Cyst-theca relationship in *Scrippsiella* (Dinophyceae) and related orthoperidinioid genera. Bot. Mar. 34: 91–106.

Lewis, J., J. D. Dodge & P. Tett, 1984. Cyst-theca relationship in some *Protoperidinium* species (Peridiniales) from Scottish sea lochs. J. Micropaleontol. 3: 25–34.

Liang, P., R. Amons, T. H. Macrae & J. S. Clegg, 1997. Purification, structure and *in vitro* molecular chaperone activity of *Artemia* P26, a small heat shock/alpha crystallin protein. Eur. J. Biochem. 243: 225–232.

Madhupratap, M., S. Nehring & J. Lenz, 1996. Resting eggs of zooplankton (Copepoda and Cladocera) from the Kiel Bay and adjacent waters (southwestern Baltic). Mar. Biol. 125: 77–87.

Marcus, N. H., 1984. Recruitment of copepod nauplii into the plankton: importance of diapause eggs and benthic processes. Mar. Ecol. Prog. Ser. 15: 47–54.

Marcus, N. H., 1990. Calanoid copepod, cladoceran, and rotifer eggs in sea-bottom sediments of northern Californian coastal waters: identification, occurrence and hatching. Mar. Biol. 105: 413–418.

Marcus, N. H., R. Lutz, W. Burnets & P. Cable, 1994. Age, viability and vertical distribution of zooplankton resting eggs from an anoxic besin: evidence of an egg bank. Limnol. Oceanogr. 39: 154–158.

Marino, D., G. Giuffré, M. Montresor & A. Zingone, 1991. An electron microscope investigation on *Chaetoceros minimum* (Levander) comb. nov. and new observations on *Chaetoceros throndsenii* (Marino, Montresor & Zingone) comb. nov. Diatom Res. 6: 317–326.

Marino, D., M. Montresor & A. Zingone, 1987. *Miraltia throndsenii* gen. nov., sp. nov., a planktonic diatom from the Gulf of Neaples. Diatom Res. 2: 205–211.

Matsuoka, K., 1985a. Organic walled dinoflagellate cysts from surface sediments of Nagasaki Bay and Senzaky Bay, West Japan. Bull. Faculty of Liberal Arts, Nagasaki Univ. (Nat. Sci.) 25: 21–115.

Matsuoka, K., 1985b. Archeopile structure in modern Gymnodiniales dinoflagellate cysts. Rev. Paleobot. Palynol. 44: 217–231.

Matsuoka, K., 1988. Cyst-theca relationships in the diplopsalid group (Peridiniales, Dinophyceae). Rev. Paleobot. Palynol. 56: 95–122.

Mc Minn, A., 1991. Recent dinoflagellate cysts from estuaries on the central coast of New South Wales, Australia. Micropaleontol. 37: 269–287.

Mc Minn, A., 1992. Recent and late quaternary dinoflagellate cyst distribution on the continental shelf and slope of southeastern Australia. Palynol. 16: 13–24.

Montresor, M., 1992. Life histories in diatoms and dinoflagellates and their relevance in phytoplankton ecology. Oebalia (Suppl.) 17: 241–257.

Montresor, M., 1995. *Scrippsiella ramonii* sp. nov. (Peridiniales, Dinophyceae), a marine dinoflagellate producing a calcareous resting cyst. Phycologia 34: 87–91.

Montresor, M., E. Montesarchio, D. Marino & A. Zingone, 1994. Calcareous dinoflagellate cysts in marine sediment of the Gulf of Naples (Mediterranean Sea). Rev. Paleobot. Palynol. 84: 45–56.

Montresor, M. & A. Zingone, 1988. *Scrippsiella precaria* sp. nov. (Dinophyceae), a marine dinoflagellate from the Gulf of Naples. Phycologia 27: 387–394.

Montresor, M., A. Zingone & D. Marino, 1993. The calcareous resting cyst of *Pentapharsodinium tyrrenicum* comb. nov. (Dinophyceae). J. Phycol. 29: 223–230.

Morey-Gaines, G. & R. H. Ruse, 1980. Encystment and reproduction of the predatory dinoflagellate *Polykrikos kofoidi* Chatton (Gymnodiniales). Phycologia 19: 230–236.

Mura, G., 1986. SEM morphological survey on the egg shell in the Italian Anostracans (Crustacea, Branchiopoda). Hydrobiologia 134: 273–286.

Mura, G., 1991. SEM morphology of resting eggs in the species of the genus *Branchinecta* from North America. J. Crust. Biol. 11: 432–436.

Mura, G., 1992. Pattern of egg shell morphology in thamnocephalids and streptocephalids of the New World (Anostraca). Crustaceana 62: 300–311.

Mura, G., F. Accordi & M. Rampini, 1978. Studies on the resting eggs of some fresh water fairy shrimps of the genus *Chirocephalus*. Biometry and Scanning Electron Microscopic Morphology (Branchiopoda, Anostraca). Crustaceana 35: 190–194.

Nehring, S., 1994. *Scrippsiella* spp. resting cyst from the german bight (North Sea): a tool for more complete check-list of dinoflagellates. Neth. J. Sea Res. 33: 57–63.

Paranjape, M., 1980. Occurrence and significance of resting cysts in a hyaline tintinnid *Helicostomella subulata* (Ehre.) Jorgensen. J. exp. mar. Biol. Ecol. 48: 23–33.

Reguera B. & S. Fraga, 1995. Autoecology and some life history stages of *Dinophysis acuta* Ehremberg. J. Plankton Res. 17: 999–1015.

Reid, P. C., 1974. Gonyaulacean dinoflagellate cyst from the British Isles. Nova Hedwigia 25: 579–637.

MIX
Papier aus verantwortungsvollen Quellen
Paper from responsible sources
FSC® C105338

If you have any concerns about our products,
you can contact us on
ProductSafety@springernature.com

In case Publisher is established outside the EU,
the EU authorized representative is:
Springer Nature Customer Service Center GmbH
Europaplatz 3, 69115 Heidelberg, Germany

Printed by Libri Plureos GmbH
in Hamburg, Germany